Lecture Notes in Computer Science   **9711**

Commenced Publication in 1973
Founding and Former Series Editors:
Gerhard Goos, Juris Hartmanis, and Jan van Leeuwen

More information about this series at http://www.springer.com/series/7407

Daming Zhu · Sergey Bereg (Eds.)

Frontiers in Algorithmics

10th International Workshop, FAW 2016
Qingdao, China, June 30 – July 2, 2016
Proceedings

 Springer

Editors
Daming Zhu
Shandong University
Jinan
China

Sergey Bereg
MS EC-31
University of Texas at Dallas
Richardson, TX
USA

ISSN 0302-9743   ISSN 1611-3349 (electronic)
Lecture Notes in Computer Science
ISBN 978-3-319-39816-7   ISBN 978-3-319-39817-4 (eBook)
DOI 10.1007/978-3-319-39817-4

Library of Congress Control Number: 2016940797

LNCS Sublibrary: SL1 – Theoretical Computer Science and General Issues

Printed on acid-free paper

This Springer imprint is published by Springer Nature
The registered company is Springer International Publishing AG Switzerland

Preface

This volume contains the papers presented at FAW 2016: The 10th International Frontiers of Algorithmics Workshop held during June 29–July 1, 2016, in QingDao, China.

FAW 2016 received 54 submissions. Each submission was reviewed by at least two Program Committee members. The committee decided to accept 25 papers.

The previous nine meetings were held during August 1–3, 2007, in Lanzhou, June 19–21, 2008, in Changsha, June 20–23, 2009, in Hefei, August 11–13, 2010, in Wuhan, May 28–31, 2011, in Jinhua, May 14–16, 2012, in Beijing, June 26–28, 2013, in Dalian, June 28–30, 2014, in Zhangjiajie, and July 3–5, 2015, in Guilin.

We had three invited plenary speakers at FAW 2016: Thomas Erlebach (Leicester University, UK), Jianer Chen (Central South University, PRC, Texas A&M University, USA), and Binhai Zhu (Montana State University, USA). We express our sincere thanks to them for their contributions to the conference and proceedings.

We would like to thank the Program Committee members and external reviewers for their hard work in reviewing and selecting papers. We are also very grateful to all the authors who submitted their work to FAW 2016. Finally, we would like to thank the editors at Springer and the local organization chairs for their hard work in the preparation of this conference.

April 2016

Daming Zhu
Sergey Bereg

Organization

Program Committee

Hee-Kap Ahn	Pohang University of Science and Technology, Korea
Sergey Bereg	University of Texas at Dallas, USA
Hans L. Bodlaender	Utrecht University, The Netherlands
Yixin Cao	Hong Kong Polytechnic University, SAR China
Zhi-Zhong Chen	Tokyo Denki University, Japan
Marek Chrobak	University of California, Riverside, USA
Daming Zhu	Shandong University, China
Jianxi Fan	Soochow University, China
Qilong Feng	Central South University, China
Henning Fernau	University of Trier, Germany
Mordecai J. Golin	Hong Kong UST, SAR China
Jiong Guo	Shandong University, China
Gregory Gutin	Royal Holloway, University of London, UK
Kun He	Huazhong University of Science and Technology
Xiuzhen Huang	Arkansas State University-Jonesboro, USA
Hiro Ito	The University of Electro-Communications, Japan
Klaus Jansen	University of Kiel, Germany
Haitao Jiang	Shandong University, China
Ming-Yang Kao	Northwestern University
Naoki Katoh	Kyoto University, Japan
Michael Langston	University of Tennessee, USA
Guojun Li	Shandong University, China
Zimao Li	South-Central University for Nationalities, China
Guohui Lin	University of Alberta, USA
Hong Liu	Shandong University, China
Tian Liu	Peking University, China
Xiaowen Liu	Indiana University-Purdue University Indianapolis, USA
Daniel Lokshtanov	UiB
Pinyan Lu	Microsoft Research Asia
Ulrike Stege	University of Victoria
Xiaoming Sun	Chinese Academy of Sciences, China
Jianxin Wang	Central South University, China
Lusheng Wang	City University of Hong Kong, SAR China
Gerhard Woeginger	TU Eindhoven, The Netherlands
Mingyu Xiao	University of Electronic Science and Technology of China
Ke Xu	Beihang University
Yinfeng Xu	Xi'an Jiaotong University

Boting Yang   University of Regina, Canada
Chee Yap   Courant, NYU, USA
Guochuan Zhang   Zhejiang University, China
Peng Zhang   Shandong University, China
Binhai Zhu   Montana State University, USA
Xiaofeng Zhu   Guangxi Normal University, China

General Chairs

John Hopcroft   Cornell University, USA
Fengjing Shao   Qingdao University, China

Program Chairs

Daming Zhu   Shandong University, China
Sergey Bereg   University of Texas at Dallas, USA

Publication Chairs

Binhai Zhu   Montana State University, USA
Zhenbo Guo   Qingdao University, China

Organization Chairs

Haodi Feng   Shandong University, China
Zhenbo Guo   Qingdao University, China

Additional Reviewers

Abu-Khzam, Faisal   Oh, Eunjin
Achlioptas, Dimitris   Paulsen, Niklas
Boucher, Christina   Phillips, Charles
Casel, Katrin   Qi, Enfeng
Hagan, Ronald   Saitoh, Toshiki
Han, Xin   Scheder, Dominik
Higashikawa, Yuya   Schweitzer, Haim
Kim, Min-Gyu   Seto, Kazuhisa
Langetepe, Elmar   Shachnai, Hadas
Li, Liang   Solis-Oba, Roberto
Li, Shuguang   Tong, Weitian
Li, Yang   Wang, Kai
Lin, Weibo   Wang, Lusheng
Liu, Juntao   Xiao, Tao
Maack, Marten   Xu, Yao
Meyer Auf der Heide, Friedhelm   Yoon, Sangduk
Mnich, Matthias   Yu, Ting
Mu, Zengchao   Zhang, Chihao

Abstracts

Algorithms for Queryable Uncertainty

Thomas Erlebach

Department of Computer Science, University of Leicester, Leicester, UK
t.erlebach@leicester.ac.uk

Abstract. Queryable uncertainty refers to settings where the input of a problem is initially not known precisely, but exact information about the input can be obtained at a cost using queries. A natural goal is then to minimize the number of the queries that are required until the precise information that has been obtained about the input is sufficient for solving the problem. The performance of an algorithm can be measured using competitive analysis, comparing the number of queries made by the algorithm to the minimum possible number of queries. We describe the witness set algorithm concept and how it yields query-competitive algorithms for minimum spanning tree and cheapest set problems under uncertainty. We also discuss the problem variant where the algorithm can make a bounded number of simultaneous queries in each round and the goal is to minimize the number of rounds.

On Relating Parameterized Tractability
and Polynomial-Time Approximability

Jianer Chen

[1]School of Information Science and Engineering, Central South University,
People's Republic of China
[2]Department of Computer Science and Engineering,
Texas A&M University, USA

Parameterized algorithms and approximation algorithms are both popular approaches to solving computationally difficult problems. A good parameterized algorithm becomes practically effective when the parameter values are small, while a good approximation algorithm provides solutions that are practically acceptable. Thus, problems with good parameterized algorithms and problems with good approximation algorithms are both regarded as "easier" problems, in particular when the problems are hard in terms of traditional complexity theory (i.e., NP-hard problems). As a consequence, there has been increasing research interests in the study of relationship between these two computational paradigms, from viewpoints of both theoretical and practical computing sciences.

We survey the major research results in the past decades on the relationship between parameterized tractability and polynomial-time approximability of NP-optimization problems, including those that are of interesting in complexity theoretical study and those that are related to algorithmic technique development. We present general techniques that enable us to show that a large class of approximable problems are parameterized tractable, and general techniques that are based on the parameterized complexity and lead to good approximation algorithms for a large class of optimization problems. In particular, we give a precise characterization of the approximation class FPTAS in terms of the parameterized complexity. Our study leads to new lower bound techniques that allow us to prove inapproximability of optimization problems based on their parameterized complexity, which makes it possible to study the practical limitations of the approximation class with polynomial-time approximation schemes (i.e., PTAS). In the study of development of algorithmic techniques, we present general techniques that translate one to the other between parameterized algorithms and approximation algorithms. In particular, our show, as an example, how the techniques allow us to develop parameterized algorithms and approximation algorithms for the MAXIMUM AGREEMENT FOREST problem on multiple phylogenetic trees, improving a long line of series results in the literature.

Genomic Scaffold Filling: A Progress Report

Binhai Zhu

Department of Computer Science,
Montana State University, Bozeman, MT 59717-3880, USA
bhz@montana.edu

Abstract. The genomic scaffold filling problem has attracted a lot of attention since 2010. The general problem is on filling an incomplete sequence (sequence scaffold) I into I', with respect to a complete reference genome G, such that the number of adjacencies between G and I' is maximized. The problem is NP-complete and APX-hard, and admits a 1.2-approximation. In this survey paper, we will first review the progress being made for this setting.

However, the sequence input I is not quite practical and does not fit most of the real datasets (where a scaffold is more often given as a list of contigs). Then, we will review the most recent progress on this new version of the genomic scaffold filling problem where, (1) a scaffold S is given, the missing genes $X = c(G) - c(S)$ can only be inserted in between the contigs, and the objective is to maximize the number of adjacencies between G and the filled S', and (2) a scaffold S is given, a subset of the missing genes $X' \subset X = c(G) - c(S)$ can only be inserted in between the contigs, and the objective is still to maximize the number of adjacencies between G and the filled S''. Some open problems will be posed for further research.

Contents

Algorithms for Queryable Uncertainty

Thomas Erlebach$^{(\boxtimes)}$

Department of Computer Science, University of Leicester, Leicester, UK
`t.erlebach@leicester.ac.uk`

Abstract. Queryable uncertainty refers to settings where the input of a problem is initially not known precisely, but exact information about the input can be obtained at a cost using queries. A natural goal is then to minimize the number of the queries that are required until the precise information that has been obtained about the input is sufficient for solving the problem. The performance of an algorithm can be measured using competitive analysis, comparing the number of queries made by the algorithm to the minimum possible number of queries. We describe the witness set algorithm concept and how it yields query-competitive algorithms for minimum spanning tree and cheapest set problems under uncertainty. We also discuss the problem variant where the algorithm can make a bounded number of simultaneous queries in each round and the goal is to minimize the number of rounds.

1 Introduction

In traditional algorithms research, one normally assumes that the precise input to a problem is given to the algorithm in advance (offline algorithms) or revealed to the algorithm over time (online algorithms). Motivated by real-world applications, one can also consider settings where the input is initially given with some uncertainty and the algorithm can obtain precise information about the input using *queries*. We refer to the uncertainty in such settings as *queryable uncertainty*. In many cases it is natural to assume that queries incur a cost, and so one wants to minimize the number of queries (or the total cost of queries) that are required until the precise information that has been obtained suffices for computing a solution to the given problem. The performance of an algorithm can then be measured using competitive analysis, comparing the number of queries made by the algorithm to the minimum possible number of queries.

There are numerous application areas where queryable uncertainty arises. For example, Kahan [1] considered a setting where the inputs are the locations of moving objects (e.g., airplanes). An uncertainty area containing the current position of a plane can easily be determined based on a known past position and the maximum speed of the plane. Determining the exact current position of a plane is possible but expensive as it involves a radio communication with the pilot. If one wants to determine the pair of airplanes with the minimum distance between each other, doing so with a minimum number of radio communications is a natural objective. Another application arising in distributed settings with

D. Zhu and S. Bereg (Eds.): FAW 2016, LNCS 9711, pp. 1–7, 2016.
DOI: 10.1007/978-3-319-39817-4_1

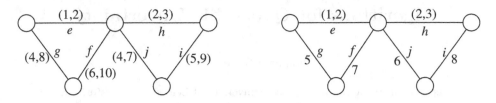

Fig. 1. Initial instance of MST-U (left) and resulting instance after four queries (right)

database caches was considered by Olston and Widom [2]. Local database caches maintain intervals of values that are guaranteed to contain the exact value. The master server needs to replicate an updated value of a data item to the caches only when the new value lies outside that interval. Here, the intervals stored in the local cache correspond to uncertainty areas, and the operation of requesting the exact value of a data item from the master server corresponds to a query. More generally, any setting where estimates of the input values are easily available and exact values can be obtained at a cost can be seen as an application of queryable uncertainty.

As a concrete problem in the context of queryable uncertainty, let us consider the minimum spanning tree (MST) problem under uncertainty, denoted by MST-U. One is given an undirected graph $G = (V, E)$ together with an uncertainty area A_e for every edge $e \in E$. The exact weight of edge $e \in E$, denoted by w_e, is initially unknown to the algorithm, but it is guaranteed that $w_e \in A_e$. The task is to output the edge set of a minimum spanning tree of $G = (V, E)$ with respect to the edge weights w_e for $e \in E$. If the algorithm queries edge e, the exact weight w_e of edge e is revealed. If only the initial uncertainty areas are known, it may be impossible to determine the edge set of an MST. Whenever the algorithm queries an edge, it receives the exact weight of that edge before querying another edge. It may not be necessary to query all edges, because it can be possible to determine that an edge is part of the MST (or not part of the MST) even if some edge weights are not known precisely.

As an example, consider the instance of MST-U shown in Fig. 1 (left), where the uncertainty area of every edge is an open interval. Without making queries it is not possible to determine the edge set of an MST, since one cannot know e.g. whether the edge g with weight in $(4, 8)$ or the edge f with weight in $(6, 10)$ has a smaller exact weight. Only the edges e and h will surely be contained in the edge set of the MST. An algorithm might now query the edges f, g, j, i in that order, producing the situation shown in Fig. 1 (right). At this point, one can determine that $\{e, g, h, j\}$ is the edge set of an MST, and the problem is solved. The algorithm has made four queries in total. In hindsight, it would have been enough to query only the edges g and i, as knowing the exact weights of these two edges would also have shown that $w_g < w_f$ and $w_j < w_i$. No single query would have been sufficient to determine the edge set of an MST, so the optimal number of queries in this example is two. The ratio between the number of queries made by the algorithm and the optimal number is 2.

while *instance is unsolved* **do**
  $W \leftarrow$ a witness set of the current instance;
  query all u in W;
end

Algorithm 1. Witness set algorithm

The rest of the paper is structured as follows. In Sect. 2 we give definitions of queryable uncertainty and query-competitive algorithms and describe witness set algorithms. Related work is discussed in Sect. 3. Results for minimum spanning tree and cheapest set problems under uncertainty are described in Sects. 4 and 5, respectively. The direction of round-competitive algorithms is introduced in Sect. 6, and we conclude in Sect. 7.

2 Preliminaries

In a setting with queryable uncertainty, an instance of a problem typically consists of some structural information S, a set U of elements with uncertain values, and a function A that maps each element $u \in U$ to a set A_u called *uncertainty area*. The exact values of the uncertain input elements, which are initially unknown to the algorithm, are represented by a function w that maps each $u \in U$ to its exact value w_u, and it is guaranteed that $w_u \in A_u$. A query of an element $u \in U$ reveals its exact value w_u. We can view a query of $u \in U$ as the operation of replacing A_u by the singleton set $\{w_u\}$.

For a given instance $I = (S, U, A, w)$ of a problem in the model of queryable uncertainty, we let $\phi(S, U, w)$ denote the set of solutions. It depends only on the exact values w_u for $u \in U$, but not on the uncertainty areas A_u. An algorithm only receives (S, U, A) as input. The goal of the algorithm is to compute a solution in $\phi(S, U, w)$ after making a minimum number of queries. Queries are made one by one, and the results of previous queries can be taken into account when determining the next query. By OPT_I (or simply OPT) we denote the minimum number of queries that provide sufficient information for computing a solution in $\phi(S, U, w)$. An algorithm that makes ALG_I queries to solve an instance I is called (strongly) ρ-*query-competitive* (or simply ρ-*competitive*) if $ALG_I \leq \rho OPT_I$ for all instances I of the problem. For randomized algorithms, ALG_I refers to the expected number of queries that the algorithm makes on instance I.

The concept of witness set algorithms was introduced by Bruce et al. [3] and stated in general form by Erlebach et al. [4]. For a given instance $I = (S, U, A, w)$ of an uncertainty problem, a set $W \subseteq U$ is called a *witness set* if it is impossible to determine a solution without querying at least one element of W. A *witness set algorithm* (see Algorithm 1) is an algorithm that repeatedly determines a witness set and queries all its elements. If it can be shown that every witness set used by the algorithm has size at most k, it follows that the algorithm is k-query-competitive [3, 4].

3   Related Work

As far as we are aware, Kahan [1] was the first to study query-competitive algorithms in the model of queryable uncertainty. He gave query-competitive algorithms with optimal competitive ratio for the problems of computing the maximum, the median and the minimum gap of n real values that are given as uncertainty areas in the form of intervals. Bruce et al. [3] studied geometric problems where the input points lie in regions represented by uncertainty areas. They introduced the concept of witness set algorithms and gave 3-query-competitive algorithms for the problems of computing maximal points or the points on the convex hull of a given set of uncertain points in Euclidean space. Query-competitive algorithms were studied for the minimum spanning tree problem in [4,5] and for cheapest set problems in [6]; we will discuss these results in more detail in Sects. 4 and 5. Gupta et al. [7] and Tseng and Kirkpatrick [8] considered variants of the model where queries yield more refined estimates instead of precise values. Goerigk et al. [9] studied the knapsack problem in a model of queryable uncertainty where the item weights are uncertain and the number of queries is bounded. The setting where the uncertain values follow known probability distributions was considered by Guha and Munagala [10], who also related adaptive query strategies to non-adaptive ones (where all queries must be specified in advance and are executed in parallel) for that setting. Non-adaptive query strategies were considered by various authors, see e.g. [2,11,12]. We refer to [13] for a more extensive survey of queryable uncertainty.

4   MST Under Uncertainty

The problem MST-U, which was defined in Sect. 1, was considered by Erlebach et al. [4]. They assume that each uncertainty area is either an open interval or a singleton set. (In fact, they note that their results hold also for a certain generalization of open intervals.)

An instance of the problem is specified by a graph $G = (V, E)$ and the uncertainty areas A_e and exact weights w_e of each edge $e \in E$. Denote the infimum and supremum of an uncertainty area A_e by L_e and U_e, respectively. For two edges e and f, define $e < f$ if $L_e < L_f$ or $L_e = L_f$ and $U_e < U_f$. Define $e \le f$ if $e < f$ or $L_e = L_f$ and $U_e = U_f$. For a cycle C, an edge $e \in C$ is called an *always maximal edge* if e is a maximum-weight edge in C for every possible choice of exact values $w_f \in A_f$ for all $f \in C$. The algorithm presented in [4], called U-RED (see Algorithm 2), processes the edges in the order defined by the $\le$ relation and adds them to an initially empty graph T. When an edge closes a cycle, there are two cases: If the cycle has an always maximal edge, that edge is removed from T. Otherwise, two edges f and g of the cycle are selected and queried. It can be shown that $\{f, g\}$ is a witness set, and therefore the algorithm is 2-query-competitive.

Erlebach et al. [4] also gave a lower bound showing that no deterministic algorithm for MST-U can achieve competitive ratio smaller than 2, implying that algorithm U-RED is best possible among deterministic algorithms.

index all edges such that $e_1 \leq e_2 \leq \ldots \leq e_m$;
let T be a graph on V without any edges;
for $i \leftarrow 1$ to m do
   add e_i to T;
   if T has a cycle C then
      if C contains an always maximal edge e then
         delete e from T;
      else
         let $f \in C$ such that $U_f = \max\{U_c | c \in C\}$;
         let $g \in C - \{f\}$ such that $U_g > L_f$;
         query f and g;
         restart the algorithm;
      end
   end
end
return T

Algorithm 2. U-RED algorithm (from [4])

They left open whether it is possible to achieve a ratio better than 2 using a randomized algorithm. Their lower bound construction only yielded a lower bound of 1.5 for the competitive ratio of randomized algorithms. Megow et al. [5] recently answered this question affirmatively and presented a randomized algorithm with competitive ratio $1 + 1/\sqrt{2} \approx 1.707$ for MST-U. In their algorithm they make use of a carefully adapted water-filling scheme from [14] for online bipartite vertex cover. They also extend their results to the case of non-uniform query costs.

5 Cheapest Set Under Uncertainty

The cheapest set problem under uncertainty (CSU) was considered by Erlebach et al. [6]. An instance of the problem consists of a set U of elements, a family $\mathcal{F}$ of subsets of U, an uncertainty area A_u for each $u \in U$, and the exact weight w_u for each $u \in U$. The task is to determine a cheapest set $S \in \mathcal{F}$, i.e., a set such that the sum of the exact weights of its elements is minimized. Many combinatorial optimization problems (including the minimum spanning tree problem) can be viewed as cheapest set problems.

Define a *robust cheapest set* to be a set C in $\mathcal{F}$ with the following property: For any choice of exact weights $w_u \in A_u$ for the elements u of $U \setminus C$, there are weights $w_u \in A_u$ for each element $u \in C$ such that C is a cheapest set. The algorithm proposed for CSU by Erlebach et al. [6] repeatedly queries all elements of a robust cheapest set until a cheapest set can be identified. They can show that all sets queried by the algorithm, except at most one, are witness sets. For the case that all sets in $\mathcal{F}$ have cardinality at most d, their algorithm therefore makes at most $dOPT + d$ queries. They also showed that this is best possible among deterministic algorithms.

For the special case of CSU where the sets correspond to edge sets that form a multi-cut in a tree with d terminal pairs, they presented an improved algorithm that makes at most $dOPT + 1$ queries, which was also shown to be optimal among deterministic algorithms.

6   Round-Competitive Algorithms

We have so far assumed that queries are made sequentially and the choice of the next query can depend on the outcomes of all previous queries. In some application scenarios, it is meaningful to consider a setting where a number of queries can be made in parallel. The queries are then made in *rounds*. The set of queries to be executed in one round must be specified at the start of that round, and the answers of the queries become available at the end of the round. One natural problem formulation then assumes that an upper bound k on the number of queries that can be executed in each round is given, and the goal is to minimize the number of query rounds that are necessary until sufficient information about the input has been obtained for determining a solution. We call this variant of an uncertainty problem the *round-style* variant of the problem. If we denote by ALG_I^r the number of query rounds required by algorithm ALG on instance I and by OPT_I^r the optimal number of query rounds, we say that the algorithm is ρ-round-competitive for a round-style problem if $ALG_I^r \leq \rho OPT_I^r$ for all instances I of the problem. Note that $OPT_I^r = \lceil OPT_I/k \rceil$, where OPT_I is the optimal number of queries for instance I.

The direction of round-style problems with queryable uncertainty does not seem to have received much attention up to now. We believe that this is an interesting direction to pursue in the future. For example, the round-style variant of MST-U could be studied. If the bound on the number of queries per round is $k = 2$, it is easy to see that algorithm U-RED can be interpreted as a round-style algorithm and that it is 2-round-competitive. For the case of arbitrary k, it is not difficult to show a constant lower bound larger than 2 on the competitive ratio of any round-style algorithm. It would be interesting to obtain tight bounds on the best possible round-competitive ratio for every value of k.

Another variant of round-style problems is obtained if one assumes that the number of rounds is fixed and the goal is to minimize the total number of queries. In the case where only a single query round is allowed, this leads to the non-adaptive query model that has been considered by various authors, see e.g. [2, 11,12]. The case where the bound on the number of rounds is larger than one does not seem to have received much attention so far.

7   Conclusions

We have discussed the model of queryable uncertainty and some of the existing work in this area, explained the concept of witness set algorithms and described query-competitive algorithms for MST-U (from [4]) and CSU (from [6]). Furthermore, we made some remarks about possible directions for research in round-competitive algorithms. Another interesting direction could be considering problems in the model of queryable uncertainty where the objective to be minimized is a combination of query cost and the cost of the solution that is being output. For example, such a setting could be relevant to scheduling problems where the processing times of tasks are uncertain and a query corresponds to performing a code analysis that runs on the same processor that executes the scheduled tasks once the schedule has been determined.

References

1. Kahan, S.: A model for data in motion. In: 23rd Annual ACM Symposium on Theory of Computing (STOC 1991), pp. 267–277 (1991)
2. Olston, C., Widom, J.: Offering a precision-performance tradeoff for aggregation queries over replicated data. In: 26th International Conference on Very Large Data Bases (VLDB 2000), pp. 144–155 (2000)
3. Bruce, R., Hoffmann, M., Krizanc, D., Raman, R.: Efficient update strategies for geometric computing with uncertainty. Theor. Comput. Syst. **38**, 411–423 (2005)
4. Erlebach, T., Hoffmann, M., Krizanc, D., Mihalák, M., Raman, R.: Computing minimum spanning trees with uncertainty. In: 25th International Symposium on Theoretical Aspects of Computer Science (STACS 2008). LIPIcs, vol. 1, pp. 277–288 (2008)
5. Megow, N., Meißner, J., Skutella, M.: Randomization helps computing a minimum spanning tree under uncertainty. In: Bansal, N., Finocchi, I. (eds.) Algorithms - ESA 2015. LNCS, vol. 9294, pp. 878–890. Springer, Heidelberg (2015)
6. Erlebach, T., Hoffmann, M., Kammer, F.: Query-competitive algorithms for cheapest set problems under uncertainty. Theor. Comput. Sci. **613**, 51–64 (2016)
7. Gupta, M., Sabharwal, Y., Sen, S.: The update complexity of selection and related problems. In: IARCS Annual Conference on Foundations of Software Technology and Theoretical Computer Science (FSTTCS 2011). LIPIcs, vol. 13, pp. 325–338 (2011)
8. Tseng, K.-C.R., Kirkpatrick, D.: Input-thrifty extrema testing. In: Asano, T., Nakano, S., Okamoto, Y., Watanabe, O. (eds.) ISAAC 2011. LNCS, vol. 7074, pp. 554–563. Springer, Heidelberg (2011)
9. Goerigk, M., Gupta, M., Ide, J., Schöbel, A., Sen, S.: The robust knapsack problem with queries. Comput. OR **55**, 12–22 (2015)
10. Guha, S., Munagala, K.: Model-driven optimization using adaptive probes. In: 18th Annual ACM-SIAM Symposium on Discrete Algorithms (SODA 2007), pp. 308–317 (2007)
11. Feder, T., Motwani, R., Panigrahy, R., Olston, C., Widom, J.: Computing the median with uncertainty. SIAM J. Comput. **32**, 538–547 (2003)
12. Feder, T., Motwani, R., O'Callaghan, L., Olston, C., Panigrahy, R.: Computing shortest paths with uncertainty. J. Algorithms **62**, 1–18 (2007)
13. Erlebach, T., Hoffmann, M.: Query-competitive algorithms for computing with uncertainty. Bulletin of the EATCS **116** (2015)
14. Wang, Y., Wong, S.C.-W.: Two-sided online bipartite matching and vertex cover: beating the greedy algorithm. In: Halldórsson, M.M., Iwama, K., Kobayashi, N., Speckmann, B. (eds.) ICALP 2015. LNCS, vol. 9134, pp. 1070–1081. Springer, Heidelberg (2015)

Genomic Scaffold Filling: A Progress Report

Binhai Zhu[✉]

Department of Computer Science,
Montana State University, Bozeman, MT 59717-3880, USA
bhz@montana.edu

Abstract. The genomic scaffold filling problem has attracted a lot of attention since 2010. The general problem is on filling an incomplete sequence (sequence scaffold) I into I', with respect to a complete reference genome G, such that the number of adjacencies between G and I' is maximized. The problem is NP-complete and APX-hard, and admits a 1.2-approximation. In this survey paper, we will first review the progress being made for this setting.

However, the sequence input I is not quite practical and does not fit most of the real datasets (where a scaffold is more often given as a list of contigs). Then, we will review the most recent progress on this new version of the genomic scaffold filling problem where, (1) a scaffold S is given, the missing genes $X = c(G) - c(S)$ can only be inserted in between the contigs, and the objective is to maximize the number of adjacencies between G and the filled S', and (2) a scaffold S is given, a subset of the missing genes $X' \subset X = c(G) - c(S)$ can only be inserted in between the contigs, and the objective is still to maximize the number of adjacencies between G and the filled S''. Some open problems will be posed for further research.

1 Introduction

Since 2001, the cost of sequencing a genome has been reduced significantly, with the current cost being around \$1 k. This results in a lot of genomes being sequenced, usually not completely finished (they are typically called *draft* genomes). On the other hand, the cost to finish these genomes completely has not been decreased as much [6]. The result is that we are having more and more draft genomes. Nonetheless, for many tools analyzing the genomic data we do need complete genomes. For instance, to compute the reversal distance between two genomes we do need two complete genomes. Hence, there is a need to turn a draft genome into a complete one.

To make the result biologically interesting, Munoz *et al.* first proposed the following *scaffold filling* problem (on multichromosomal genomes with no gene repetition) as follows [24]. Given a complete (permutation) genome R and an incomplete scaffold S, fill the missing genes in $R - S$ into S to have S' such that the genomic distance (or DCJ distance [26]) between R and S' is minimized. It was shown that this problem can be solved in polynomial time. In [16], Jiang *et al.* considered the case for singleton genomes without gene repetition

© Springer International Publishing Switzerland 2016
D. Zhu and S. Bereg (Eds.): FAW 2016, LNCS 9711, pp. 8–16, 2016.
DOI: 10.1007/978-3-319-39817-4_2

(i.e., permutations), using the simplest *breakpoint* distance as the similarity measure. It was not surprising that this problem was shown to be polynomially solvable; in fact, even for the two-sided case when both the input scaffolds, being a reference to each other, are incomplete permutations.

When the genomes and scaffolds contain gene repetitions, the problem becomes harder. (That should not be considered as a surprise as even computing certain similarity measure between two complete genomes is NP-complete, for instance, with the exemplar breakpoint distance [2,4,7,9,19], exemplar adjacency number [8,10], or the minimum common string partition [11].) The similarity measure adopted for the scaffold filling problem is the *number of common (string) adjacencies*, which can be computed in polynomial time [2,15,16]. In [15,16], it was shown by Jiang *et al.* that filling a scaffold to maximize the number of common string adjacencies (SF-MNSA) is NP-hard. (Formally, the problem is to fill an incomplete sequence scaffold I into I', with respect to a complete reference genome G, such that the missing letters in $G - I$ are inserted back to I and the number of common adjacencies between G and I' is maximized.) A factor-1.33 approximation was designed in [15,16], and this bound has been improved to 1.25 [21], and to 1.20 [17]. For the corresponding two-sided case, i.e., when two sequence scaffolds are references to each other, the problem admits a factor-1.5 approximation with the number of common adjacencies between the filled scaffolds being maximized [22]. Using the number of common adjacencies as a parameter, it was shown that this problem is also fixed-parameter tractable (FPT) — this only handles that case when G and I' are not very similar so it is only of a theoretical meaning [5].

Recently, a practical factor is seriously considered [18]. Firstly, the 'scaffold' used in most of these papers is an incomplete sequence, i.e., a missing gene can be inserted anywhere in such a 'scaffold'. In practice, most of the real datasets are not in this format; in fact, a scaffold in a real dataset is usually composed of a sequence of contigs, where a contig is usually computed with mature tools like Celera Assembler [1], hence should not be arbitrarily altered. This case was only briefly considered a few years ago [16,24]. Secondly, take a complete reference genome G and a scaffold S, there is no guarantee that the filled scaffold S' is of the same length as that of G; in fact, sometimes we could know roughly the length of the target genome S^* (S' should be as close to S^* as possible). Then, we might only need to insert a subset of letters in $G - S$ into S (to obtain S').

Formally, the above two problems are called One-sided Scaffold Filling (One-sided-SF-max), and One-sided Subset Scaffold Filling (One-sided-SF-max($\subset$)) respectively. (For the important practical case when a gene can only appear at most d times in G, we call the corresponding problems One-sided-SF-max(d) and One-sided-SF-max($\subset, d$) respectively.) The objective function in both cases are to maximize the number of common adjacencies between the reference and the filled scaffold.

The paper is organized as follows. In Sect. 2, we give the preliminaries. In Sect. 3, we review the results in three categories: when genomes have no gene duplications, approximation results for the one-sided cases, and FPT results for the one-sided cases. We conclude the paper in Sect. 4 by giving directions for further research.

2   Preliminaries

Throughout this paper we focus only on singleton genomes (i.e., each is a sequence). But the results can be easily generalized to multichromosomal or circular genomes, with minor changes.

At first, we review some necessary definitions, which are also defined in [16, 27]. We assume that all genes and genomes are unsigned, and it is straightforward to generalize the result to signed genomes. Given a gene set Σ, a string P is called *permutation* if each element in Σ appears exactly once in P. We use $c(P)$ to denote the set of elements in permutation P. A string A is called *sequence* if some genes appear more than once in A, and $c(A)$ denotes genes of A, which is a multi-set of elements in Σ. For example, $\Sigma = \{a, b, c, d\}$, $A = abcdacd$, $c(A) = \{a, a, b, c, c, d, d\}$. A *sequence scaffold* is an incomplete sequence, typically obtained by some sequencing and assembling process. A substring with m genes (in a sequence) is called an *m-substring*, and a 2-substring is also called a *pair*; as the genes are unsigned, the relative order of the two genes of a pair does not matter, i.e., the pair xy is equal to the pair yx. Given an incomplete sequence (or sequence scaffold) $A = a_1a_2a_3 \cdots a_n$, let $P_A = \{a_1a_2, a_2a_3, \ldots, a_{n-1}a_n\}$ be the set of pairs in A.

Definition 1. *Given two sequence scaffolds $A = a_1a_2 \cdots a_n$ and $B = b_1b_2 \cdots b_m$, if $a_ia_{i+1} = b_jb_{j+1}$ (or $a_ia_{i+1} = b_{j+1}b_j$), where $a_ia_{i+1} \in P_A$ and $b_jb_{j+1} \in P_B$, we say that a_ia_{i+1} and b_jb_{j+1} are matched to each other. In a maximum matching of pairs in P_A and P_B, a matched pair is called an* **adjacency***, and an unmatched pair is called a* **breakpoint** *in A and B respectively.*

It follows from the definition that sequence scaffolds A and B contain the same set of adjacencies but distinct breakpoints. The maximum matched pairs in B (or equally, in A) form the *(common) adjacency set* between A and B, denoted as $a(A, B)$. We use $b_A(A, B)$ and $b_B(A, B)$ to denote the set of breakpoints in A and B respectively. We illustrate the above definitions in Fig. 1.

For a sequence A and a multi-set of elements X, let $A + X$ be the set of all possible resulting sequences after filling all the elements in X into A. We define a contig as a string over a gene set Σ whose contents should not be altered. A *scaffold* S is simply a sequence of contigs $\langle C_1, ..., C_m \rangle$. We define $c(S) = c(C_1) \cup \cdots \cup c(C_m)$. Now, we define the problems on scaffolds formally.

Given two incomplete sequences (or sequence scaffolds) $A = a_1a_2 \cdots a_n$ and $B = b_1b_2 \cdots b_m$, as we can see, each gene except the four ending ones is involved in two adjacencies or two breakpoints or one adjacency and one breakpoint. To get rid of this imbalance, we add "#" to both ends of A and B.

Definition 2. *Scaffold Filling to Maximize the Number of (String) Adjacencies (SF-MNSA).*

Input*: two sequence scaffolds A and B over a gene set Σ and two multi-sets of elements X and Y, where $X = c(B) - c(A)$ and $Y = c(A) - c(B)$.*

Question*: Find $A^* \in A + X$ and $B^* \in B + Y$ such that $|a(A^*, B^*)|$ is maximized.*

$$sequence\ scaffold\ A = \langle c\ b\ c\ e\ d\ a\ b\ a\ \rangle$$
$$sequence\ scaffold\ B = \langle a\ b\ a\ b\ d\ c \rangle$$
$$P_A = \{cb, bc, ce, ed, da, ab, ba\}$$
$$P_B = \{ab, ba, ab, bd, dc\}$$
$$matched\ pairs\ :\ (ab \leftrightarrow ba), (ba \leftrightarrow ab)$$
$$a(A, B) = \{ab, ba\}$$
$$b_A(A, B) = \{cb, bc, ce, ed, da\}$$
$$b_B(A, B) = \{ab, bd, dc\}$$

Fig. 1. An example for adjacency and breakpoint definitions.

The one-sided SF-MNSA problem is a special instance of the SF-MNSA problem where one of X and Y is empty. We formally define it as follows.

Definition 3. *One-sided SF-MNSA.*

Input: *a complete sequence G and an incomplete sequence scaffold I over a gene set Σ, a multi-set $X = c(G) - c(I) \neq \emptyset$ with $c(I) - c(G) = \emptyset$.*

Question: *Find $I^* \in I + X$ such that $|a(I^*, G)|$ is maximized.*

Note that while the two-sided SF-MNSA problem is more general and more difficult, the One-Sided SF-MNSA problem is more practical as a lot of genome analysis are based on some reference genome [24]. We next consider the usual scaffolds composed of sequences of contigs.

Definition 4. *One-Sided-SF-max.*

Input: *a complete genome G and a scaffold $S = \langle C_1, C_2, ..., C_m \rangle$ where G and the contig C_i's are over a gene set Σ, a multiset $X = c(G) - c(S) \neq \emptyset$.*

Question: *Find $S^* \in S + X$ such that $|a(S^*, G)|$ is maximized.*

One-Sided-SF-max($\subset$) is exactly the same as One-Sided-SF-max except that only a subset $X' \subset X$ need to be inserted into S. When a gene can appear at most d times in G, the two versions of problems are abbreviated as One-Sided-SF-max(d) and One-Sided-SF-max($\subset, d$) respectively.

3   Current Status

We review the current status for the research on the genomic scaffold filling problems.

3.1   Results on Filling Permutation Scaffolds

When the genomes contain no duplicated genes, the genomic scaffold filling problems are all known to be polynomially solvable. For the one-sided case, when the distance measure is DCJ (double-cut-and-join) and the scaffolds are composed of contigs, Munoz *et al.* used the *breakpoint graph* to obtain a polynomial-time solution (in fact, a linear-time solution after the breakpoint graph is constructed). Later, the result was generalized to the two-sided case by Jiang *et al.* [16]. When the distance measure is the breakpoint distance, and when the scaffolds are incomplete permutations (meaning missing genes can be inserted anywhere), Jiang *et al.* showed that both of the one-sided and two-sided problems can be solved in $O(n^2)$ time, where n is the total number of genes in the input [16].

Recently, Liu et al. [23] studied the one-sided scaffold filling problem when the scaffold S is given as a set of contigs, i.e., $S = \langle C_1, C_2, \cdots, C_m \rangle$. Let the complete reference genome (permutation) be R and let $X = c(R) - c(S)$. Let $\alpha(C_i), \beta(C_i)$ be the first and last letter of C_i respectively. Then $\langle \beta_i, \alpha_{i+1} \rangle$ (or simply $\beta_i \alpha_{i+1}$) constitutes a *slot* where missing genes can inserted between β_i and α_{i+1}. We write $\beta_0 = -\infty$ and $\alpha_{m+1} = +\infty$, where $\langle -\infty, \alpha_1 \rangle$ and $\langle \alpha_m, +\infty \rangle$ are the leftmost and rightmost (open) slot respectively.

Define a type-1 (resp. type-2) substring s of length $\ell \geq 1$, over X, as one which can be inserted in the slot $\langle \beta_i, \alpha_{i+1} \rangle$, for some i, to increase the total number of adjacencies by $\ell + 1$ (resp. ℓ).

Note that if $\beta_i \alpha_{i+1}$ is already an adjacency with respect to R, then in general it is possible that s is inserted in the slot to generate $|s| + 1$ adjacencies (while destroying the adjacency $\beta_i \alpha_{i+1}$). However, when the genome contains no gene duplication it can be shown that there exists an optimal solution which always preserves such an existing adjacency $\beta_i \alpha_{i+1}$.

Similarly, define a type-3 substring s of length $\ell \geq 1$, over X, as one which can be inserted in the slot $\langle \beta_i, \alpha_{i+1} \rangle$, for some i, to increase the number of adjacencies by $\ell - 1$. Note that a type-3 substring can only form adjacencies internally, hence it does not matter where we insert s — provided that it does not destroy any existing adjacency.

Clearly, due to that there is no gene duplication in R, a type-i substring with length ℓ must be a substring (or the reversal of a substring) of R. (This property does not hold when there are gene duplications.) We show an example as follows:

$$R = \langle 1, 2, 3, 4, 5, 6, 7, 8, 9, 10, 11, 12, 13, 14, 15 \rangle,$$

$$S = \langle \boxed{1,3}, \boxed{5,6}, \boxed{15,2}, \boxed{14,13,10,9} \rangle.$$

We have $\alpha_1 = 1, \beta_1 = 3, \alpha_2 = 5, \beta_2 = 6, \alpha_3 = 15, \beta_3 = 2, \alpha_4 = 14, \beta_4 = 9$. Then, $X = \{4, 7, 8, 11, 12\}$ are missing from S. The optimal solution is

$$S^* = \langle \boxed{1,3}, 4, \boxed{5,6}, 7, 8, \boxed{15,2}, 11, 12, \boxed{14,13,10,9} \rangle.$$

In this case, 4 is type-1, $\langle 7, 8 \rangle$ is type-2, and $\langle 11, 12 \rangle$ is type-3.

Then, Liu et al. [23] tried to first identify type-i substrings, and then fill them in the other of $i = 1, 2, 3$. The running time is dominated by the computation of a maximum matching in a bipartite graph, which takes $O(n^{2.5})$ time. We summarize the results in following Table 1.

Table 1. Results on scaffold filling when the genomes contain no gene duplications. In the first column, 'contigs' means a scaffold is composed of a list of contigs, 'permutation' means a scaffold is an incomplete permutation.

Problem	Distance/Similarity measure	Status
One-sided, singleton, contigs	breakpoint/adjacency number	P [23]
One-sided, multichromosome, contigs	DCJ	P [24]
two-sided, singleton, permutation	breakpoint/adjacency number	P [16]
two-sided, multichromosome, contigs	DCJ	P [16]

3.2   Results on One-Sided Scaffold Filling

When the genomes contain duplicated genes, the genomic scaffold filling problems become NP-hard. The past effort has been mainly focused on the one-sided case. (We comment that the two-sided SF-MNSA can be approximated with a factor 1.5 [22].) So we will focus on One-sided SF-MNSA and One-sided-SF-max, representing that the scaffold is an incomplete sequence and a list of contigs respectively.

For One-sided SF-MNSA, Jiang *et al.* first showed that it is NP-hard by a reduction from Exact Cover by 3-Sets (X3C) [15,16]. Subsequently, a factor-1.33 approximation was given [15,16]. The main idea is that when a scaffold is a sequence (meaning missing genes can be inserted anywhere), there is no type-3 substrings. Then, the idea is to maximize the inserted type-1 substrings of length one and two, using a greedy method. Finally, for the remaining genes, it can be done so that each inserted one contributes at least one adjacency. (This last step is in fact not trivial, the details were filled later [21,27].)

In [21], the approximation factor was improved to 1.25. The idea was to insert i-type-1, $i = 1, 2, 3$, substrings using a mixture of greedy search, maximum matching and local search. Then, for the remaining genes, it can be done so that each inserted one contributes at least one adjacency. Recently, the approximation factor for One-sided SF-MNSA was improved to 1.2 and the problem was shown to be APX-hard [17]. The method was based on non-oblivious local search [20]. (As of this writing, it seems there was a small bug in the proof and a paper was devoted in this proceeding to fix that.)

While the research on the One-sided SF-MNSA has been fruitful, it does not help much on solving the practical problem. The main reason is that a scaffold in reality is usually computed with mature tools, like Celera Assembler [1]. Hence, in real datasets a scaffold is usually given as a list of contigs, each should not be altered arbitrarily. Recently, research on this version, formally called One-sided-SF-max, has been started.

In [18], a simple reduction from Hamiltonian Path to One-sided-SF-max was constructed. Thus, the problem is NP-hard. Then, a factor-2 approximation was presented using greedy search on type-1 substrings of length one and two, also on type-2 substrings of length one. Finally, a maximum matching method was used to make sure that for each pair *useful* genes, at least one adjacency can be

Table 2. Approximation results on One-sided SF-MNSA, Two-sided SF-MNSA, and One-sided-SF-max.

Problem	Similarity measure	Approximation ratio
One-sided SF-MNSA	adjacency number	1.33 [15, 16]
One-sided SF-MNSA	adjacency number	1.25 [21]
One-sided SF-MNSA	adjacency number	1.20 [17]
Two-sided SF-MNSA	adjacency number	1.50 [22]
One-sided-SF-max	adjacency number	2.00 [18]

computed. (Formally, a gene is useful if it can contribute some adjacency in an optimal solution.)

We summarize the approximation results for One-sided SF-MNSA, Two-sided SF-MNSA, and One-sided-SF-max in following Table 2.

3.3   Parameterized Results on Scaffold Filling

In this subsection, we briefly review the FPT results on scaffold filling. (Readers are referred to [12,13,25] for standard FPT concepts and definitions.) In fact, as scaffold filling is a maximization problem, any FPT algorithm parameterized on the solution size is only of theoretical meaning.

In [5], Bulteau *et al.* showed that SF-MNSA is FPT, and the running times are $O^*(2^{O(k)})$ for the one-sided case and $O^*(2^{O(k \log k)})$ for the two-sided case respectively. The technique is based on color-coding [3] and subset enumeration.

In [18], it was shown that One-sided-SF-max(d) is FPT and the running time is $O^*((2d)^{O(k)})$ and when d is unbounded whether it is FPT is still open. On the other hand, it was shown that One-sided-SF-max($\subset$), parameterized by the number of genes inserted, is W[1]-hard. The reduction is from the Partial Vertex Cover (PVC) problem (via the standard FPT-reduction from Independent Set to PVC) [14]. Again, we list the corresponding results in following Table 3.

4   Future Research Directions

In this paper, we have thoroughly reviewed the current research on scaffold filling problems. While many theoretically interesting results have been obtained, a lot

Table 3. FPT results/status on SF-MNSA, One-sided-SF-max and One-sided-SF-max($\subset$).

Problem	Parameter k	FPT status
SF-MNSA	adjacency number	FPT [5]
One-sided-SF-max	adjacency number	open [18]
One-sided-SF-max($\subset$)	number of genes inserted	W[1]-hard [18]

still need to be done to make practical impact on the problem. We list some of these problems as follows.

1. For One-sided-SF-max, is it possible to design an approximation algorithm with a factor less 2? less than 1.5?
2. For One-sided-SF-max, is it possible to design an FPT algorithm parameterized by the solution size? As a positive answer will not solve the practical problem, how about using the number of breakpoints as a parameter? Is the problem FPT then?
3. For One-sided-SF-max($\subset$), is it possible to design an FPT approximation with a factor less than 2?

Acknowledgments. I would like to thank my collaborators for this series of research: Chenglin Fan, Haitao Jiang, Nan Liu, David Sankoff, Boting Yang, Chunfang Zheng, Farong Zhong, Daming Zhu and Peng Zou.

References

1. http://wgs-assembler.sourceforge.net/
2. Angibaud, S., Fertin, G., Rusu, I., Thevenin, A., Vialette, S.: On the approximability of comparing genomes with duplicates. J. Graph Algorithms Appl. **13**(1), 19–53 (2009)
3. Alon, N., Yuster, R., Zwick, U.: Color-coding. J. ACM **42**(4), 844–856 (1995)
4. Blin, G., Fertin, G., Sikora, F., Vialette, S.: The exemplar breakpoint distance for non-trivial genomes cannot be approximated. In: Das, S., Uehara, R. (eds.) WALCOM 2009. LNCS, vol. 5431, pp. 357–368. Springer, Heidelberg (2009)
5. Bulteau, L., Carrieri, A.P., Dondi, R.: Fixed-parameter algorithms for scaffold filling. Theoret. Comput. Sci. **568**, 72–83 (2015)
6. Chain, P.S., Grafham, D.V., Fulton, R.S., et al.: Genome project standards in a new era of sequencing. Science **326**, 236–237 (2009)
7. Chen, Z., Fu, B., Zhu, B.: The approximability of the exemplar breakpoint distance problem. In: Cheng, S.-W., Poon, C.K. (eds.) AAIM 2006. LNCS, vol. 4041, pp. 291–302. Springer, Heidelberg (2006)
8. Chen, Z., Fu, B., Xu, J., Yang, B., Zhao, Z., Zhu, B.: Non-breaking similarity of genomes with gene repetitions. In: Ma, B., Zhang, K. (eds.) CPM 2007. LNCS, vol. 4580, pp. 119–130. Springer, Heidelberg (2007)
9. Chen, Z., Fu, B., Fowler, R., Zhu, B.: On the inapproximability of the exemplar conserved interval distance problem of genomes. J. Comb. Optim. **15**(2), 201–221 (2008)
10. Chen, Z., Fu, B., Goebel, R., Lin, G., Tong, W., Xu, J., Yang, B., Zhao, Z., Zhu, B.: On the approximability of the exemplar adjacency number problem of genomes with gene repetitions. Theoret. Comput. Sci. **550**, 59–65 (2014)
11. Cormode, G., Muthukrishnan, S.: The string edit distance matching problem with moves. In: Proceedings of the 13th ACM-SIAM Symposium on Discrete Algorithms (SODA 2002), pp. 667–676 (2002)
12. Downey, R., Fellows, M.: Parameterized Complexity. Springer, New York (1999)
13. Flum, J., Grohe, M.: Parameterized Complexity Theory. Springer, Heidelberg (2006)

14. Guo, J., Niedermeier, R., Wernicke, S.: Parameterized complexity of vertex cover variants. Theor. Comput. Syst. **41**(3), 501–520 (2007)
15. Jiang, H., Zhong, F., Zhu, B.: Filling scaffolds with gene repetitions: maximizing the number of adjacencies. In: Giancarlo, R., Manzini, G. (eds.) CPM 2011. LNCS, vol. 6661, pp. 55–64. Springer, Heidelberg (2011)
16. Jiang, H., Zheng, C., Sankoff, D., Zhu, B.: Scaffold filling under the breakpoint, related distances. IEEE/ACM Trans. Comput. Biol. Bioinform. **9**(4), 1220–1229 (2012)
17. Jiang, H., Ma, J., Luan, J., Zhu, D.: Approximation and nonapproximability for the one-sided scaffold filling problem. In: Xu, D., Du, D., Du, D. (eds.) COCOON 2015. LNCS, vol. 9198, pp. 251–263. Springer, Heidelberg (2015)
18. Jiang, H., Fan, C., Yang, B., Zhong, F., Zhu, D., Zhu, B.: Genomic scaffold filling revisited. In: Proceedings of the 27th Annual Symposium on Combinatorial Pattern Matching (CPM 2016), Tel Aviv, Israel, 27–29 June 2016 (2016)
19. Jiang, M.: The zero exemplar distance problem. In: Tannier, E. (ed.) RECOMB-CG 2010. LNCS, vol. 6398, pp. 74–82. Springer, Heidelberg (2010)
20. Khanna, S., Motwani, R., Sudan, M., Vazirani, U.: On syntactic versus computational views of approximability. SIAM J. Comput. **28**(1), 164–191 (1998)
21. Liu, N., Jiang, H., Zhu, D., Zhu, B.: An improved approximation algorithm for scaffold filling to maximize the common adjacencies. IEEE/ACM Trans. Comput. Biol. Bioinform. **10**(4), 905–913 (2013)
22. Liu, N., Zhu, D., Jiang, H., Zhu, B.: A 1.5-approximation algorithm for two-sided scaffold filling. Algorithmica **74**(1), 91–116 (2016)
23. Liu, N., Zou, P., Zhu, B.: A polynomial time solution for permutation scaffold filling. Technical report, Department of Computer Science, Montana State University (2016)
24. Muñoz, A., Zheng, C., Zhu, Q., Albert, V., Rounsley, S., Sankoff, D.: Scaffold filling, contig fusion and gene order comparison. BMC Bioinform. **11**, 304 (2010)
25. Niedermeier, R.: Invitation to Fixed-Parameter Algorithms. Oxford University Press, New York (2006)
26. Yancopoulos, S., Attie, O., Friedberg, R.: Efficient sorting of genomic permutations by translocation, inversion and block interchange. Bioinformatics **21**, 3340–3346 (2005)
27. Zhu, B.: A retrospective on genomic preprocessing for comparative genomics. In: Chauve, C., et al. (eds.) Models and Algorithms for Genome Evolution, pp. 183–206. Springer, London (2013)

Better Approximation Algorithms
for Scaffolding Problems

Zhi-Zhong Chen[1($\boxtimes$)], Youta Harada[1], Eita Machida[1], Fei Guo[2],
and Lusheng Wang[3]

[1] Division of Information System Design, Tokyo Denki University, Saitama,
Hatoyama 350-0394, Japan
zzchen@mail.dendai.ac.jp
[2] School of Computer Science and Technology, Tianjin University, Tianjin, China
guofeieileen@163.com
[3] Department of Computer Science, City University of Hong Kong,
Tat Chee Avenue, Kowloon, Hong Kong SAR
lwang@cs.cityu.edu.hk

Abstract. Scaffolding is one of the main stages in genome assembly. During this stage, we want to merge contigs assembled from the paired-end reads into bigger chains called *scaffolds*. For this purpose, the following graph-theoretical problem has been proposed: Given an edge-weighted complete graph G and a perfect matching D of G, we wish to find a Hamiltonian path P in G such that all edges of D appear in P and the total weight of edges in P but not in D is maximized. This problem is NP-hard and the previously best polynomial-time approximation algorithm for it achieves a ratio of $\frac{1}{2}$. In this paper, we design a new polynomial-time approximation algorithm achieving a ratio of $\frac{5-5\epsilon}{9-8\epsilon}$ for any constant $0 < \epsilon < 1$. Several generalizations of the problem have also been introduced in the literature and we present polynomial-time approximation algorithms for them that achieve better approximation ratios than the previous bests. In particular, one of the algorithms answers an open question.

Keywords: Approximation algorithms · Randomized algorithms · Scaffolding · Matchings

1 Introduction

Sequencing the whole genome of an organism is a vital component for detailed molecular analysis of the organism, and genome projects are now underway or complete [6]. Unfortunately, with current genome-sequencing technologies, it is impossible to continuously read from one end of a long chromosome to the other. So, a commonly used method for sequencing a chromosome is to first randomly shear multiple copies of the chromosome into many small fragments of varying sizes, then accurately sequence the fragments to obtain *reads*, and further assemble the reads into a sequence of the whole chromosome. The assembling process

© Springer International Publishing Switzerland 2016
D. Zhu and S. Bereg (Eds.): FAW 2016, LNCS 9711, pp. 17–28, 2016.
DOI: 10.1007/978-3-319-39817-4_3

typically consists of two steps. The first step is called *contigging*, where we use confident overlaps between the reads to piece together larger segments of continuous sequences called *contigs* (each of which consists of two strands, namely, the *forward strand* and the *reverse strand*). The second step is called *scaffolding*, where we linking contigs together into *scaffolds* by using longer fragments of a known length whose ends are sequenced (called *paired-end reads*). A recent comprehensive evaluation of available software tools shows that scaffolding is still computationally intractable [4].

The scaffolding problem can be formulated as the problem of finding a special Hamiltonian path in an edge-weighted complete graph G as follows [5]. For each contig c, G has two vertices f_c and r_c, where f_c corresponds to the forward strand of c while r_c corresponds to the reverse strand of c. The edge $\{f_c, r_c\}$ is assigned a weight of 0 in G and is called a *dummy edge* for convenience. Each non-dummy edge $\{u, v\}$ of G corresponds to a bundle of paired-end reads each of which connects strand u (of a contig c) with strand v (of another contig c'), and the weight of $\{u, v\}$ in G is equal to the size of the corresponding bundle. In case no such bundle exists for $\{u, v\}$, then the weight of $\{u, v\}$ in G is 0. Note that the set D of dummy edges is a perfect matching of G. A Hamiltonian path P in G is *D-valid* if P contains all dummy edges of G. Given G and D, the objective is to compute a D-valid Hamiltonian path in G such that the total weight of edges in P is maximized over all D-valid Hamiltonian paths in G.

The scaffolding problem is NP-hard [1,2]. Indeed, the problem is APX-hard because the maximum asymmetric traveling salesman problem is APX-hard [7] and can be reduced to the scaffolding problem as follows. Given an edge-weighted digraph G, we construct an (undirected) graph H from G by splitting each vertex u of G into two vertices u_{in} and u_{out} so that (1) $\{u_{in}, u_{out}\}$ is an edge (of weight 0) in both H and D and (2) each arc (u, v) in G is transformed into an edge $\{u_{out}, v_{in}\}$ (of the same weight as (u, v)) in H. Mandric and Zelikovsky [5] propose two heuristics for the scaffolding problem. One of them is based on maximum-weight matching and the other is based on the greedy method. Although their heuristics perform well in practice, the heuristics are not shown to have a worst-case performance guarantee.

In order to take the desired structure of the genome (namely, the number of circular or linear chromosomes) into consideration, Chateau and Giroudeau [1,2] generalizes the scaffolding problem as follows. In addition to G and D, we are also given two nonnegative integers σ_p and σ_c. Instead of a single D-valid Hamiltonian path in G, we want to find a collection of exactly σ_p paths and exactly σ_c cycles such that the paths and cycles are disjoint and contain all edges of D. For convenience, we refer to such a collection as a *D-valid (σ_p, σ_c)-cover* of G. Note that a D-valid Hamiltonian path in G is just a D-valid $(1, 0)$-cover of G. Moreover, G has a D-valid (σ_p, σ_c)-cover if and only if $\sigma_p + \sigma_c \geq 1$ and $|D| \geq \sigma_p + 2\sigma_c$ [2]. So, we can hereafter assume that the input $(G, D, \sigma_p, \sigma_c)$ always satisfies $\sigma_p + \sigma_c \geq 1$ and $|D| \geq \sigma_p + 2\sigma_c$. The new objective is to compute a D-valid (σ_p, σ_c)-cover C of G such that the total weight of edges in C is maximized over all D-valid (σ_p, σ_c)-covers of G. We call this generalization the *generalized scaffolding problem* (GSP for short).

In the special case of GSP where the input satisfies $|D| = \sigma_p + 2\sigma_c$, a D-valid (σ_p, σ_c)-cover of G is simply a collection of disjoint edges and cycles with 4 edges and hence can be found by computing a maximum-weight matching in a suitably constructed graph [2]. Moreover, in the special case where $(\sigma_p, \sigma_c) = (0, 1)$, a very simple $O(n^3)$-time approximation algorithm achieving a ratio of $\frac{1}{2}$ can be designed [1,2], where n is the number of vertices in the input graph. This algorithm is also applicable to the scaffolding problem, i.e., the special case of GSP where $(\sigma_p, \sigma_c) = (1, 0)$. Furthermore, in the special case where the input satisfies $|D| \geq 2(\sigma_p + 2\sigma_c)$, an $O(n^3)$-time approximation algorithm achieving a ratio of $\frac{1}{3}$ can be designed [2]. However, the approximability of the remaining case where $\sigma_p + 2\sigma_c < |D| < 2(\sigma_p + 2\sigma_c)$ was left as an open question in [2].

In this paper, we improve the algorithmic results in [1,2] and answer the above open question in [2]. More specifically, we first design a new $O(n^3)$-time approximation algorithm for the scaffolding problem that achieves a ratio of $\frac{5-5\epsilon}{9-8\epsilon}$ for any constant $0 < \epsilon < 1$. This is done by first designing a randomized algorithm and then derandomizing it. The randomized algorithm finds two D-valid Hamiltonian paths and outputs the better one between the two. The randomized algorithm is inspired by the algorithm in [3] for the maximum traveling salesman problem. We also show that our analysis is almost tight. We then design an $O(n^3)$-time approximation algorithm for GSP that *always* achieves a ratio of $\frac{1}{3}$. A simple crucial idea behind the algorithm is to first compute a maximum-weight matching M in the input graph G such that $M \cap D = \emptyset$ and $|M| = |D| - \sigma_p$. With a minor modification, the algorithm achieves a ratio of $\frac{1}{2}$ for the special case of GSP where $|D| \geq \sigma_p + 3\sigma_c$. With another minor modification, the algorithm achieves a ratio of $\min\left\{\frac{2}{5}, \frac{1+2\epsilon}{3}\right\}$ for the special case where $|D| \geq \sigma_p + (2 + \epsilon)\sigma_c$ for any constant $0 < \epsilon < 1$.

We also modify the approximation algorithm for the scaffold problem so that it works for two special cases of GSP. For the special case of GSP where the input satisfies $|D| \geq 9(\sigma_p + \sigma_c)$ (respectively, $|D| \geq 6(\sigma_p + \sigma_c)$), the modified algorithm runs in $O\left(\left(\sigma_c^2 + 1\right) n^3\right)$ time and achieves a ratio of $\frac{5-4\epsilon}{9}$ (respectively, $\frac{7-6\epsilon}{13}$) for any constant $0 < \epsilon < 1$.

Weller *et al.* [8] define a different generalization of the scaffolding problem as follows. The input is the same as to GSP but without the condition $|D| \geq \sigma_p + 2\sigma_c$, and the objective is to find a D-valid (σ_p', σ_c')-cover C of G with $\sigma_p' \leq \sigma_p$ and $\sigma_c' \leq \sigma_c$ such that the total weight of edges in C is maximized over all D-valid (σ_p'', σ_c'')-covers of G with $\sigma_p'' \leq \sigma_p$ and $\sigma_c'' \leq \sigma_c$. We call this generalization the *loosely generalized scaffolding problem* (LGSP for short). The previously best approximation algorithm for LGSP achieves a ratio of $\frac{1}{2}$ [8] and runs in $O(n^3)$ time. In this paper, we show that the approximation algorithm for the scaffold problem can be modified to approximate LGSP as well without altering the approximation ratio and the time complexity.

The remainder of this paper is organized as follows. Section 2 gives basic definitions that will be used in the remainder of the paper. Section 3 presents approximation algorithms for the scaffolding problem, Sect. 4 presents approximation algorithms for LGSP, and Sect. 5 presents approximation algorithms for GSP and its special cases. Due to lack of space, the proofs of all lemmas and theorems are omitted here and will be given in the journal version.

2 Basic Definitions

Throughout this paper, a graph means an undirected graph without parallel edges or self-loops. Let G be a graph. We denote the vertex set of G by $V(G)$, and denote the edge set of G by $E(G)$. For a subset U of $V(G)$, $G[U]$ denotes the graph obtained from G by removing the vertices in $V(G) - U$ (together with the edges incident to them). For a subset F of $E(G)$, $G - F$ denotes the graph obtained from G by removing the edges in F. The *length* of a cycle or path C is the number of edges in C and is denoted by $|C|$. A *k-cycle* is a cycle of length k, while a *k-path* is a path of length k. A *path component* of G is a connected component of G that is a path.

For a matching M of G, a path or cycle C of G is *M-alternating* if $E(C)$ can be partitioned into two matchings M_1 and M_2 such that $M_1 \subseteq M$ and $M_2 \cap M = \emptyset$. A *2-matching* in G is a subgraph H of G with $V(H) = V(G)$ in which the degree of each vertex is at most 2. For two nonnegative integers σ_p and σ_c, a (σ_p, σ_c)-*cover* of G is a 2-matching of G in which there are exactly σ_c cycles and exactly σ_p path components, while a (σ_p^-, σ_c^-)-*cover* of G is a (k_p, k_c)-cover of G with $k_p \leq \sigma_p$ and $k_c \leq \sigma_c$. For a perfect matching D in G, a (σ_p, σ_c)- or (σ_p^-, σ_c^-)-cover H of G is D-*valid* if $D \subseteq E(H)$. It is easy to see that G has a D-valid (σ_p, σ_c)-cover if and only if $|D| \geq \sigma_p + 2\sigma_c$ [1,2].

Suppose that each edge of G has a nonnegative weight. The *weight* of an $F \subseteq E(G)$ is the total weight of edges in F and is denoted by $w(F)$. The *weight* of a subgraph H of G is $w(E(H))$ and is denoted by $w(H)$. A *maximum-weight matching* in G is a matching in G whose weight is maximized over all matchings in G. A *maximum-weight D-valid (σ_p, σ_c)-cover* (respectively, (σ_p^-, σ_c^-)-*cover*) of G is a D-valid (σ_p, σ_c)-cover (respectively, (σ_p^-, σ_c^-)-cover) of G whose weight is maximized over all D-valid (σ_p, σ_c)-covers (respectively, (σ_p^-, σ_c^-)-cover) of G.

In the *generalized scaffolding problem* (GSP for short), we are given a quadruple $(G, D, \sigma_p, \sigma_c)$, where G is a complete graph whose edges have nonnegative weights, D is a perfect matching in G with $w(D) = 0$, and σ_p and σ_c are two nonnegative integers with $\sigma_p + \sigma_c \geq 1$ and $|D| \geq \sigma_p + 2\sigma_c$. The objective of GSP is to compute a maximum-weight D-valid (σ_p, σ_c)-cover of G. The input to the *loosely generalized scaffolding problem* (LGSP for short) is the same as to GSP but without the condition $|D| \geq \sigma_p + 2\sigma_c$, and the objective is to compute a maximum-weight D-valid (σ_p^-, σ_c^-)-cover of G. The *scaffolding problem* is the special case of GSP and LGSP where $(\sigma_p, \sigma_c) = (1, 0)$.

For a random variable X, $\mathcal{E}[X]$ denotes the expected value of X.

3   Algorithms for the Scaffolding Problem

Throughout this section, let (G, D) be an instance of the scaffolding problem, Opt be a maximum-weight D-valid Hamiltonian path in G, and $n = |V(G)|$. Since D is a perfect matching of G, $n = 2|D|$ vertices. We may assume that $|D| \geq 3$, because otherwise the problem is trivially solved in $O(1)$ time.

We first design a randomized approximation algorithm achieving an expected ratio of $\frac{5-5\epsilon}{9-8\epsilon}$ for any constant $0 < \epsilon < 1$. In other words, the randomized algorithm finds a D-valid Hamiltonian path P in G such that $\mathcal{E}[w(P)] \geq \frac{5-5\epsilon}{9-8\epsilon} \cdot w(Opt)$. We then derandomize the algorithm.

3.1   The Randomized Algorithm

The algorithm starts by computing a maximum-weight matching M in $G - D$. It holds that $w(M) \geq w(Opt)$, because $E(Opt) \setminus D$ is a matching in $G - D$. Moreover, the graph $H = (V(G), D \cup M)$ is a collection of disjoint cycles and paths. If H is a single path, then it is a maximum-weight D-valid Hamiltonian path in G and hence we are done by outputting H. Similarly, if H is a single cycle, then we are done by outputting the path P obtained from H by deleting one edge in M whose weight is minimized among the edges in M. Since $|M| \geq 3$, $w(P) \geq \frac{2}{3}w(M) \geq \frac{2}{3}w(Opt)$. So, we hereafter assume that H is neither a single path nor a single cycle. Then, H has at least two connected components.

Let Opt_{int} denote the set of all edges $\{u, v\}$ of Opt such that some cycle C in H contains both u and v. Let Opt_{ext} denote the set of edges in Opt but not in Opt_{int}. Let $\beta = w(Opt_{\text{int}})/w(Opt)$.

Our algorithm then computes two D-valid Hamiltonian paths P_1 and P_2 of G, outputs the heavier one between them, and stops. P_1 is computed by modifying H as follows. Fix a parameter $0 < \epsilon < 1$. A cycle C of H is *long* if $|E(C) \cap M| > \epsilon^{-1}$; otherwise, C is *short*. For each cycle C in H, if C is long, then we delete the minimum-weight edge in $E(C) \cap M$ from H; otherwise, we replace C by a maximum-weight D_C-valid Hamiltonian path of $G[V(C)]$, where $D_C = D \cap E(C)$. Then, H becomes a collection of disjoint paths and hence can be transformed into a D-valid Hamiltonian path P_1 of G by adding more edges.

Lemma 1. $w(P_1) \geq (1 - \epsilon) \cdot w(Opt_{\text{int}}) = (1 - \epsilon)\beta \cdot w(Opt)$.

When $w(Opt_{\text{ext}})$ is large, $w(Opt_{\text{int}})$ is small and $w(P_1)$ may be small, too. P_2 is aimed at the case where $w(Opt_{\text{ext}})$ is large. Its computation is given below.

3.2   Computation of P_2

To compute P_2, we first compute a maximum-weight matching M' in an auxiliary graph K, where $V(K) = V(G)$ and $E(K)$ consists of those $\{u, v\} \in E(G)$ such that u and v belong to different connected components of H. Since $D \subseteq E(H)$ and Opt is D-valid, Opt_{ext} is a matching in K and we have the next lemma:

Lemma 2. $w(M') \geq w(Opt_{\text{ext}})$.

In order to obtain P_2, we next use M' to modify H as follows.

1. For each connected component C of H with $|E(C) \cap M| = 1$, delete the edge in $E(C) \cap M$ from H with probability $\frac{1}{2}$. (*Comment:* C is a 3-path.)
2. For each connected component C of H such that $|E(C) \cap M| \geq 2$, perform the following substeps:
   (a) Partition $E(C) \cap M$ into two nonempty subsets $M_{C,1}$ and $M_{C,2}$.
   (b) For a random $i \in \{1, 2\}$, delete the edges in $M_{C,i}$ from H. (*Comment:* Each connected component of H is now a D-alternating path.)
3. For each $e \in M'$ such that both endpoints of e are of degree at most 1 in H, add e to H. (*Comment:* After this step, each connected component of H is a D-alternating path or cycle.)
4. For each cycle C in H, select one edge $e \in E(C) \cap M'$ uniformly at random, and delete e from H. (*Comment:* After this step, each connected component of H is a D-alternating path.)
5. Connect the connected components of H into a D-valid Hamiltonian path P_2 of G by adding edges.

Lemma 3. *For each $e \in M$, the probability that e remains in P_2 is $\frac{1}{2}$. Moreover, for each $e \in M'$, the probability that e remains in P_2 is at least $\frac{1}{8}$.*

3.3 Analysis of the Algorithm

Our algorithm is clearly correct. As for its approximation ratio, we can use Lemma 3 to show the next bound on $\mathcal{E}[w(P_2)]$ and in turn the next theorem.

$$\mathcal{E}[w(P_2)] \geq \frac{1}{2}w(M) + \frac{1}{8}w(Opt_{\text{ext}}) \geq \left(\frac{1}{2} + \frac{1}{8}(1 - \beta)\right) \cdot w(Opt), \qquad (1)$$

Theorem 1. *For any fixed $0 < \epsilon < 1$, there is a randomized approximation algorithm for the scaffolding problem that runs in $O(n^3)$ time and achieves an expected approximation ratio of $\frac{5-5\epsilon}{9-8\epsilon}$.*

We can construct a graph G to show that our above analysis is almost tight.

3.4 Derandomization

The above randomized algorithm makes random choices only in Steps 1, 2b, and 4. To derandomize Step 4, we just modify it as follows:

5. For each cycle C in H, delete one edge $e \in E(C) \cap M'$ from H such that $w(e)$ is minimized over all edges in $E(C) \cap M'$.

When processing each C in Step 1, we need one random bit. Similarly, when processing each C in Step 2b, we need one random bit. So, Steps 1 and 2 need r random bits in total, where r is the number of connected components in $(V(G), D \cup M)$. In the above analysis of the randomized algorithm, only the proof of Lemma 3 is based on the mutual independence between these random bits. Indeed, by inspecting the proof, we can easily see that the proof is still valid even if the random bits are only pairwise independent. So, we can derandomize it via conventional approaches. Therefore, we have the following theorem:

Theorem 2. *There is an approximation algorithm for the scaffolding problem that runs in $O(n^3)$ time and achieves a ratio of $\frac{5-5\epsilon}{9-8\epsilon}$.*

4   An Approximation Algorithm for LGSP

Throughout this section, let $(G, D, \sigma_p, \sigma_c)$ be an instance of LGSP, Opt be a maximum-weight D-valid (σ_p^-, σ_c^-)-cover of G, and $n = |V(G)|$. Our goal is to modify the algorithm in Sect. 3 (without altering its approximation ratio and time complexity) so that it becomes an approximation algorithm for LGSP.

Fix a parameter $\epsilon > 0$. We proceed almost in the same way as in Sect. 3. More specifically, we define Opt_{int}, Opt_{ext}, and β as before. However, we need to modify the construction of P_1 and P_2 so that they become D-valid (σ_p^-, σ_c^-)-covers of G with $w(P_1) \geq (1-\epsilon) \cdot w(Opt_{\text{int}})$ and $\mathcal{E}[w(P_2)] \geq \frac{1}{2}w(M) + \frac{1}{8}w(Opt_{\text{ext}})$. Indeed, if $\sigma_p \geq 1$, then we construct P_2 as before; otherwise, we first construct P_2 as before but further transform P_2 into a cycle by (adding the edge of G between the endpoints of P_2). In either case, P_2 is clearly a D-valid (σ_p^-, σ_c^-)-cover of G with $\mathcal{E}[w(P_2)] \geq \frac{1}{2}w(M) + \frac{1}{8}w(Opt_{\text{ext}})$.

To construct P_1, we start with the graph $H = (V(G), D \cup M)$ and then modify it into a D-valid (σ_p^-, σ_c^-)-cover of G as follows.

1. Let s be the number of short cycles in H. If H has neither long cycles nor path components, let $\chi = 0$; otherwise, let $\chi = 1$.
2. For each long cycle C in H, delete the minimum-weight edge in $E(C) \cap M$ from H.
3. If H has two or more path components, connect them into a single path component by adding edges.
4. If $s < \sigma_c$ and $\chi = 1$, then transform the unique path component of H into a cycle (by adding the edge of G between the endpoints of the path).
5. For each short cycle C in H, compute a maximum-weight D_C-valid Hamiltonian path P_C of $G[V(C)]$, where $D_C = D \cap E(C)$.

Obviously, if either $s < \sigma_c$, or $s = \sigma_c$ and $\chi \leq \sigma_p$, then after Step 4, H is a D-valid (σ_p^-, σ_c^-)-cover of G with $w(H) \geq (1 - \epsilon) \cdot w(M) \geq (1 - \epsilon) \cdot w(Opt)$. So, we hereafter assume that either $s > \sigma_c$, or $s = \sigma_c$ and $0 = \sigma_p < \chi = 1$.

Let $C_1, \ldots, C_s$ be the short cycles in H. We may assume that $w(C_i) - w(P_{C_i}) \geq w(C_j) - w(P_{C_j})$ for all $1 \leq i < j \leq s$. For each $i \in \{1, \ldots, s\}$, let $V_i = V(C_i)$, $E_i = E(G[V_i])$, and $Opt_{\text{int},i} = Opt_{\text{int}} \cap E_i$. A simple crucial point is that if the graph $(V_i, Opt_{\text{int},i})$ has a (respectively, no) cycle, then C_i (respectively, P_{C_i}) is at least as good as $Opt_{\text{int},i}$ in the following sense:

- $w(C_i) \geq w(Opt_{\text{int},i})$ (respectively, $w(P_{C_i}) \geq w(Opt_{\text{int},i})$).
- Like the graph $(V_i, Opt_{\text{int},i})$, C_i (respectively, P_{C_i}) can be transformed into a D-valid (σ_p^-, σ_c^-)-cover of G by adding edges of G (not contained in E_i).

Now, we finish constructing P_1 by further modifying H as follows:

6. If $\sigma_p \geq 1$, then replace C_i by P_{C_i} for all i with $\sigma_c < i \leq s$, and further connect the path components of H into a single path by adding edges of G.

7. If $\sigma_p = 0$, then replace C_i by P_{C_i} for all i with $\sigma_c \le i \le s$, connect the path components of H into a path by adding edges of G, and transform the path into a cycle by adding the edge of G between the endpoints of the path.

Lemma 4. $w(P_1) \ge (1 - \epsilon) \cdot w(Opt_{int})$.

Theorem 3. *Let ϵ be a constant with $0 < \epsilon < 1$. There is an approximation algorithm for* LGSP *that runs in $O\left(n^3\right)$ time and achieves a ratio of $\frac{5-5\epsilon}{9-8\epsilon}$.*

5   Approximation Algorithms for GSP

Throughout this section, let $(G, D, \sigma_p, \sigma_c)$ be an instance of GSP, Opt be a maximum-weight D-valid (σ_p, σ_c)-cover of G, and $n = |V(G)|$. Since D is a perfect matching of G, $n = 2|D|$. If $\sigma_p + \sigma_c \le 1$, then Theorem 3 shows that we can find a D-valid (σ_p, σ_c)-cover $\mathcal{C}$ of G in $O(n^3)$ time such that $w(\mathcal{C}) \ge \frac{5-5\epsilon}{9-8\epsilon} \cdot w(Opt)$ for any constant $0 < \epsilon < 1$. So, we hereafter assume that $\sigma_p + \sigma_c \ge 2$.

In the sequel, we first design an $O(n^3)$-time approximation algorithm for GSP that achieves a ratio of $\frac{1}{3}$. By refining the algorithm, we then obtain $O(n^3)$-time approximation algorithms that achieve ratios better than $\frac{1}{3}$ for two special cases. Moreover, by refining the algorithm for the scaffolding problem, we further obtain polynomial-time approximation algorithms that achieve even better ratios for two even more special cases.

For a 2-matching H in G, we use $\#_c(H)$ (respectively, $\#_p(H)$) to denote the number of cycles (respectively, path components) in H. The number of connected components of H equals $\#_c(H) + \#_p(H)$.

Lemma 5. *Assume that $|D| \ge \sigma_p + k\sigma_c$ for some integer $k \ge 2$. Further suppose that K is a 2-matching of G such that $D \subseteq E(K)$, $\#_c(K) + \#_p(K) \ge \sigma_c + \sigma_p$, $\#_c(K) \le \sigma_c$, each path component P of K satisfies $|E(P) \cap M| < k$, and each cycle C in K satisfies $|E(C) \cap M| \le k$. Then, we can transform K into a D-valid (σ_p, σ_c)-cover $\mathcal{C}$ of G with $w(\mathcal{C}) \ge w(K)$.*

5.1   A Ratio-3 Approximation Algorithm

Our algorithm starts by computing a matching M in $G - D$ such that $|M| = |D| - \sigma_p$ and $w(M) \ge w(N)$ for all matchings N in $G - D$ with $|N| = |D| - \sigma_p$. M can be computed in $O(n^3)$ time as follows. First, we construct an auxiliary graph G' from $G - D$ by adding $2\sigma_p$ new vertices and connecting each of them to each original vertex by a new edge of weight $w(G) + 1$. Note that each maximum-weight matching in G' must contain $2\sigma_p$ edges incident to the new vertices. So, we compute a maximum-weight matching N in G' in $O(n^3)$ time. By deleting the edges incident to the new vertices from N, we obtain M.

Since $E(Opt) \setminus D$ is a matching in G and $|E(Opt) \setminus D| = |D| - \sigma_p$, $w(M) \ge w(Opt)$. Moreover, $H = (V(G), D \cup M)$ is a collection of disjoint cycles and paths. Indeed, exactly σ_p connected components of H are paths. If $\#_c(H) = \sigma_c$, then we are done. Moreover, if $\#_c(H) > \sigma_c$, then we can obtain a D-valid (σ_p, σ_c)-cover $\mathcal{C}$ of G with $w(\mathcal{C}) \ge \frac{1}{2}w(M) \ge \frac{1}{2}w(Opt)$ by modifying H as follows.

1. Arbitrarily select $\#_c(H) - \sigma_c + 1$ cycles in H.
2. For each cycle C selected in Step 1, transform C into a path P_C by removing a minimum-weight edge in $E(C) \cap M$ from H.
3. Connect the paths P_C obtained in Step 2 into a cycle by adding edges of G.

So, we hereafter assume that $\#_c(H) < \sigma_c$. It remains to show that we can modify H into a 2-matching K such that $w(K) \geq \frac{1}{3}w(M)$ and K together with $k = 2$ satisfies the conditions in Lemma 5. To this end, we modify H in two stages. In the first stage, we modify each connected component C of H such that C is *either* a cycle containing an even number of edges of M *or* a path, as follows.

1. Partition $E(C) \cap M$ into two (possibly empty) subsets N_1 and N_2 such that for each $i \in \{1, 2\}$, each connected components of $C - N_i$ is of length 1 or 3.)
2. For an $i \in \{1, 2\}$ with $w(N_i) \leq w(N_{3-i})$, delete the edges in N_i from H.

In the second stage, we modify each cycle C in H such that $|E(C) \cap M|$ is odd, as follows.

1. Partition $E(C) \cap M$ into three (possibly empty) subsets N_1, N_2, and N_3 such that for each $i \in \{1, 2, 3\}$, each connected components of $C - N_i$ is of length 1 or 3.)
2. Choose an $i \in \{1, 2, 3\}$ such that $w(N_i) \leq w(N_j)$ for all $j \in \{1, 2, 3\}$. Delete the edges in N_i from H.

After the second stage, H is now a 2-matching K such that $w(K) \geq \frac{1}{3}w(M)$ and K together with $k = 2$ satisfies the conditions in Lemma 5. So, we have:

Theorem 4. *There is an approximation algorithm for* GSP *that runs in* $O(n^3)$ *time and achieves a ratio of* $\frac{1}{3}$*.*

5.2   The Special Case Where $|D| \geq \sigma_p + 3\sigma_c$

Throughout this subsection, we assume that $|D| \geq \sigma_p + 3\sigma_c$.

We want to show that by modifying the algorithm in Sect. 5.1, we can achieve a ratio of $\frac{1}{2}$. The idea is to modify the second stage by keeping all 6-cycles in H intact and transforming each $(4\ell + 2)$-cycle C in H with $\ell \geq 2$ into a collection $\mathcal{C}_C$ of paths and cycles with $w(\mathcal{C}_C) \geq \frac{\ell+1}{2\ell+1}w(C)$ as follows.

1. Select an edge $e \in E(C) \cap M$ such that $w(e) \geq w(e')$ for all $e' \in E(C) \cap M$.
2. Partition $N = E(C) \setminus \{e\}$ into two subsets N_1 and N_2 such that for each $i \in \{1, 2\}$, one of the connected components of $C - N_i$ is a 5-path and the others are 3-paths.)
3. For an $i \in \{1, 2\}$ with $w(N_i) \leq w(N_{3-i})$, delete the edges in N_i from H.

Let K be the graph obtained from $H = (V(G), D \cup M)$ by the above modified algorithm. Clearly, K is a 2-matching in G, $w(K) \geq \frac{1}{2}w(M)$, and K together with $k = 3$ satisfies the conditions in Lemma 5. So, we finally have:

Theorem 5. *For the special case of* GSP *where the input* $(G, D, \sigma_p, \sigma_c)$ *satisfies* $|D| \geq \sigma_p + 3\sigma_c$*, there is an approximation algorithm that runs in* $O(n^3)$ *time and achieves a ratio of* $\frac{1}{2}$*.*

5.3 The Special Case Where $|D| \geq \sigma_p + (2 + \epsilon)\sigma_c$

Throughout this subsection, we assume that $|D| \geq \sigma_p + (2 + \epsilon)\sigma_c$ for some constant $0 < \epsilon < 1$. Since $|D| - \sigma_p - 2\sigma_c$ is an integer, $|D| \geq \sigma_p + 2\sigma_c + \lceil \epsilon \sigma_c \rceil$. Let $r = \min\left\{ \frac{2}{5}, \frac{1+2\epsilon}{3} \right\}$.

We want to show that by modifying the algorithm in Sect. 5.1, we can achieve a ratio of r. Let b_6 be the number of 6-cycles in $H = (V(G), M)$, and X be the set of the heaviest $\lceil \epsilon b_6 \rceil$ 6-cycles in H. The idea is to modify the second stage by keeping the cycles in X intact and further transforming each $(4\ell+2)$-cycle $C \notin X$ in H with $\ell \geq 1$ into a collection $\mathcal{C}_C$ of paths and cycles with $w(\mathcal{C}_C) \geq \frac{\ell}{2\ell+1} w(C)$ as follows.

1. Select an edge $e \in E(C) \cap M$ such that $w(e) \leq w(e')$ for all $e' \in E(C) \cap M$.
2. Partition $N = E(C) \setminus \{e\}$ into two subsets N_1 and N_2 such that for each $i \in \{1, 2\}$, one of the connected components of $C - (N_i \cup \{e\})$ is a 1-path and the others are 3-paths.)
3. Let i be an integer in $\{1, 2\}$ with $w(N_i) \leq w(N_{3-i})$. Delete the edges in $N_i \cup \{e\}$ from H.

Let K be the graph obtained from $H = (V(G), D \cup M)$ by the above modified algorithm. Clearly, K is a 2-matching in G, $D \subseteq E(K)$, $\#_c(K) + \#_p(K) \geq \sigma_c + \sigma_p$, $\#_c(K) \leq \sigma_c$, each path component P of K satisfies $|E(P) \cap M| \leq 1$, exactly $\lceil \epsilon b_6 \rceil$ cycles in K are 6-cycles, and the other cycles in K are 4-cycles. Since at least a fraction $\frac{1+2\epsilon}{3}$ of the total weight of 6-cycles in the graph $(V(G), M)$ still remains in K, we have $w(K) \geq r \cdot w(M)$.

We claim that K can be transformed into a D-valid (σ_p, σ_c)-cover of G by adding edges. The proof is similar to that of Lemma 5. In more details, as long as $\#_c(K) < \sigma_c$ and K has a path component P with $|E(P) \cap M| \geq 1$, we repeatedly modify K by adding the edge of G between the endpoints of P so that P is transformed into a cycle. If now $\#_c(K) = \sigma_c$, then we can add $\#_p(K) - \sigma_p$ edges to K so that $\#_p(K) - \sigma_p + 1$ path components of K are connected into a single path. So, assume that $\#_c(K) < \sigma_c$. Then, each path component P of K satisfies $|E(P) \cap D| = 1$, exactly $\lceil \epsilon b_6 \rceil$ cycles in K are 6-cycles, and the other cycles in K are 4-cycles. Thus, $|D| = \#_p(K) + 3\lceil \epsilon b_6 \rceil + 2(\#_c(K) - \lceil \epsilon b_6 \rceil)$. Since $|D| \geq \sigma_p + 2\sigma_c + \lceil \epsilon \sigma_c \rceil$ and $b_6 < \sigma_c$, we now have $\#_p(K) - \sigma_p \geq 2\sigma_c + \lceil \epsilon \sigma_c \rceil - \lceil \epsilon b_6 \rceil - 2\#_c(K) \geq 2(\sigma_c - \#_c(K))$. So, as in the proof of Lemma 5, we can transform K into a D-valid (σ_p, σ_c)-cover of G by adding edges of G.

Theorem 6. *Let ϵ be a constant with $0 < \epsilon < 1$. For the special case of GSP where the input $(G, D, \sigma_p, \sigma_c)$ satisfies $|D| \geq \sigma_p + (2 + \epsilon)\sigma_c$, there is an approximation algorithm that runs in $O(n^3)$ time and achieves a ratio of $\min\left\{ \frac{2}{5}, \frac{1+2\epsilon}{3} \right\}$.*

5.4 The Special Case Where $|D| \geq 9(\sigma_p + \sigma_c)$

Throughout this subsection, we assume that $|D| \geq 9(\sigma_p + \sigma_c)$. Our goal is to modify the algorithm in Sect. 3 so that it becomes an approximation algorithm for the special case achieving a ratio of $\frac{5-4\epsilon}{9}$.

Fix a parameter $\epsilon > 0$, and let $h = \lceil \epsilon^{-1} \rceil$. We first preprocess M by performing the following steps for each connected component C of the graph $(V(G), D \cup M)$ with $|E(C) \cap M| > h$:

1. Partition $E(C) \cap M$ into h subsets $M_{C,1}, \ldots, M_{C,h}$ so that for each $i \in \{1, \ldots, h\}$, $C - M_{C,i}$ is a collection of paths each with at most h edges of M.
2. Choose an $i \in \{1, \ldots, h\}$ such that $w(M_{C,i}) \leq w(M_{C,j})$ for all $j \in \{1, \ldots, h\}$. Delete the edges in $M_{C,i}$ from M. (*Comment:* $w(M_{C,i}) \leq \epsilon w(C)$.)

Now, the graph $H = (V(G), D \cup M)$ is a collection of disjoint cycles and paths each containing at most h edges in M. Moreover, during the preprocessing, $w(M)$ is decreased by at most a fraction ϵ of its original value. So, after the processing, $w(M) \geq (1 - \epsilon) \cdot w(Opt)$.

We then proceed as in Sect. 3. In more details, we define Opt_{int}, Opt_{ext}, and β as before. However, we need to modify the construction of P_1 and P_2 so that they become D-valid (σ_p, σ_c)-covers of G with $w(P_1) \geq w(Opt_{int})$ and $\mathcal{E}[w(P_2)] \geq \frac{1}{2}w(M) + \frac{1}{8}w(Opt_{ext})$.

To modify the construction of P_2 in Sect. 3.2, we first replace Steps 2a and 5 in Sect. 3.2 by Steps 2a, 5 and 6 below, respectively.

2̆a. Partition $E(C) \cap M$ into two nonempty subsets $M_{C,1}$ and $M_{C,2}$ such that for each $i \in \{1, 2\}$, each connected component of $C - M_{C,i}$ contains at most two edges of M.
5̆. For each path component P of H with $|E(P) \cap M'| \geq 3$, do the following:
   (a) Partition $E(P) \cap M'$ into 3 nonempty subsets $M'_{P,1}$, $M'_{P,2}$, and $M'_{P,2}$ such that for each $i \in \{1, 2, 3\}$, each connected component of $P - M'_{P,i}$ contains at most two edges of M'.
   (b) For a random $i \in \{1, 2, 3\}$, delete the edges in $M'_{P,i}$ from H. (*Comment:* After this step, each connected component of H is a path Q with $|E(Q) \cap (M \cup M')| \leq 8$. So, $K = H$ and $k = 9$ satisfy the conditions in Lemma 5, if we replace M in the lemma by $M \cup M'$.)
6̆. Transform H into a D-valid (σ_p, σ_c)-cover P_2 of G as in the proof of Lemma 5.

Lemma 6. *For each $e \in M'$, the probability that e remains in P_2 is at least $\frac{1}{8}$.*

By Lemma 6, we have the following inequality:

$$\mathcal{E}[w(P_2)] \geq \frac{1}{2}w(M) + \frac{1}{8}w(Opt_{ext}) \geq \left(\frac{1}{2}(1 - \epsilon) + \frac{1}{8}(1 - \beta)\right) \cdot w(Opt), \quad (2)$$

To modify the construction of P_1 in Sect. 3.2, we use dynamic programming. The details are omitted due to lack of space. The resulting P_1 is a D-valid (σ_p, σ_c)-cover of G such that $w(P_1) \geq w(Opt_{int}) = \beta \cdot w(Opt)$.

Theorem 7. *Let ϵ be a constant with $0 < \epsilon < 1$. For the special case of GSP where the input $(G, D, \sigma_p, \sigma_c)$ satisfies $|D| \geq 9(\sigma_p + \sigma_c)$, there is an approximation algorithm that runs in $O\left((\sigma_c^2 + 1)n^3\right)$ time and achieves a ratio of $\frac{5-4\epsilon}{9}$.*

5.5   The Special Case Where $|D| \geq 6(\sigma_p + \sigma_c)$

Throughout this subsection, we assume that $|D| \geq 6(\sigma_p + \sigma_c)$. Suppose that we modify the algorithm in Sect. 5.4 by replacing Step 5̌ with the following step:

5̌. For each path component P of H with $|E(P) \cap M'| \geq 2$, do the following:
   (a) Partition $E(P) \cap M'$ into two nonempty subsets $M'_{P,1}$ and $M'_{P,2}$ such that for each $i \in \{1, 2\}$, each connected component of $P - M'_{P,i}$ contains at most one edge of M'.
   (b) For a random $i \in \{1, 2\}$, delete the edges in $M'_{P,i}$ from H. (*Comment:* After this step, each connected component of H is a path P with $|E(P) \cap (M \cup M')| \leq 5$. So, $K = H$ and $k = 6$ satisfy the conditions in Lemma 5, if we replace M in the lemma by $M \cup M'$.)

Lemma 7. *For each $e \in M'$, the probability that e remains in P_2 is at least $\frac{1}{12}$.*

By Lemma 7, we have the following inequality:

$$\mathcal{E}[w(P_2)] \geq \frac{1}{2}w(M) + \frac{1}{12}w(Opt_{\text{ext}}) \geq \left(\frac{1}{2}(1 - \epsilon) + \frac{1}{12}(1 - \beta)\right) \cdot w(Opt), \quad (3)$$

Theorem 8. *Let ϵ be a constant with $0 < \epsilon < 1$. For the special case of* GSP *where the input $(G, D, \sigma_p, \sigma_c)$ satisfies $|D| \geq 6(\sigma_p + \sigma_c)$, there is an approximation algorithm that runs in $O\left((\sigma_c^2 + 1)n^3\right)$ time and achieves a ratio of $\frac{7-6\epsilon}{13}$.*

References

1. Chateau, A., Giroudeau, R.: Complexity and polynomial-time approximation algorithms around the scaffolding problem. In: Dediu, A.-H., Martín-Vide, C., Truthe, B. (eds.) AlCoB 2014. LNCS, vol. 8542, pp. 47–58. Springer, Heidelberg (2014)
2. Chateau, A., Giroudeau, R.: A complexity and approximation framework for the maximization scaffolding problem. Theor. Comput. Sci. **595**, 92–106 (2015)
3. Hassin, R., Rubinstein, S.: An approximation algorithm for the maximum traveling salesman problem. Inf. Process. Lett. **67**, 125–130 (1998)
4. Hunt, M., Newbold, C., Berriman, M., Otto, T.D.: A comprehensive evaluation of assembly scaffolding tools. Genome Biol. **15**, R42 (2014)
5. Mandric, I., Zelikovsky, A.: ScaffMatch: scaffolding algorithm based on maximum weight matching. Bioinformatics **31**, 2632–2638 (2015)
6. Pagani, I., Liolios, K., Jansson, J., Chen, I.-M., Smirnova, T., Nosrat, B., Markowitz, V.M., Kyrpides, N.C.: The genomes on-line database (GOLD) v. 4: status of genomic and metagenomic projects and their associated metadata. Nucleic Acids Res. **40**, D571–D579 (2012)
7. Papadimitriou, C.H., Yannakakis, M.: The traveling salesman problem with distances one and two. Math. Oper. Res. **18**, 1–11 (1993)
8. Weller, M., Chateau, A., Giroudeau, R.: On the complexity of scaffolding problems: from cliques to sparse graphs. In: Lu, Z., et al. (eds.) COCOA 2015. LNCS, vol. 9486, pp. 409–423. Springer, Heidelberg (2015). doi:10.1007/978-3-319-26626-8_30

How to Block Blood Flow by Using Elastic Coil

Zihe Chen[✉], Danyang Chen, Xiangyu Wang, Jianping Xiang, Hui Meng,
and Jinhui Xu

State University of New York, Buffalo, USA
{zihechen,danyangc,xiangyuw,jxiang2,huimeng,jinhui}@buffalo.edu

Abstract. Endovascular coiling is a primary treatment for intra-cranial
aneurysm, which deploys a thin and detachable metal wire inside the
aneurysm so as to prevent its rupture. Emerging evidence from medical
research and clinical practice has suggested that the coil configuration
inside the aneurysm plays a vital role in properly treating aneurysm and
predicting its outcome. In this paper, we propose a novel virtual coiling technique, called *Ball Winding*, for generating a coil configuration
with ensured blocking ability. It can be used as an automatic tool for
virtually simulating coiling before its implantation and thus optimizes
such treatments. Our approach is based on integer linear programming
and computational geometry techniques, and takes into consideration
the packing density and coil distribution as the performance measure-
ments. Experimental results on both random and real aneurysm data
suggest that our proposed method yields near optimal solution. *abstract
environment.*

Keywords: Endovascular coiling · Optimization · Computational
geometry

1 Introduction

Intracranial aneurysms affect up to 5 % of the US population [13]. The rupture
of the intra-cranial aneurysms triggers the sub-arachnoid hemorrhage which can
lead to a high mortality rate [16]. The risk of rupture is increased by the velocities
and volume of the blood flowing into the aneurysm. In recent years, a treatment
called *endovascular coiling* (or simply *coiling*; see Fig. 1) becomes one of the
most important and common surgical methods to prevent the rupture [7,8,10].
This treatment is performed by deploying a very thin and detachable metal wire,
called coil, within the aneurysmal sac through the arteries. Once the aneurysmal
sac is occluded with coils, a blood clot is formed around the coil due to the
response of human body, thus blocking off the aneurysmal inflow and significantly
reducing the risk of rupture. In practice, the coils can be pre-structured with a
specific shape in order to achieve an optimized occlusion rate in aneurysm.

This work was supported in part by NIH through grant NIH 1R01NS091075 and
NSF through grant IIS-1422591.

D. Zhu and S. Bereg (Eds.): FAW 2016, LNCS 9711, pp. 29–40, 2016.
DOI: 10.1007/978-3-319-39817-4_4

Packing density is the ratio between the volume of the inserted coils and the volume of the aneurysm. Currently, it is the only coil deployment parameter which has been clinically correlated with aneurysmal occlusion [1]. However, many studies show that there are recurrence at high packing density and occlusion at low packing density [12], which suggests that there exist other factors (*e.g.*, coil distribution) affecting the coil treatment.

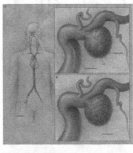

(a) Coils in aneurysms.

Several techniques have previously been developed for coil embolization inside the aneurysms [1,4,9,15]. Due to their heuristic nature, they all suffer from various types of limitations when applied in clinical practice. Roughly speaking, there are two kinds of limitations. (*a*) Techniques in [9,15] focus only on obtaining the desired packing density, but ignore other factors that may influence the occlusion. (*b*) Existing virtual coiling methods often neglect the physical properties of coil. For example, the technique in [9] deploys coil in a random-walk fashion and does not view coil as elastic-rod-like material, making it difficult to be applied in practice.

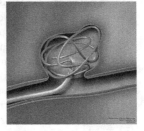

(b) Simulation of coil.

Fig. 1. Aneurysm and coil [2].

In this paper, we develop a new virtual method, called *Ball Winding*, to provide a rapid simulation tool for coiling. Combining with computational flow dynamics tool to simulate the blood flow, such a virtual tool could potentially provide critical information for doctors to optimize their coiling treatments. Our proposed method has the flexibility in controlling packing density, and can also provide accurate geometric features of the coils for its deployment. Our method considers three major issues critical to the performance of coiling: (*a*) Blocking ability, (*b*) Deployability, and (*c*) Scalability.

Our method first approximates the 3D shape of the aneurysm sac by one (or more) convex region(s). Then we carefully place a set of 3D blocking points inside the convex region(s). The blocking points are locations where the coil is expected to pass through. To avoid packing overly dense coil inside the aneurysm, our method minimizes the total number of blocking points by using an integer linear programming technique. After that, we generates a coil curve to pass through the neighborhood of each blocking point. To ensure that the resulting coil curve is scalable, we partition the space into layers, starting from the center to the boundary of the aneurysm, where each layer has a ball or ellipsoid shape. Then we deploy a smooth coil curve, in a ball winding fashion, from inner layers to outer layers with minimized torsion and bending energy.

2 r-blocking

In this section, we build an integer linear programming model to generate the blocking points inside the aneurysm sac. We view the aneurysm sac as a

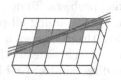

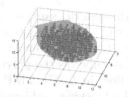

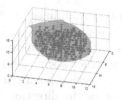

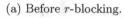

(a) Before r-blocking.   (b) After r-blocking.

Fig. 2. Select a (red) line from the lines intersecting the same set of grid cells. (Color figure online)

Fig. 3. Selecting blocking points by r-blocking.

polyhedron $\mathcal{P}$. To ensure the desired blocking ability, the blocking points are expected to be placed in such a way that any line l across the aneurysm intersects at least r balls centered at these blocking points and with radius δ for some constant r and δ, where line l represents a possible direction of the blood flow (note that since the blood flow could be reflected by the wall of the aneurysm, the direction of its velocity could be arbitrary), and the ball centered at a blocking point p represents the blocking region of p (which is the maximum region that a point of the coil can affect or slow down the blood flow). Thus, we have the following r-blocking problem.

Definition 1 (r-blocking). *Let $\mathcal{P}$ be a polyhedron in $\mathbb{R}^3$ with diameter D. A r-blocking of $\mathcal{P}$ is a set of points $\mathcal{B} = \{b_1, b_2, \cdots, b_m\}$ inside $\mathcal{P}$ with minimum cardinality such that for every line l intersecting with $\mathcal{P}$, there are at least rD_l/D points $\{b_{l_1}, b_{l_2}, \cdots, b_{l_{\lceil rD_l/D \rceil}}\} \subset \mathcal{B}$ satisfying the condition of*

$$d(b_{l_i}, l) \leq \delta, 1 \leq i \leq \lceil rD_l/D \rceil, \tag{1}$$

where D_l is the length of intersecting segment $l \cap \mathcal{P}$, $d(b_{l_i}, l)$ is the Euclidean distance between point b_{l_i} and line l, and δ is a predefined constant (i.e., radius of a blocking region of a coil point).

To solve the above r-blocking problem, we first investigate its hardness, and consider the following minimum line cover problem.

Definition 2 (Minimum Line Cover (MLP)). *Given a set L of lines, the minimum line cover problem is to find a minimum-cardinality set of points P such that there is a least one point in P on each line in L.*

Theorem 1. *The r-blocking problem is NP-hard.*

Proof. To prove this theorem, we first notice that the MLP problem is NP-hard [3,6]. Thus, our main idea is to reduce the MLP problem to our r-blocking problem. For this purpose, we first consider a restricted version of the r-blocking problem called direction-fixed r-blocking. It is obtained by first building a grid of edge length $2\delta/\sqrt{3}$ inside $\mathcal{P}$, and then finding a set of lines in the following way: for all lines passing through the same set of grid cells, randomly select one of them (see Fig. 2). This will give us a finite number of lines. Then, we reduce the MLP problem to this restricted version of r-blocking problem. To do this, we construct a sufficiently large cube (as the polyhedron $\mathcal{P}$) containing all possible intersection points of the lines in L. The MLP problem then becomes a special case of the direction-fixed r-blocking problem with $r = 1$. This means that the direction-fixed r-blocking problem is NP-hard. Since direction-fixed r-blocking problem is a restricted version of the r-blocking problem, we can easily reduce the restricted version to the unrestricted version. Thus, the r-blocking problem is also NP-hard.   □

After understanding the hardness of the problem, we now generate the set of blocking points inside the aneurysm sac $\mathcal{P}$. We first perform the following pre-processing.

1. Build a grid in $\mathcal{P}$ with grid points $\mathcal{G} = \{g_1, g_2, \cdots, g_n\}$ (see Fig. 3a). The distance between every neighboring pair of points is no more than 2δ.
2. Select a set of lines representing all possible directions of the blood flow in the following way: For all lines within δ distance to the same set of points, we pick exactly one of them as their representative (see Fig. 2). Let the resulting set of lines be $\mathcal{L} = \{l_1, l_2, \cdots, l_\lambda\}$.

Clearly, our goal is to find a set of blocking points $\mathcal{B} = \{g_1^B, g_2^B, \cdots, g_m^B\}$ with minimum cardinality from $\mathcal{G}$ such that each line l in $\mathcal{L}$ is within δ distance to at least $\lceil rD_l/D \rceil$ points in $\mathcal{B}$.

Next, we introduce an integer linear programming model for finding the desired set of blocking points $\mathcal{B}$.

For each point $g_i \in \mathcal{G}$, we introduce an indicator variable x_i. x_i equals 1 if $g_i \in \mathcal{B}$, and 0 otherwise. For the line set $\mathcal{L}$, we generate a coefficient matrix $A \in \mathbb{R}^{\lambda \times n}$ in the following way. The i-th row of A corresponds to line l_i and entry a_{ij} (*i.e.*, the entry in the i-th row and j-th column) encodes the distance information between line l_i and point $g_j \in \mathcal{G}$. a_{ij} equals 1 if the distance between l_i and g_j is smaller than (or equal to) δ, and 0 otherwise.

Integer Linear Programming

$$\min \sum_{i=1}^{n} x_i \tag{2}$$

$$s.t.\ x_i \in \{0, 1\}, \forall 1 \leq i \leq n \tag{3}$$

$$\sum_{i=1}^{n} a_{ji} x_i \geq \lceil rD_{l_j}/D \rceil, \forall 1 \leq j \leq \lambda \tag{4}$$

Theorem 2. *The optimal solution of the above integer linear programming is equivalent to the optimal solution of the r-blocking problem in aneurysm sac $\mathcal{P}$.*

Since solving an integer linear program is in general NP-hard [11,14], we first relax the above integer linear programming formulation to a **Linear Programming** by replacing the integrality constraint (Eq. (3)) with

$$0 \le x_i \le 1, \forall 1 \le i \le n. \tag{5}$$

Solving the above linear programming, we obtain a non-negative vector $X = (x_1, x_2, \cdots, x_n)^T$. Since X is not necessarily a 0/1 vector, to yield a feasible solution to the integer linear programming problem, we need to perform a rounding procedure on X. Our rounding algorithm is motivated by the idea that the r-blocking problem can be viewed as the union of r carefully selected 1-blocking problem (*i.e.*, the case of $r = 1$). Our approach is thus to first solve the 1-blocking problem by using a multi-level rounding technique. After $\log \lambda$ rounds, we obtain an integral solution to the 1-blocking problem. To yield a solution to the r-blocking problem, we then repeatedly solve the 1-blocking problem until a feasible solution to the r-blocking problem is achieved. At each iteration, we "cancel" the solution to the 1-blocking problem obtained in previous iterations so that the number of blocking points increases for each line in $\mathcal{L}$.

Below are the main steps of our rounding algorithm, where the vector multiplication of $U = (u_1, u_2, \cdots, u_n)$ and $V = (v_1, v_2, \cdots, v_n)$ is defined as follows

$$U \otimes V = (u_1 v_1, u_2 v_2, \cdots, u_n v_n). \tag{6}$$

Algorithm 1. Rounding Algorithm

1: Initialize an n index set $\{Ind_1, Ind_2, \cdots, Ind_n\}$ with each $Ind_i = 0$. Set $\bar{\lambda} = \lambda$.
2: Solve the linear programming of 1-blocking and output solution vector X.
3: Let A_i denote the i-th ($1 \le i \le n$) row of A and $A_i' = A_i \otimes X^T$. Select the top $\frac{\bar{\lambda}}{2}$ rows based on the value of $\max(A_i')$. For each selected row A_i, if $a_{ij} x_j = \max(A_i')$, set $Ind_j = 1$. Note that if there are more than one x_j satisfying $a_{ij} x_j = \max(A_i')$ in the i-th row, randomly set one index Ind_j to 1.
4: Add the following constraint to the linear programming: for $1 \le i \le n$, if $Ind_i = 1$, then set $x_i = 1$. Let $\bar{\lambda} = \frac{\bar{\lambda}}{2}$.
5: Repeat Step 2-4 until $\bar{\lambda} < 1$.
6: Let X be the solution vector after Step 5, and $Y = (y_1, y_2, \cdots, y_n)^T = AX$. Take out all such rows A_i of A satisfying $y_i < \lceil rD_{l_i}/D \rceil$ to form a new matrix $\bar{A}$.
7: Set $A = \bar{A}$ and let $\bar{\lambda}$ be the number of rows of A. For all i, if $Ind_i = 1$, set entries in i-column of A to 0.
8: Repeat Step 2-7 until every y_i is no less than $\lceil rD_{l_i}/D \rceil$.

The running time of the above rounding algorithm is $O(T(n, \lambda) r \log \lambda)$, where $T(n, \lambda)$ is the time needed for solving the linear programming problem. Experiments in Sect. 4 show that the above algorithm yields near optimal solution.

3   Ball Winding Approach for Blocking Points

Once obtaining the set $\mathcal{B}$ of blocking points, we need to compute a coil curve to pass through these points or their neighborhoods to achieve the desired blocking performance. In order to do this, we have to resolve two main issues, deployability and scalability. Deployability is to determine whether a computed coil curve configuration is able to be deployed inside the aneurysm sac, and scalability concerns whether we can uniformly deal with aneurysms with different sizes. For deployability, we consider two major factors: (1) whether the coil curve will encounter any deadlock (a drawback of some exist methods [9]) during its deployment process, and (2) whether the coil curve is stable or, in other words, whether its energy (including both torsion and bending energy) is minimized. To resolve these issues, we propose the following ball winding approach, motivated from the concept of winding a yarn into a ball.

We first compute a minimum enclosing ball $B(\mathcal{P})$ of $\mathcal{P}$, and evenly partition $\mathcal{P}$ into L layers using L-1 spheres concentric with $B(\mathcal{P})$. Then, starting from a blocking point s in the innermost layer, we compute a narrow slab D_s (*i.e.*, a region bounded by two parallel planes with width 2ϵ for some small $\epsilon \leq \delta$; in Sect. 3.1) to contain s and a maximum number of blocking points G_s inside the current or neighboring layer (for the purpose of minimizing the coil length).

We use the algorithm in Sect. 3.3 to find a coil curve passing through all points in G_s with minimum bending energy and almost 0 torsion energy. Therefore, the computed coil curve for G_s is stable (*i.e.*, major factor (2)).

After obtaining the coil curve for G_s, we then start from the last point s'' of the coil curve and repeat the above procedure until all points are finished (in a layer-by-layer fashion). During the winding process, the two consecutive slabs change their orientation. To avoid causing large torsion energy, we assume that the coil has been pre-shaped at the starting point s of each slab. Thus the whole coil curve will have minimized energy. Also since the coil curve is winded in a layer-by-layer fashion, it will not be blocked by existing coil curve and thus avoids the deadlock problem (*i.e.*, major factor (1)). This means that the resulting coil curve is deployable. Clearly, the winding procedure can handle aneurysms with different sizes in the same way, thus it is also scalable.

3.1   Finding a Narrow Slab

Let s be a starting point, and B_s be the set of points in the same or neighboring layers of s which have not yet been processed. To find the aforementioned slab D_s, we first notice that D_s has to satisfy the condition that D_s maximizes the set $G_s = \{p \in B_s | p$ is inside $D_s\}$ among all possible slabs with width 2ϵ and containing s on its center plane. Figure 4a gives an example of starting point s (green point) and its slab D_s.

Denote $\boldsymbol{n_s}$ as the normal vector of D_s. Since s lies in the center plane of D_s, D_s can be uniquely determined by s and $\boldsymbol{n_s}$. Therefore, finding slab D_s can be reduced to searching for a vector $\boldsymbol{n_s}$ which maximizes the cardinality of the following set (see Fig. 4b in which point p is in the set of (7) but q is not)

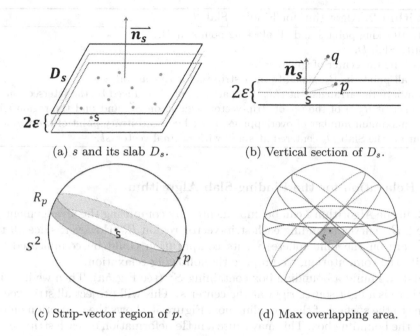

(a) s and its slab D_s.   (b) Vertical section of D_s.

(c) Strip-vector region of p.   (d) Max overlapping area.

Fig. 4. Finding normal vector to determine D_s.

$$\{p \in B_s : |\vec{ps} \cdot \boldsymbol{n}_s| \le \epsilon\}. \tag{7}$$

Dividing both sides of Eq. (7) by $\|\vec{ps}\|$ and since $\frac{\vec{ps}}{\|\vec{ps}\|}$ is a unit vector, our problem of finding $\boldsymbol{n}_s$ in the Euclidean Space can be converted to a problem in sphere S^2.

Definition 3 (Strip-Vector Region of p). *For any $p \in B_s$, a strip-vector region of p is defined as*

$$R_p = \{t \in S^2 : |\frac{\vec{ps}}{\|\vec{ps}\|} \cdot \vec{ts}| \le \frac{\epsilon}{\|\vec{ps}\|}\}. \tag{8}$$

The colored strip in Fig. 4c is R_p. Similar to vector $\boldsymbol{n}_s$, the vector $\vec{ts}$ formed by any point t in the strip-vector region of p also satisfies inequality (7). Thus, $\vec{ts}$ can be viewed as a candidate for $\boldsymbol{n}_s$. This implies the following lemma.

Lemma 1. *Finding a vector $\boldsymbol{n}_s$ that maximizes the cardinality of the set in (7) is equivalent to finding a vector $\vec{ts}$ such that t lies in the common intersection of a maximum number of R_p's for all $p \in B_s$.*

The colored area in Fig. 4d is the common intersection of a maximum number of strip-vector regions. The corresponding vector $\vec{st}$ of any point t in this area can be used as the desired slab's normal vector $\boldsymbol{n}_s$.

Algorithm 2. Algorithm for Finding-Slab D_s

Input: Blocking point s and all blocking points in B_s.
Output: Slab D_s.
1: Let s be the center of sphere S^2.
2: For all points in B_s, generate their strip-vector regions on S^2.
3: Compute the arrangement (*i.e.,* the partition of S^2 induced by the intersections of the set of R_p's) of the set of strip-vector regions on S^2, and find the region with the maximum number of overlappings. Let t be an arbitrary point of the region.
4: Output the Slab D_s centered at s and with normal vector $\overrightarrow{st}$.

3.2 Relaxation for the Finding-Slab Algorithm

Step 3 in the Algorithm 2 can be implemented by computing the arrangement [5] of the $2|B_s|$ curves bounding each strip-vector region R_p. However, since all the curves are on the unit sphere S^2, its computation could be complicated. To simplify the computation, we propose the following relaxation.

First, we build a bounding box containing S^2 (see Fig. 5a). Then we imagine that there is a light source right at the center s. This will project all strip-vector regions to the surface of the bounding box. Figure 5b shows the projection on one side of the bounding box. This may cause a little deformation on each strip-vector region, but the topological relationship of overlapping stays the same. Thus, our problem of finding overlapping areas on S^2 can be converted to the problem of finding overlapping areas on surface of the bounding box. Since each side of the bounding box is a 2D shape, the computation can be greatly simplified. Once the region with the maximum number of overlapping is determined on the 6 sides of the bounding box, we randomly select a point t' from the region, and project it back to S^2 to obtain the desired point t.

Algorithm 3. Algorithm for Coil Curve Generation

1: Project all the points in G_s to the center plane of D_s along the direction of the normal vector. Let G'_s be the projection of G_s and p' be the projection of $p \in G_s$.
2: Compute the circumcircle C of G'_s and let c denote the center of C. Use a set of concentric circles inside C to divide C into layers. Note that by adjusting the radius or the number of concentric circles, we can make s either in the innermost layer or in the outermost layer (see Fig. 6a).
3: Calculate all angles between $\overrightarrow{cp'}$ and $\overrightarrow{cq}$ ($\forall q \in G'_p$). In each layer of C, sort all points based on their angular order. If s is in the innermost layer, add points in the arranged order from inner layers to outer layers into a queue Q; otherwise, add points from outer layers to inner layers. Figure 6b shows an example of ordering all points in G_s where s is in the innermost layer.
4: Use a cubic spline to connect all points in G'_s in the order of Q as the coil curve (Fig. 6c).

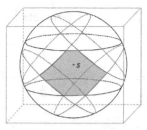

(a) Bounding box of S^2.   (b) Projection to one side.

Fig. 5. Projecting strip-vector regions onto the bounding box.

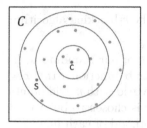

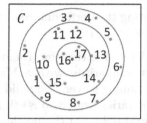

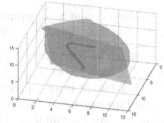

(a) Partition points into layers.   (b) Sort points by their angular order.   (c) A segment of coil curve for points in one slab.

Fig. 6. Steps to generate coil curve for G_s.

3.3 Coil Curve Generation

Once obtaining the set G_s of blocking points inside a slab D_s for a starting point s, we need to generate a coil curve to pass through the neighborhoods of all points in G_s. To ensure the generated coil curve has low bending energy, we use Algorithm 3.

4 Evaluations

To evaluate the performance of our proposed ball winding approach, we implement our algorithms on a Windows PC with a CPU i5-3570, 3.4 Ghz and 32 GB memory. We consider two types of data, one is randomly generated data set with ground truth and certain level of noise, and the other is in vivo aneurysm data.

4.1 Performance of the r-blocking Algorithm

We validate our r-blocking algorithm using two experiments. The first experiment is for comparing the performance of our rounding algorithm with a Matlab built-in binary integer programming function `bintprog` on real data set. The

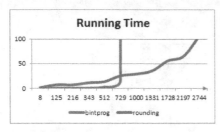
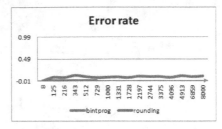

(a) Running time comparison.   (b) Error rate comparison.

Fig. 7. Rounding vs `bintprog`. (Color figure online)

second experiment is for evaluating the performance of the r-blocking algorithm on random data sets.

Rounding vs `bintprog`: To determine the performance of our rounding technique (in Sect. 2), we show its comparison with a Matlab function `bintprog` which is a binary integer programming solver. We build a grid with n grid points in a real aneurysm sac, where n varies from 8 to 8000. r is chosen to be around half of the average number of grid points in the neighborhood of each line.

Figure 7a shows that the running time of our rounding algorithm is a growing (slightly super-linear) function of the number (denoted as n) of grid points, which is consistent with its time complexity. The running time is acceptable even for large n. The running time for `bintprog` is short when n is small. However, when n is larger than 800, it takes more than 20000 seconds (≥ 5 h) without generating any result. Figure 7b shows the error rate comparison, where error rate is the percentage of the number of redundant blocking points on each line comparing to its optimal solution. Experiments suggest that the error rate of our rounding algorithm is around 0.08, which is near optimal.

Random Data: We first randomly generate n points in the space as ground truth (base points). Then randomly choose pairs of the base points to generate lines (*i.e.*, each line passes a pair of base points). Every base point is on at least two lines and each line intersects with r other lines. We also add k noise points and each noise point lies on at most one line. The expected point chosen rate for the r-blocking problem on the above data set is $\frac{n}{n+k}$.

To show the robustness of our method, we test it on different level of noise, where n varies from 10 to 1000 and r varies from 1 to 10. For each data size, we

Table 1. r-blocking on random data.

Noise level	10 %	20 %	30 %	40 %	50 %
Expect	0.8929	0.8000	0.6993	0.5988	0.5000
Output	0.8929	0.8133	0.7016	0.6168	0.5250

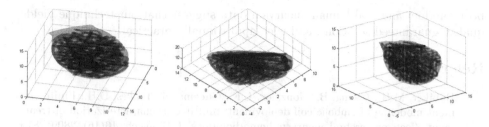

Fig. 8. Generated coil configurations for three aneurysm datasets.

Table 2. Performance comparison

Methods	Blocking ability	Deployable	Packing density	Coil distribution	Real time generation	Cause deadlocks
Ball winding	Guaranteed	Yes	Flexible	Evenly	Yes	No
DPP	No guarantee	No	Around 30 %	No consideration	Yes	Yes
SLB	No guarantee	Yes	Not mentioned	No consideration	Yes	Not mentioned

run 6 times and take the average. Table 1 shows the results, which also suggests that our algorithm yields near optimal solutions.

4.2 Performance of Coil Generation for Real Aneurysm Data

We also validate our ball winding approach using several in vivo aneurysm 3-D meshed images from patients. Figure 8 shows the results.

Comparison with Previous Work: We first note that no existing method is capable of generating coils that guarantee occlusion and ensure no recurrence. Also, since there is no ground truth for "optimal" coil, it is difficult to determine the coil related parameters. In order to validating the performance of our proposed approach, we compare our result with two existing results in several aspects (Table 2). One of the previous result is a dynamic path planning (DPP) approach [9] and the other one is based on serially-linked beams (SLB) [16].

5 Conclusion

In this paper, we propose an effective ball-winding method for generating coils inside aneurysm. Our technique is based on a number of interesting ideas, and has several unique advantages over existing techniques. Experimental results on both random and real human aneurysm data suggest that our technique yields quality guaranteed solutions, and is computationally practical.

References

1. Babiker, M.H., Chong, B., Gonzalez, L.F., Cheema, S., Frakes, D.H.: Finite element modeling of embolic coil deployment: multifactor characterization of treatment effects on cerebral aneurysm hemodynamics. J. Biomech. **46**(16), 2809–2816 (2013)
2. Brettler, S.: Endovascular coiling for cerebral aneurysms. AACN Adv. Crit. Care **16**(4), 515–525 (2005)
3. Cannon, S., Fai, T.G., Iwerks, J., Leopold, U., Schmidt, C.: Np-hardness of the minimum point and edge 2-transmitter cover problem (2014)
4. Damiano, R., Xiang, J., Levy, E., Meng, H.: A new virtual coiling method and its use in evaluation of combining coiling and flow-diversion treatment in a patient-specific aneurysm. In: ASME 2014 International Design Engineering Technical Conferences and Computers and Information in Engineering Conference, p. V003T12A017. American Society of Mechanical Engineers (2014)
5. De Berg, M., Schwarzkopf, O.C., Van Kreveld, M., Overmars, M.: Computational Geometry. Springer, Heidelberg (2008)
6. Dumitrescu, A., Jiang, M.: On the approximability of covering points by lines, related problems (2013). arXiv preprint arXiv: 1312.2549
7. Guglielmi, G.: History of the genesis of detachable coils: a review. J. Neurosurg. **111**(1), 1–8 (2009)
8. Molyneux, A.: International Subarachnoid Aneurysm Trial (ISAT) of neurosurgical clipping versus endovascular coiling in 2143 patients with ruptured intracranial aneurysms: a randomised trial. Lancet **360**(9342), 1267–1274 (2002)
9. Morales, H.G., Larrabide, I., Geers, A.J., San Roman, L., Blasco, J., Macho, J.M., Frangi, A.F.: A virtual coiling technique for image-based aneurysm models by dynamic path planning. IEEE Trans. Med. Imaging **32**(1), 119–129 (2013)
10. Muschenborn, A.D., Ortega, J.M., Szafron, J.M., Szafron, D.J., Maitland, D.J.: Porous media properties of reticulated shape memory polymer foams and mock embolic coils for aneurysm treatment. Biomed. Eng. Online **12**(1), 103 (2013)
11. Papadimitriou, C.H.: On the complexity of integer programming. J. ACM (JACM) **28**(4), 765–768 (1981)
12. Piotin, M., Spelle, L., Mounayer, C., Salles-Rezende, M.T., Giansante-Abud, D., Vanzin-Santos, R., Moret, J.: Intracranial aneurysms: treatment with bare platinum coilsaneurysm packing, complex coils, and angiographic recurrence 1. Radiology **243**(2), 500–508 (2007)
13. Rinkel, G.J., Djibuti, M., Algra, A., Van Gijn, J.: Prevalence and risk of rupture of intracranial aneurysms a systematic review. Stroke **29**(1), 251–256 (1998)
14. Schrijver, A.: Theory of Linear and Integer Programming. Wiley, Amsterdam (1998)
15. Sorteberg, A., Sorteberg, W., Aagaard, B.D., Rappe, A., Strother, C.M.: Hemodynamic versus hydrodynamic effects of Guglielmi detachable coils on intraaneurysmal pressure and flow at varying pulse rate and systemic pressure. Am. J. Neuroradiol. **25**(6), 1049–1057 (2004)
16. Wei, Y., Cotin, S., Dequidt, J., Duriez, C., Allard, J., Kerrien, E., et al.: A (near) real-time simulation method of aneurysm coil embolization. Aneurysm **8**(29), 223–248 (2012)

Linear Time Algorithm for 1-Center in $\Re^d$ Under Convex Polyhedral Distance Function

Sandip Das[1]([⊠]), Ayan Nandy[1], and Swami Sarvottamananda[2]

[1] Indian Statistical Institute, Kolkata 700 108, India
sandipdas@isical.ac.in
[2] Ramakrishna Mission Vivekananda University, Belur Math, Howrah, WB, India

Abstract. In this paper we present algorithms for computing 1-center of a set of points for convex polyhedral distance function in $\Re^d$ for any d. Given polyhedral P of size m, the running time of our algorithm for computing 1-center of n points in $\Re^2$ for convex polygonal distance function d_P is $O(nm \log^2 m)$. For $d > 2$, we present an $O(3^{3d^2} nm^2 \log^d m)$ algorithm to compute 1-center of n points in $\Re^d$ for convex polyhedral distance function d_P, $|P| = m$. Both the algorithms are linear time for fixed d and fixed polyhedron P.

1 Introduction

In 1857, Sylvester [11] posed *facility location problem* of finding optimal location of facility for n customers represented by points in the plane. Geometrically, this problem is equivalent to finding the smallest circle that encloses the given set of n points where optimal facility location is the center of this circle. This circle is called *minimum enclosing circle*, *minimum spanning circle* or *minimum $\mathcal{P}$-circle*. The problem is also known as *Euclidean 1-center problem* or *intersection radius problem*. For geometric objects other than points the facility location problem is called *minimum stabbing circle* in plane and *minimum stabbing ball* in $\Re^d$. This problem has been studied for various types of facilities, clients as well as various distance metrics. In 1983, Megiddo [8,9] settled the problem for Euclidean distance metric in $\Re^d$ by giving an optimal $O(n)$ algorithm.

Chew and Dyrsdale [4] introduced convex polyhedral distance function, d_P. Let C_1 and C_2 be two convex sets of points in $\Re^d$. Let P be a convex polyhedra of size m in $\Re^d$ with the origin o in its interior. The *convex polyhedral distance function* $d_P(C_1, C_2)$ from C_1 to C_2 is defined as follows

$$d_P(C_1, C_2) = \inf\{t \geq 0 \mid t \in \Re, \text{ there exists } p \in C_1, q \in C_2, \text{ such that } q \in (tP + p)\}$$

where $tP + p$ is the polyhedra translated to p and scaled by a factor $t \geq 0$. Chew and Dyrsdale [4] presented an asymptotically optimal algorithm for computing Voronoi diagrams based on d_P. They solved all-nearest neighbors, minimum spanning trees, largest empty convex shape, motion planning for a convex shape

© Springer International Publishing Switzerland 2016
D. Zhu and S. Bereg (Eds.): FAW 2016, LNCS 9711, pp. 41–52, 2016.
DOI: 10.1007/978-3-319-39817-4_5

using Voronoi Diagram for d_P. It seems their construction of Voronoi diagram for d_P may be used to compute farthest Voronoi diagram for d_P which may then be used implicitly to compute 1-center for convex distance functions in super-linear time. Icking and Ma [6] showed that the size of the Voronoi diagram of a set of n points for d_P is tightly bounded by $\theta(n^2)$. This bound holds for farthest Voronoi diagram for d_P too. Any method for the computation of 1-center using farthest Voronoi diagram for d_P will have $\Omega(n^2)$ complexity for n points. Alonso et al. [1,2] showed how we can compute minimum enclosing balls, circumballs and circumcentres of simplices in normed planes.

In this paper we show how to compute the 1-center of n points for convex polyhedral distance function in any dimension optimally in $O(n)$ time. We believe that this is the first attempt to solve the 1-center problem in linear time for any non-metric distance function. Observe that d_P is not symmetric and hence it is not a metric. Conventionally, we call the "polyhedral distance function" as the "polygonal distance function" in $\Re^2$.

Sharir and Welzl [10] formulated a framework for *LP-type* problems to solve a class of optimization problem using randomization techniques. Matousek et al. [7] gave a randomized algorithm to solve a general LP-type problem in expected $O(n)$ time. Chazelle and Matousek [3] gave a linear deterministic algorithm to solve LP-type problems with some additional computational constraints using derandomization. We observe that the problem discussed here is in the frame-work of LP-type problem, so it may be solved in expected linear time. But proposing a deterministic linear time algorithm in n is the main challenge of this paper. We are also able to keep the dependence of running time complexity on m to a poly-logarithmic factor of m in $\Re^2$ and m^2 in $\Re^d$, where m is the size of convex polyhedra in the convex distance function.

In Sect. 2, we give some useful concepts and an algorithm to compute bisectors of two objects for convex polyhedral distance function. In Sect. 3 we present a linear time algorithm for computing 1-center of point in $\Re^2$ for convex polygonal distance function constraint on a given line. In Sect. 4 we present an $O(nm \log^2 m)$ algorithm for computing 1-center of n points in $\Re^2$ for a convex polygonal distance function $d_P, |P| = m$. In Sect. 5 we present a $O(3^{3d^2} nm^2 \log^d m)$ to compute 1-center of n points in $\Re^d$ for convex polyhedral distance function $d_P, |P| = m$.

2   Concepts and Primitives

Note that the distance function d_P is not a metric as d_P is not symmetric. Let S be a finite set of n convex sets in $\Re^d$, $S = \{S_i : 1 \le i \le n\}$. The *1-center* of S for convex polyhedral distance function d_P is the point $c \in \Re^d$ such that the maximum of the distances from sets S_i's to c is minimized. See Fig. 1. This problem is equivalent to the following min-max optimization problem:

$$\min_{c \in \Re^d} \max_{1 \le i \le n} d_P(S_i, c)$$

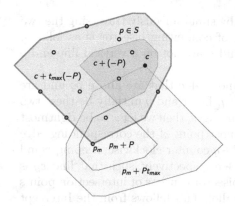
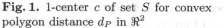
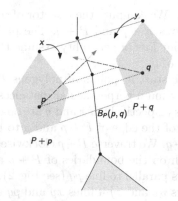

Fig. 1. 1-center c of set S for convex polygon distance d_P in $\Re^2$

Fig. 2. Bisector $B_P(p,q)$ of points p and q in $\Re^2$

Due to asymmetric nature of d_P, there is another similar min-max optimization problem:

$$\min_{c \in \Re^d} \max_{1 \leq i \leq n} d_P(c, S_i)$$

This problem can be reduced to the earlier problem by replacing the convex polyhedron P by the polyhedra $-P$. $-P$ is the polyhedra $\{-x \mid x \in P\}$. The origin o will be in the interior of $-P$ since o is in the interior of P. $-P$ is also termed as the point reflection of P about o. We can show that $d_P(c, S_i) = d_{-P}(S_i, c)$. Therefore, solving one of the above problems leads to solving the other.

The problem of 1-center for convex polyhedral distances is equivalent to the problem of finding the *minimum enclosing polyhedron* with a fixed shape and orientation if the sets S_i's are points. If the sets S_i's are convex sets, the problem is equivalent to the problem of finding the *minimum intersecting/stabbing polyhedron* with a fixed shape and orientation.

Our algorithm hinges on computing bisectors of geometric objects in S efficiently. The prune and search technique that we employ does not drop a fraction of input size in every iteration. Instead we use a weighted prune and search to drop a fraction of weight in each iteration.

The *bisector* of two points p and q, $B_P(p,q)$, is the set of points that are equidistant from both of them. Set theoretically, the *bisector* is set of points given by $\{x \mid d_P(p, x) = d_P(q, x)\}$.

In $\Re^2$ the bisector of two points consists of piecewise non-intersecting line segments, with non-intersecting unbounded rays or *wedges* at both ends. See Fig. 2. We refer these line segments, rays and wedges as *segments* of the bisector. The following lemma states the computational complexity of the bisector of two points.

Lemma 1. *The bisector of two points p and q in $\Re^2$ for a convex polygonal distance function d_P can be computed in $O(m)$ time, where $|P| = m$.*

Proof. We compute the bisector of p and q by simultaneously traversing the two polygons $P + p$ and $P + q$. The procedure of computing bisector is as follows. Without loss of generality assume that p and q are not on a vertical line and p is left of q.

We start at points of polygons $P + p$ and $P + q$ that are above pq and are at maximum perpendicular distances from pq. Let x and y initially be these two points respectively. If x and y are on collinear edges, then we let x to be rightmost point of the edge of $P + p$ and y to be leftmost point of the corresponding edge of $P+q$. We traverse $P+p$ clockwise and $P+q$ counter-clockwise, keeping x and y both on the boundaries of $P + p$ and $P + q$ respectively, such that line xy is always parallel to line pq (see Fig. 2). The bisector consists of intersection points of rays $\vec{px}$ and $\vec{qy}$ if lines xy and pq are parallel. This follows from the Intercept Theorem. Observe that the bisector is piecewise linear except at the wedges. Therefore we compute the bisector as the set of points of inflection when either x or y is at a vertex of P. We terminate the procedure when x and y reach the maximum perpendicular distance points from pq below pq. It should be noted if P has an edge parallel to pq, then we get wedges at the corresponding end of the bisector, which we compute separately. This procedure implies an $O(m)$ algorithm.   □

Observe that the bisector, by the construction of Lemma 1, is monotonic in the direction perpendicular to pq except at wedges, that is, the bisector in non-wedge portions, intersects every parallel line to pq at one and only one point. At the wedges, the bisector intersects every parallel line to pq in an interval.

When p and q are points in $\Re^d$, the algorithm to construct the bisector is similar traversal algorithm, except we need to traverse the surface of polyhedra $P + p$ and $P + q$ which consists of higher dimension faces. We can still compute the bisector efficiently in linear time of the size of the bisector. We use the bisectors in the algorithms of Sects. 3, 4 and 5.

3   Computing 1-Center Constrained on a Line Segment for Points in $\Re^2$ Under Convex Polygonal Distance Function

Let S be set of n points in $\Re^2$. Let L be a given line segment on the plane. We show in this section how we can compute 1-center of S that is constrained to lie on line segment L for convex polygonal distance function d_P. This is equivalent to solving the following optimization problem

$$\min_{c \in L} \max_{p \in S} d_P(p, c).$$

Let c_L be the solution of the min-max optimization problem which is the 1-center constrained on the line L.

We compute c_L using weighted prune and search technique as given below. We use weights to improve the efficiency of algorithm. Instead of point removal, weight trimming leads to decreased number of search operations. After every iteration we update S whenever points, that do not change 1-center, are dropped.

First we arbitrarily pair the points in S to get $\lfloor n/2 \rfloor$ point-pairs. We compute the bisectors of every point-pair. Each bisector has at most $m - 2$ points of inflection.

Let B be the set of bisectors in consideration in any iteration, and let I be the interval inside line segment L where the constrained 1-center c_L is known to lie. Initially, B is the set of all $\lfloor n/2 \rfloor$ bisectors of above point pairs, and I is whole line segment L. We also associate a weight to each bisector $b \in B$. In each iteration of the algorithm we drop a fraction of total weight of bisectors of B as well as shorten the possible interval I where an optimal 1-center c_L lies. We may also drop some of the bisectors in B and then add some other bisectors in each iteration. However, we may not necessarily drop bisectors in each and every iteration of the algorithm.

We assume that B contains only those bisectors that have segments crossed by I. If a bisector does not satisfy this property, then we can drop the nearer point of the defining points and re- pair with some other unpaired point.

We assign a weight to a bisector b in such a way that in every iteration we drop at least a fraction of total weight. The weight is dropped due to the reduction of weights of some bisectors, and dropping of some other bisectors. We give higher weights to bisectors intersecting I more number of times. We determine the weight of a bisector b by counting the number of proper crossings of the interior of the interval I with segments of bisector b. The crossing at any wedge, ray or line-segment of b is called *proper crossing* only if I crosses the bisector fully. We fix w as $2^{1/(\lfloor \log_2(m-1) \rfloor + 1)}$. We assign weights $1, w, w, w^2, w^2, w^2, w^2, w^3, \ldots, w^{\lfloor \log_2 k \rfloor}, \ldots, w^{\log_2(m-1)}$ to bisectors for which I crosses $1, 2, 3, 4, 5, 6, 7, 8, \ldots, k, \ldots (m-1)$ times respectively.

We begin every iteration of the algorithm by recomputing crossings and weights of the bisectors b in B.

For all the bisectors $b \in B$, we compute the number of crossings of b with interior of the interval I. Suppose there are k crossings ($k \geq 1$). Let the $\lceil k/2 \rceil$-th crossing of b with the interior of I be $p(b)$. For a crossed wedge $p(b)$ is any point in the crossing. We assign $p(b)$ same weight as the weight assigned to the bisector b, that is, $w^{\lfloor \log_2 k \rfloor}$. We compute the weighted median x_m of all $p(b), b \in B$.

We wish to determine which side of x_m in the interval I, an optimal 1-center c_L lies. To do this, we determine the points in S such that $d_P(p, x_m)$ is maximum. Equivalently, we find those points $p \in S$ for which $d_{-P}(x_m, p)$ is maximum. We call these *extreme* points, and let the set of extreme points be S_{max}. To find all extreme points, we compute a scaled polygon $-P$ translated to x_m with minimum positive scale that contains all points of S. Let this scaling factor be t_{max}. Naturally, $t_{max} \geq d_P(p, c_L)$ as c_L minimizes the distance for all points $p \in S$ and specifically for the extreme points in S_{max}. Thus c_L lies inside the polygon P translated to an extreme point $p_m \in S_{max}$ and scaled to touch x_m, that is, polygon $(p_m + t_{max}P)$. So, x_m lies in the interior of I and on the boundary of polygon $(p_m + t_{max}P)$, that is, $c_L \in I \cap (p_m + t_{max}P)$ for an

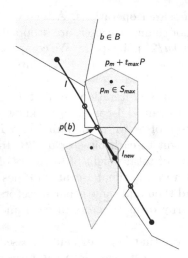

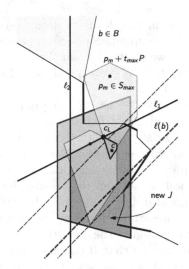

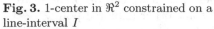

Fig. 3. 1-center in $\Re^2$ constrained on a line-interval I

Fig. 4. 1-center in $\Re^2$ lies in region J

extreme point $p_m \in S_{max}$. See Fig. 3. This not only gives us the side of x_m in I that 1-center c_L lies but also a shortened interval where c_L is known to lie. We compute the intersection of all such intervals. We call this interval I_{new}. Two cases arise. One, the intersection I_{new} is the single point x_m in which case we terminate our algorithm and return x_m as an optimal 1-center c_L for the convex distance function d_P constrained on L.

In the other case, the intersection I_{new} in not empty. For those bisectors b that do not cross interval I_{new}, the whole of line segment I lies in one of the regions partitioned by the bisector b. This means that one of the defining points of bisector b is always nearer than other defining point for convex polygonal distance d_P. We drop the nearer point, as the farther point will always be farther from I_{new} than the dropped point. We preserve the farther point for the next step. We also remove these bisectors from B and from further computation.

We arbitrarily pair the preserved points from the previous step, compute their bisectors and include these bisectors in B after ensuring that the bisectors intersect I_{new} (otherwise we drop a point and consider next point for pairing). There might be a single left over point for odd number of points.

If I_{new} is smaller than I then we repeat the iteration with I as I_{new} and the new B. Otherwise, if I_{new} is same as I we use any straight forward algorithm to compute c_L and terminate the algorithm.

Briefly the steps of algorithm are as follows:

1. We compute the crossings of $b \in B$ with I and compute its weight.
2. We compute middle most crossing $p(b)$ of bisector $b \in B$.
3. We compute weighted median x_m of the middle most crossings $p(b)$'s, for $b \in B$.

4. We identify the side of the weighted median x_m w.r.t. I in which the constrained 1-center c_L lies. I is restricted to I_{new} accordingly.
5. If I_{new} is a single point x_m then we terminate the algorithm and output x_m as c_L.
6. We drop the nearer points for bisectors in B, that do not cross I_{new}, and preserve the farther points. We remove these bisectors from B.
7. We re- pair the preserved points of the broken pairs. We compute their bisectors and include them into B.
8. We repeat from step 1 until B remains unchanged.
9. We compute c_L using any straight forward algorithm.

3.1   Algorithm Analysis

Theorem 1. *The 1-center of a set S of n points, constrained on a line segment L, for a convex polygonal distance function d_P, with P of size m, can be computed in $O(nm \log m)$ time.*

Proof. We can compute the initial B in $O(nm)$ time.

Let W be weight of bisectors in any iteration. Initially we start with total weight $W \le n \cdot w^{\lfloor \log(m-1) \rfloor}$. At every iteration we drop at least a fraction of the weight W.

We can compute the bisector crossing of b with I in $O(m)$ time from Lemma 1. The time complexity of each iteration is $O(|B| \cdot m)$, where $|B|$ is the cardinality of B. Also,
$W \ge |B| \ge W/(w^{\lfloor \log(m-1) \rfloor}) \ge W/2$, since all weights are between 1 and 2.

At the end of every iteration, a subset of bisectors in B with at least half of the weight have middle most segment that is earlier crossed by I in the interior, now outside of restricted I, I_{new}. When any bisector in this subset do not have any segments crossed by I_{new}, then a defining point is dropped and the remaining point is paired again in next iteration. This means that two bisectors with weight at least 1 are removed and one bisector with weight at most $w^{\lfloor \log_2(m-1) \rfloor} = 2/w$ is added. This also leads to a weight reduction by a factor of w. Note that even if we have a single remaining unpaired point, it does not change the analysis much.

If any bisector in the subset above has segments crossed by I_{new}, and the number of such crossed segments is k, then this bisector will have at least $2k+1$ segments crossed by I at the beginning of the iteration. Thus the weight of this bisector is reduced from at least $w^{\lfloor \log_2(2k+1) \rfloor}$ to $w^{\lfloor \log_2 k \rfloor}$, i.e., the weight is reduced by a factor of w.

Thus the recurrence relation for worst case time complexity $F(W) : \Re^+ \to \mathbb{N}$, when the weight is W, is given by:
$F(W) \le F(W/2 + W/2w) + O(|B| \cdot m)$, if $W > W_0$
$F(W) = O(1)$, if $W \le W_0$, for some fixed small weight W_0.
This equation reduces to,
$F(W) \le F(W(1/2 + 1/2w)) + O(\lfloor W \rfloor \cdot m)$
$\Rightarrow F(W) = O(\lfloor W/(1/2 - 1/2w) \rfloor \cdot m)$.

This gives us the worst cast time complexity $T(n)$ of the algorithm.

$$T(n) = F(n/2 \cdot w^{\lfloor log_2(m-1) \rfloor})$$
$$= O(\lfloor n \cdot w^{\lfloor log_2(m-1) \rfloor} \cdot m/(1/2 - 1/2w) \rfloor)$$
$$= O(\lfloor n \cdot w^{\lfloor log_2(m-1)+1 \rfloor} \cdot m \cdot 2/(w-1) \rfloor)$$

Since $w = 2^{1/(\log_2(m-1)+1)}$ and $1/(e^{1/x} - 1) = x - 1/2 + O(1/x)$ for $x > 1$, it follows that $T(n) = O(\lfloor n \cdot m \cdot \log_2(m-1) \rfloor)$ □

4   Computing 1-Center for Points in $\Re^2$ for a Given Convex Polygonal Distance Function

Let S be set of n points in $\Re^2$. S will be updated as we iterate through our algorithm by dropping points such that 1-center is unchanged. We show how we can compute 1-center of S for a given convex polygonal distance function using the algorithm of Sect. 3 and the prune and search technique. In every iteration we update S by dropping some points that do not change 1-center. In this section the problem of computing 1-center is equivalent to the min-max optimization problem $\min_{c \in \Re^2} \max_{p \in S} d_P(p, c)$. Let c be the solution of the min-max optimization problem which is 1-center.

We can use 1-center c_L of S for the convex polygonal distance function d_P that is constrained on line L, to determine whether it is also the unconstrained 1-center, and if it is not, we can determine the side of L the unconstrained 1-center of S for the convex polygonal distance function d_P lies. We compute the smallest positively scaled polygon $-P$ translated to c_L that contains all the points of S. Let the scaling factor be t_L. Then for all points $p \in S$, $d_{-P}(c_L, p) \leq t_L$, and therefore, $d_P(p, c_L) \leq t_L$. It is obvious that optimal scaling factor $t_{max} \leq t_L$, that is, $d_P(p, c) \leq t_L$, and therefore $c \in (p + t_L P)$. We just need to check only the *extreme points* of S that are on the boundary of scaled $-P$ translated to c_L. Let the set of these extreme points be S_{max}. Then c lies in region $\cap_{p_m \in S_{max}} (p_m + t_L P)$. Let this region be R_L. Computing R_L in the immediate neighborhood of c_L enables us to determine if c_L is optimal, and if it is not, then the side of L that c lies. If c_L is optimal then R_L will be a point or a line segment. If c_L is not optimal then R_L will be a polygon touching c_L but not crossing it, which lets us determine the side of L that c lies.

Lemma 2. *Given a line L, we can determine in linear time, whether the 1-center c for points S in $\Re^2$ for a given polygonal distance function d_P, lies on line L, and if it does not, then we can determine in linear time the side of line L that c lies.*

As before, we compute c by a weighted modification of prune and search technique. First we arbitrarily pair the points in S to get $\lfloor n/2 \rfloor$ point pairs. We compute bisector of each of these point pairs. Each bisector has at most $m - 2$ points of inflection.

Initially, let B be the set of all $\lfloor n/2 \rfloor$ bisectors. In every iteration, some bisectors of B get dropped and some new bisectors get added. Also, let J be a vertical trapezoidal or triangular region where we determine the 1-center to lie.

Initially J is whole plane $\Re^2$. In every iteration of the algorithm the region J gets truncated. We ensure that at the end of each iteration the region J remains a vertical trapezoid or triangle. Set B of bisectors and region J where 1-center is known to lie are two important entities in the algorithm.

We give weights to a bisector b depending on number of segments of the bisector b properly crossed by the interior of J. See Fig. 4. Let this number be k and let $w = 2^{1/(\lceil \log_2(m-1)\rceil+1)}$. We give weight $w^{\lceil \log_2 k \rceil}$ to the bisector b. In the algorithm given below, in every iteration, any bisector b is represented by a line not necessarily one of its segments. The representative line for any bisector b, denoted by $l(b)$ is given as following: (1) if k for bisector b is 1, line $l(b)$ is the extended line of either the relevant line segment or ray that is crossed by J in the interior, or any one of the side of the relevant wedge which is crossed by interior of J, (2) if k for bisector b is greater than 1, then $l(b)$ is a line orthogonal to direction of monotonicity of b separating $\lfloor k/2 \rfloor$-th and $(\lfloor k/2 \rfloor +1)$-th crossed segment of b by interior of J. In brief, if b is crossed by interior of J more than once, $l(b)$ is the partitioning line that partitions the crossed segments of b roughly in half.

In every iteration, in a way somewhat similar to the method proposed by Megiddo [8], we compute three lines l_1, l_2 and l_3 and determine which side of l_1, l_2 does c lie. We refer this as oracle $\mathcal{O}$. However, unlike Megiddo, we do not ensure that we drop a fraction of points in every iteration. Instead, we ensure that a fraction of weight is dropped in each iteration, while also ensuring that the reduced region where c is known to lie remains trapezoidal.

All the bisectors in B have segments that are crossed by interior of J. We compute the representative lines $l(b)$ for $b \in B$. We compute the median slope s_m of $l(b)$'s and pair any line of slope $\geq s_m$ with another line of slope $\leq s_m$. We also compute their intersections, and store them in a set X. We compute a line l_1 that has slope s_m and partitions the weighted intersections of X in half. We use the algorithm presented in Sect. 3 as oracle $\mathcal{O}$ to restrict J if above or below l_1. We compute a line l_2 which is vertical and partitions the weighted intersections of X on the other side of restricted J in half. We again use the oracle in Sect. 3 to restrict J if l_2 intersects J. The restricted J may not be a vertical trapezoid. If J is not a vertical trapezoid or a triangle, then we select a line l_3 that partitions J into two vertical trapezoids. We restrict J with respect to l_3 using the oracle of Sect. 3 once more. The final restricted J will be a vertical trapezoid or a triangle.

If J is not reduced from what it was in the beginning of the iteration then we terminate the algorithm and compute c by any straight forward method. Also, if the oracle itself gives an optimal 1-center then there too we terminate the algorithm and output the 1-center that oracle reports.

If J is reduced from what it was before, we check the bisectors b in the set B. If b does not intersect the new restricted J, then we drop the nearer of the defining points of the bisector b and keep the farther defining point of the bisector b. We also remove such bisectors b from B.

After removing all the bisectors that have a point dropped in the previous step, we re- pair the kept points ensuring that the new bisectors have segments that are crossed by the interior of restricted J. Then we repeat the iteration.

Briefly the steps of algorithm are as follows:

1. For each $b \in B$, we compute the segments of b that are crossed by J. Let number of crossed segments be k.
2. We give weight $w^{\lceil \log_2 k \rceil}$ to the representative line $l(b)$ of bisectors in B.
3. We compute lines l_1, l_2 and optionally l_3, for the set of $l(b)$'s, and run the oracle $\mathcal{O}$ on them. Each succeeding line depending on the output of oracle on preceding line. We restrict J accordingly. If any run of oracle outputs c then we exit the algorithm.
4. We drop the nearer point for bisectors, whose segments are not crossed by J, and preserve the farther point. We remove these bisectors from B.
5. We re- pair the kept points of the broken pairs. We compute their bisectors and include them into B.
6. We repeat from step 1 until B remains unchanged.
7. We compute c using any straight forward algorithm.

4.1   Algorithm Analysis

Theorem 2. *The 1-center of a set of n points for convex polyhedral function d_P can be computed in $O(nm \log^2 m)$ time using the algorithm in Sect. 4.*

Proof. We compute the initial set of bisectors B in $O(nm)$ time. Let W be the total weight of all bisectors in B in any iteration. Initially we start with weight $W \leq n \cdot w^{\lceil \log_2 (m-1) \rceil}$ which is reduced in every iteration as B gets modified. At the end of every iteration bisectors corresponding to $W/4$ have their representative line $l(p)$ not intersecting the reduced J.

If for the corresponding bisector b, no segment of b is crossed by reduced J then we drop the bisector b and introduce a new bisector for every two dropped bisectors. Then a weight of at least 2 is dropped and a weight of at most $w^{\lceil \log_2 (m-1) \rceil}$ is added. There is weight reduction by a factor of w for the bisectors not intersecting the reduced J.

If for the corresponding bisector b, some segment of b is crossed by reduced new J, and number of such intersections and crossings with reduced new J is k then b would have at least $2k - 1$ of segments crossed by the interior of non-reduced old J. Then the weight of corresponding bisector is reduced from at least $w^{\lceil \log_2 (2k-1) \rceil}$ to $w^{\lceil \log_2 k \rceil}$, that is, weight is reduced by at least a factor w.

The recurrence relation for worst case time complexity $F(W)$, when the weight is W is given by:
$$F(W) \leq F(3W/4 + W/4w) + O(|B| \cdot m \log_2 m), W > W_0$$
$$F(W) = O(1), W \leq W_0.$$
This recurrence implies that the time complexity of the algorithm is $O(\lfloor nm \log_2^2 (m-1) \rfloor)$. Here $|B|$ is the cardinality of B, which is atmost $\frac{n}{2}$. $\square$

5   Computing 1-Center for Points in $\Re^d$ for Convex Polyhedral Distance

Let S be a set of n points in $\Re^d$. Let P be a fixed polyhedra of size m that contains origin o in the interior. We call a k-dimensional face as a k-face, a k-dimensional affine space as a k-space, a $(k-1)$-space, say A_{k-1}, in a k-space, say A_k, as a k-hyperplane of A_k, for any dimension k.

In this section we present the algorithm for computing in linear time the 1-center of S with respect to convex polyhedral distance function d_P. The method that we propose is a search method in spirit of recursive multidimensional search method proposed by Dyer in [5]. We observe that the bisector of two points can be computed in $O(m^2)$ time. To simplify the structure of bisector and also the presentation of algorithm we remove wedges in bisector by redefining bisector $B_P(p,q)$ of two points p and q as boundary between closed sets $\{x \mid d_P(p,x) \le d_P(q,x)\}$ and closure of open set $\{x \mid d_P(q,x) < d_P(p,x)\}$. The complexity and correctness of the algorithm does not change by this modification. Points p and q can be arbitrarily interchanged in the presentation of the algorithm.

First we note that the bisector $B_P(p,q)$ of two points p and q in $\Re^d$ for convex polyhedral distance function d_P is of size $O(m^2)$ and can be constructed in time $O(m^2 d^3)$ where $|P| = m$. Also, the bisector $B_P(p,q)$ intersects any line parallel to pq only at an equidistant single point to p and q and each subface of the bisector $B_P(p,q)$ is convex. If the number of k-subfaces of polyhedra P is m_k and the number of k-subfaces of the bisector $B_P(p,q)$ of p and q is b_k, then
$$b_k \le m_k m_{d-1} + m_{k+1} m_{d-2} + \cdots + m_{\lfloor (d+k-1)/2 \rfloor} m_{\lceil (d+k-1)/2 \rceil}, \ 0 \le k \le d-1$$
$$\Rightarrow b_0 + b_1 + \cdots + b_{d-1} \le m^2.$$

We call the $(d-1)$-faces of a bisector b as segments of the bisector b. The computation and storage of a bisector, say b, of two points in S, say p and q, that is $b = B_P(p,q)$, requires $O(m^2)$ time and space respectively. We can extend $(d-2)$-subfaces parallel to pq to get $O(m^2)$ $(d-1)$-hyperplanes for each $(d-2)$-subface of b. These hyperplanes will create a subdivision of $\Re^d$. We can prove this by using the fact that b is monotonic in hyperplane parallel to line pq. We shall call this subdivision $S_{div}(b)$ henceforth.

We construct an oracle $\mathcal{O}_k$ for 1-center of points in $\Re^d$ for convex polyhedral distance function d_P constrained on any given k-space recursively for $0 \le k \le d$. Note that the points are in $\Re^d$ but 1-center is constrained on a k-space. $\mathcal{O}_0$ is straight forward. Let us assume that an oracle, $\mathcal{O}_{k-1}$, for 1-center for convex polyhedral distance function d_P for set of points in $\Re^d$ constrained in $(k-1)$-space is available. Then this oracle can be used to compute 1-center for convex polyhedral distance function d_P for set of points in $\Re^d$ constrained in any k-space. With each bisector b we keep a set of some of the $(k-1)$-spaces corresponding to $(k-1)$-faces of subdivision $S_{div}^k(b)$ of A_k. We denote this set by $H(b)$. Initially it consists of all the $(k-1)$-spaces corresponding to $(k-1)$-faces of the subdivision $S_{div}^k(b)$ of A_k. Also let the initial size of $H(b)$ be $M(b)$.

The steps of algorithm for oracle $\mathcal{O}_k$:
Step 1: Let initial B be the $\lfloor n/2 \rfloor$ bisectors, for each $b \in B$ we compute $S_{div}(b)$, $S_{div}^k(b)$, $H(b)$, $K(b)$ and representative $h(b)$, and assign it weight $w(b)^{K(b)}$ where

$K(b) = \lfloor \log_2 M(b) \rfloor 9^{k-1} + c, 1 \leq M(b) \leq m^2$.

Step 2: Solve 9^{k-1} queries O_{k-1} for representatives $h(b)$'s for $b \in B$. We also determine which side of A_k with respect to the query space c lies. If any oracle computes c then output c and terminate the algorithm.

Step 3: Bisector representatives corresponding to at least half of the weight, will have new representatives or will be dropped which corresponds to a dropped point.

Step 4: Remove the bisectors with dropped points from B.

Step 5: Re-pair the dropped points and introduce new bisectors for these pairs in B along with computing $S_{div}(b)$, $S_{div}^k(b)$, $H(b)$, $K(b)$, representative $h(b)$, and weight $w(b)^{K(b)}$.

Step 6: If no points are dropped and no representative of bisector is replaced then terminate the algorithm and compute c by any straight forward method. Otherwise repeat from step 2.

Hence, the 1-center for a set of n points in $\Re^d$ can be computed in $O(3^{3d^2} nm^2 \log_2^d m)$ for convex polyhedral distance function d_P for polyhedra P of size m.

References

1. Alonso, J., Martini, H., Spirova, M.: Minimal enclosing discs, circumcircles, and circumcenters in normed planes (part I). Comput. Geom. **45**(5–6), 258–274 (2012)
2. Alonso, J., Martini, H., Spirova, M.: Minimal enclosing discs, circumcircles, and circumcenters in normed planes (part II). Comput. Geom. **45**(7), 350–369 (2012)
3. Chazelle, B., Matousek, J.: On linear-time deterministic algorithms for optimization problems in fixed dimension. J. Algorithms **21**(3), 579–597 (1996)
4. Chew, L.P., Dyrsdale, R.L.: Voronoi diagrams based on convex distance functions. In: Proceedings of the First Annual Symposium on Computational Geometry, Baltimore, Maryland, USA, 5–7 June 1985, pp. 235–244 (1985)
5. Dyer, M.E.: On a multidimensional search technique and its application to the euclidean one-centre problem. SIAM J. Comput. **15**(3), 725–738 (1986)
6. Icking, C., Klein, R., Ma, L., Nickel, S., Weißler, A.: On bisectors for different distance functions. In: Milenkovic, V. (ed.) Proceedings of the Fifteenth Annual Symposium on Computational Geometry, Miami Beach, Florida, USA, 13–16 June 1999, pp. 291–299. ACM (1999)
7. Matousek, J., Sharir, M., Welzl, E.: A subexponential bound for linear programming. In: Proceedings of the Eighth Annual Symposium on Computational Geometry, Berlin, Germany, 10–12 June 1992, pp. 1–8 (1992)
8. Megiddo, N.: Linear-time algorithms for linear programming in R^3 and related problems. SIAM J. Comput. **12**(4), 759–776 (1983)
9. Megiddo, N.: Linear programming in linear time when the dimension is fixed. J. ACM **31**(1), 114–127 (1984)
10. Sharir, M., Welzl, E.: A combinatorial bound for linear programming and related problems. In: Jantzen, M., Finkel, A. (eds.) STACS 1992. LNCS, vol. 577, pp. 567–579. Springer, Heidelberg (1992)
11. Sylvester, J.J.: A question in the geometry of situation. Q. J. Math. **1**, 79 (1857)

Positive Zero Forcing and Edge Clique Coverings

Shaun Fallat[1], Karen Meagher[1], Abolghasem Soltani[1], and Boting Yang[2]([⊠])

[1] Department of Mathematics and Statistics, University of Regina, Regina, Canada
[2] Department of Computer Science, University of Regina, Regina, Canada
boting@uregina.ca

Abstract. Zero forcing parameters, associated with graphs, have been studied for over a decade, and have gained popularity as the number of related applications grows. In this paper, we investigate positive zero forcing within the context of certain edge clique coverings. A key object considered here is the compressed cliques graph. We study a number of properties associated with the compressed cliques graph, including: uniqueness, forbidden subgraphs, connections to Johnson graphs, and positive zero forcing.

1 Introduction

Suppose that G is a simple finite graph with vertex set $V = V(G)$ and edge set $E = E(G)$. We use $\{u, v\}$ to denote an edge with endpoints u and v. Further, for a graph $G = (V, E)$ and $v \in V$, the vertex set $\{u : \{u, v\} \in E\}$ is the *neighbourhood* of v, denoted as $N_G(v)$, and the size of the neighbourhood of v is called the *degree* of v. For $V' \subseteq V$, the vertex set $\{x : \{x, y\} \in E, x \in V \setminus V'$ and $y \in V'\}$ is the *neighbourhood* of V', denoted as $N_G(V')$. Also, we let the set $N_G[v] = \{v\} \cup N_G(v)$ denote the *closed neighbourhood* of the vertex v. We use $G[V']$ to denote the subgraph induced by V', which consists of all vertices of V' and all of the edges in G that contain only vertices from V'. We let K_n denote the complete graph on n vertices. We will also refer to a complete graph on n vertices as a *clique* on n vertices.

Our interest in this work is to consider how positive zero forcing sets are related to cliques in a graph and, further, clique intersection, and edge clique coverings. Zero forcing on a graph was originally designed to be used as a tool to bound the maximum nullity associated with collections of symmetric matrices derived from a graph G [1,2]. Independently, this parameter has been studied in conjunction with control of quantum systems [4,14], and it has also been studied within the context of fast-mixed searching [15]. Positive zero forcing was

---

S. Fallat—Research supported in part by an NSERC Discovery Research Grant, Application No.: RGPIN-2014-06036.
K. Meagher—Research supported in part by an NSERC Discovery Research Grant, Application No.: RGPIN-341214-2013.
B. Yang—Research supported in part by an NSERC Discovery Research Grant, Application No.: RGPIN-2013-261290.

© Springer International Publishing Switzerland 2016
D. Zhu and S. Bereg (Eds.): FAW 2016, LNCS 9711, pp. 53–64, 2016.
DOI: 10.1007/978-3-319-39817-4_6

an adaptation of conventional zero forcing to play a similar role for positive semidefinite matrices [3,7–9,16].

Zero forcing in general is a graph colouring problem in which an initial set of vertices are coloured black, while the remaining vertices are coloured white. Using a designated colour rule, the objective is to change the colour of as many white vertices to black as possible. There are two common rules, which are known as zero forcing and positive zero forcing. The process of a black vertex u changing the colour of a white vertex v to black is usually referred to as "u forces v". The size of the smallest initial set of black vertices that will "force" all vertices black is called either the zero forcing number or the positive zero forcing number of G depending on which rule is used.

Here we are more interested in the behaviour of the positive zero forcing number in connection with cliques and edge clique coverings in a graph. In particular, we consider the positive zero forcing number of G, when maximal cliques of G satisfy certain intersection properties. Consequently, we now carefully review some basic terminology associated with positive zero forcing in a graph.

The positive zero forcing rule is based on a colour change rule similar to the zero forcing colour change rule [3,7,8]. Suppose G is a graph and B a subset of vertices; we initially colour all of the vertices in B black, while all remaining vertices are designated white. Let $W_1, \ldots, W_k$ be the sets of vertices in each of the connected components of $G - B$ (removing the vertices of B from G). If u is a vertex in B and w is the only white neighbour of u in the induced subgraph $G[W_i \cup B]$, then u can force the colour of w to black. This rule is called the *positive colour change rule*. If all vertices of G are black after repeatedly applying the positive colour change rule, then the set of initial black vertices is called a *positive zero forcing set*. The size of the smallest positive zero forcing set of a graph G is called the *positive zero forcing number* of G, denoted by $Z_+(G)$. If a subset S of $V(G)$ is a positive zero forcing set with $|S| = Z_+(G)$, then we refer to S as an *optimal* positive zero forcing set for G.

It is known that by following the sequence of forces throughout the conventional zero forcing process, a path covering of the vertices is derived (see [3, Proposition 2.10] for more details). When the positive colour change rule is applied, two or more vertices can perform forces at the same time, and a vertex can force multiple vertices from different components at the same time. This implies that the positive colour change rule produces a partitioning of the vertices into sets of vertex disjoint induced rooted trees, which we will refer to as *forcing trees*, in the graph.

Cliques in a graph play an important role in determining both zero forcing and positive zero forcing sets. We explore this correspondence further in this paper. As an example, consider chordal graphs. For chordal graphs it is known that the positive zero forcing number is equal to the number of vertices minus the fewest number of cliques that contain all of the edges (see, for example, [9]).

2   Simply Intersecting Edge Clique Coverings

Recall that a *clique* in a graph is a subset of vertices which induces a complete subgraph. A clique in a graph is *maximal* if no vertex in the graph can be added to it to produce a larger clique. An *edge clique covering* of a graph is a set of cliques with the property that every edge is contained in at least one of the subgraphs induced by one of the cliques in the set. (Note that unless the graph has isolated points, an edge clique covering that contains every edge also contains every vertex.) The *size* of an edge clique covering is the number of cliques in the covering. For a graph G, we denote the size of a smallest edge clique covering by $cc(G)$. Further, we call a given edge clique covering *minimal* if the number of cliques in this covering is equal to $cc(G)$. An edge clique covering for a graph G is called a *min-max clique covering* if its size is $cc(G)$ and every clique in it is maximal.

Observation. *For any graph G, there exists an edge clique covering with size $cc(G)$ in which every clique is maximal, that is, there is always a min-max clique covering of G.*

Let $\mathcal{C} = \{C_1, \ldots, C_\ell\}$ be an edge clique covering for G. If for any set of distinct triples $i, j, k \in \{1, \ldots, \ell\}$ it is the case that $C_i \cap C_j \cap C_k = \emptyset$, we say the clique covering has *simple intersection*. If the edge clique covering has simple intersection, then a vertex that is in $C_i \cap C_j$ (where $i \neq j$) does not belong to any other clique in $\mathcal{C}$. There are many examples of graphs with an edge clique covering with simple intersection. Throughout this paper, we will only consider the property of simple intersection for min-max clique coverings.

3   Non-unique Edge Clique Coverings

Let G be a graph with n vertices, which are labeled $0, 1, \ldots, n-1$, and let $S = \{k_1, k_2, \ldots, k_\ell\}$ be a set of positive integers such that $k_1 < k_2 < \cdots < k_\ell < (n+1)/2$. The *circulant graph*, denoted by $\text{circ}(n, S)$, has each vertex i in $\{0, \ldots, n-1\}$ adjacent to $i \pm k_1$, $i \pm k_2$, $\ldots, i \pm k_\ell$ (mod n). The graph $\text{circ}(6, \{1, 2\})$ plays a key role in identifying graphs that possess a unique min-max clique covering satisfying simple intersection.

This graph is isomorphic to the left graph in Fig. 1. This graph has one vertex for each subset of $\{1, 2, 3, 4\}$ of size two, and two vertices are adjacent if the sets intersect. This graph is also known as the Johnson graph $J(4, 2)$ [6].

The graph $\text{circ}(6, \{1, 2\})$ has two min-max clique coverings. One edge clique covering of this graph is $\{C_1, C_2, C_3, C_4\}$ where C_i is the set of all vertices with a label that contains an i. A second edge clique covering is formed by the following sets of vertices of $\text{circ}(6, \{1, 2\})$:

$$\{\{1,2\}, \{1,3\}, \{2,3\}\}, \quad \{\{1,2\}, \{1,4\}, \{2,4\}\},$$
$$\{\{1,3\}, \{1,4\}, \{3,4\}\}, \quad \{\{2,3\}, \{2,4\}, \{3,4\}\}.$$

The next fact is a key to characterizing the graphs that possess more than one min-max clique covering.

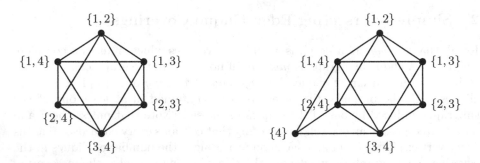

Fig. 1. The graph $\mathrm{circ}(6,\{1,2\})$ (left), and a graph that contains $\mathrm{circ}(6,\{1,2\})$ (right).

Theorem 1. *If a graph G has two distinct min-max clique covers that both satisfy simple intersection, then G contains $\mathrm{circ}(6,\{1,2\})$ as an induced subgraph.*

Unfortunately, the converse to Theorem 1 is false in general. A simple example can be derived from the graph $\mathrm{circ}(6,\{1,2\})$ by adding an additional vertex (see the right graph in Fig. 1). It is easy to verify that this graph has a unique min-max clique covering that satisfies simple intersection, but certainly contains $\mathrm{circ}(6,\{1,2\})$ as an induced subgraph.

4   Compressed Cliques Graphs

If a graph has a min-max clique covering that satisfies simple intersection, then the graph can be simplified in a way that allows us to determine both the edge clique covering number and the positive zero forcing number of the original graph from the simplified graph.

Let G be a graph and $S = \{v_1, v_2, \ldots, v_k\}$ a set of vertices in G. The *contraction of S in G* is the graph formed by replacing the vertices in S by a single vertex v_S, where v_S is adjacent to any vertex in $V(G)\backslash S$ that is adjacent to a vertex in S.

Let G be a graph with a min-max clique covering $\mathcal{C} = \{C_1, C_2, \ldots, C_\ell\}$ that has simple intersection. We can construct a new graph that is related to G called a *compressed cliques graph*. For distinct $i, j \in \{1, \ldots, \ell\}$ define the sets

$$C_{i,j} = C_i \cap C_j, \quad C_{i,i} = C_i \backslash \bigcup_{\substack{j \in \{1,\ldots,\ell\} \\ j \neq i}} C_j,$$

(these sets may be empty). First, for each pair of distinct $i, j \in \{1, \ldots, \ell\}$, if $C_{i,j}$ is non-empty, then contract all the vertices in $C_{i,j}$ to a single vertex labeled $v_{i,j}$. Second, if $C_{i,i}$ is non-empty, then contract all the vertices in $C_{i,i}$ to a single vertex; label this vertex $v_{i,i}$. In this graph the vertices $v_{i,j}$ and $v_{i',j'}$ are adjacent if and only if the sets $\{i, j\}$ and $\{i', j'\}$ have non-empty intersection.

We have seen that a graph may have multiple min-max clique covers that satisfy simple intersection, for example, the graph $\mathrm{circ}(6, \{1, 2\})$. However, for this graph, the compressed cliques graphs, for either clique cover are isomorphic. In fact, both are isomorphic to $\mathrm{circ}(6, \{1, 2\})$. Thus, an obvious question is to verify that the associated compressed cliques graphs are isomorphic in the presence of distinct min-max cliques covers satisfying simple intersection (see the next section). Assuming this fact, we denote *the* compressed cliques graph of G by $\mathcal{C}(G)$ and give some simple examples. For any integer n, $\mathcal{C}(K_n) = K_1$, $\mathcal{C}(K_n \backslash \{e\}) = P_3 (n \geq 3)$, $\mathcal{C}(C_n) = C_n$, and $\mathcal{C}(P_n) = P_n$ (where P_n is the path with n vertices).

Theorem 2. *Let G be a graph with n vertices and $\mathcal{C} = \{C_1, C_2, \ldots, C_k\}$ be a min-max clique covering of G. If $\mathcal{C}$ satisfies simple intersection, then $\mathcal{C}(G)$ isomorphic to G if and only if all of the sets $C_{i,i}$ and $C_{i,j}$ for $i, j \in \{1, \ldots, k\}$ contain no more than one vertex.*

Corollary 1. *Suppose G is a graph that possesses a min-max clique covering having simple intersection. Then $\mathcal{C}(\mathcal{C}(G)) = \mathcal{C}(G)$.*

We will make use of the following map $\phi : V(G) \longrightarrow V(\mathcal{C}(G))$ defined as

$$\phi(v) = \begin{cases} v_{i,j} & \text{if } v \in C_i \cap C_j, \\ v_{i,i} & \text{if } v \in C_i \text{ and no other cliques.} \end{cases} \quad (1)$$

We will need the following fact in Theorem 4.

Lemma 1. *If there is a path from u to v in G, then either $\phi(u) = \phi(v)$ or there is a path from $\phi(u)$ to $\phi(v)$ in $\mathcal{C}(G)$.*

5 Uniqueness of Compressed Cliques Graphs

We have seen that the circulant graph $\mathrm{circ}(6, \{1, 2\})$ has two distinct min-max clique covers that both satisfy simple intersection. Furthermore, if we expand each vertex of $\mathrm{circ}(6, \{1, 2\})$ to a clique of any positive size and expand edges correspondingly (join the vertices of these cliques if the corresponding vertices were adjacent in $\mathrm{circ}(6, \{1, 2\})$), all of the resulting graphs have two distinct min-max clique covers that both satisfy simple intersection. In order to ensure the concept of a compressed cliques graph well-defined, we need to consider the uniqueness of the compressed cliques graph up to isomorphism. We will show that if a graph has multiple min-max clique covers that satisfy simple intersection, the corresponding compressed cliques graphs are unique, up to isomorphism.

Lemma 2. *For a graph G, let $\mathcal{C}$ be a min-max clique cover of G that satisfies simple intersection. For a vertex $v \in V(G)$ and two distinct cliques $C_1, C_2 \in \mathcal{C}$, we have $v \in C_1 \cap C_2$ if and only if $N_G[v] = C_1 \cup C_2$.*

Lemma 3. *For a graph G, let $\mathcal{A}$ and $\mathcal{B}$ be two distinct min-max clique covers of G that both satisfy simple intersection. Then for any vertex $v \in V(G)$, exactly one of the following two possibilities occurs. (1) If only one clique A in $\mathcal{A}$ contains the vertex v, then there is a clique $B \in \mathcal{B}$ such that $A = B$. (2) If two distinct cliques $A, A' \in \mathcal{A}$ contain the vertex v, then there are two distinct cliques $B, B' \in \mathcal{B}$ such that $v \in B \cap B'$.*

Lemma 4. *For a graph G, let $\mathcal{A}$ and $\mathcal{B}$ be two distinct min-max clique covers of G that both satisfy simple intersection. If there are two distinct cliques A and A' in $\mathcal{A}$ and two distinct cliques B and B' in $\mathcal{B}$ such that $A \cap A' \cap B \cap B' \neq \emptyset$, then $A \cap A' = B \cap B'$.*

Now we can verify that the compressed cliques graph for a graph with a min-max clique covering with simple intersection is well-defined.

Theorem 3. *If a graph G has a min-max clique cover that satisfies simple intersection, then the compressed cliques graph of G is unique.*

Corollary 2. *Suppose G has two distinct min-max clique covers that both satisfy simple intersection. Then $\mathcal{C}(G) = \mathrm{circ}(6, \{1, 2\})$.*

Let $\mathcal{G}$ be the set of all graphs that have a min-max clique cover satisfying simple intersection. Let $\mathcal{R}$ be a binary relation on $\mathcal{G}$. We say two graphs $G_1, G_2 \in \mathcal{G}$ have the relation $\mathcal{R}$ if $\mathcal{C}(G_1) = \mathcal{C}(G_2)$. It is easy to see that $\mathcal{R}$ is an equivalence relation, and hence induces a partition of the set $\mathcal{G}$. For any equivalence class that contains a graph H, it is easy to see that the graph $\mathcal{C}(H)$ is the minimum element in the class. Thus, the compressed cliques graph of $\mathcal{C}(H)$ is $\mathcal{C}(H)$ itself.

6 Min-Max Clique Coverings of Compressed Cliques Graphs

Before we begin our analysis on positive zero forcing in compressed cliques graphs, we consider min-max clique coverings of compressed cliques graphs. We begin with the following useful lemma. Let $\phi : V(G) \to V(\mathcal{C}(G))$ be the map defined in (1).

Lemma 5. *Let G be a graph with a min-max clique covering C that has simple intersection. If C is a clique in $\mathcal{C}(G)$, then the preimage of C under ϕ is a clique in G.*

Theorem 4. *Let G be a graph in which there is a min-max clique covering with simple intersection. Let $\mathcal{C}(G)$ be the compressed cliques graph of G. Then*

$$\mathrm{cc}(G) = \mathrm{cc}(\mathcal{C}(G)).$$

We now have the following interesting consequence.

Corollary 3. *Assume that G is a graph with a min-max clique covering with simple intersection. Let $\{D_1, \ldots, D_\ell\}$ be the set of cliques in $\mathcal{C}(G)$ such that D_i is the set of vertices in $\mathcal{C}(G)$ that are labeled $v_{i,j}$ from some $j \in \{1, 2, \ldots, \ell\}$. Then the following statements hold: (1) The set $\{D_1, D_2, \ldots, D_\ell\}$ forms a min-max clique cover of $\mathcal{C}(G)$; and (2) this clique cover of $\mathcal{C}(G)$ has simple intersection.*

7 Positive Zero Forcing Sets in Compressed Cliques Graphs

Using the strong connections to edge clique coverings, we now explore positive zero forcing sets and the positive zero forcing number of compressed cliques graphs. This is one of our main motivations for developing this derived graph. Our results in this section are related to the following simple observation about positive zero forcing sets.

Lemma 6. *Let G be a graph and u and v be distinct vertices in G. If $N_G[u] = N_G[v]$, then any positive zero forcing set for G will contain at least one of u and v.*

This can be generalized to a subset of vertices as in the following fact.

Corollary 4. *Let G be a graph and $\{u_1, u_2, \ldots, u_k\}$ be a subset of k vertices from G. If $N_G[u_1] = N_G[u_2] = \cdots = N_G[u_k]$, then any positive zero forcing set for G will contain at least $k - 1$ of the vertices $\{u_1, u_2, \ldots, u_k\}$.*

We can apply the previous result to cliques in a min-max clique covering with simple intersection.

Lemma 7. *Assume that G is a graph and $\{C_1, C_2, \ldots, C_\ell\}$ is a min-max clique covering with simple intersection. If S is a positive zero forcing set, then the sets $(V(G)\backslash S) \cap C_{i,j}$ and $(V(G)\backslash S) \cap C_{i,i}$ for any $i, j \in \{1, \ldots, n\}$ have size at most one.*

Theorem 5. *Let G be a graph (that is connected, but not a clique) in which the maximal cliques have simple intersection and let $\mathcal{C}(G)$ be the compressed cliques graph of G. Then*

$$|V(G)| - Z_+(G) = |V(\mathcal{C}(G))| - Z_+(\mathcal{C}(G))$$

and there exist forcing trees for G and $\mathcal{C}(G)$ that differ only in that these forcing trees for G are isolated vertices.

8 Connections with Johnson Graphs

The compressed cliques graph is related to the *Johnson graph* $J(m, 2)$ [6]. The graph $J(m, 2)$ has the set of pairs from $\{1, 2, \ldots, m\}$ as its vertex set and two vertices are adjacent if and only if they have non-empty intersection. This graph is the line graph of the complete graph.

We will use a generalization of this graph that we denote by $J'(m, 2)$. The vertices of $J'(m, 2)$ are the set $\{i, j\}$ and $\{i\}$ where $i, j \in \{1, 2, \ldots, m\}$ and two sets are adjacent if and only if they have non-empty intersection. These graphs play a major role in the theory of compressed cliques graphs.

Lemma 8. *If G is a graph with $\mathrm{cc}(G) > 1$ and G has a min-max clique covering that satisfies simple intersection, then the compressed cliques graph of G is an induced subgraph of $J'(\mathrm{cc}(G), 2)$.*

There is a simple clique covering for this generalization of the Johnson graph. Define C_i to be the set of all vertices in $J'(m, 2)$ that contain i. Then C_i is a maximal clique and $\mathcal{C} = \{C_1, \ldots, C_m\}$ is a min-max clique covering that has simple intersection.

Lemma 9. *For any $m > 3$, $J'(m, 2)$ has the following properties:*
(1) $|V(J'(m, 2))| = \binom{m}{2} + m$, (2) $\mathrm{cc}(J'(m, 2)) = m$, (3) $Z_+(J'(m, 2)) = \binom{m}{2}$, and (4) $Z_+(J'(m, 2)) = |V(J'(m, 2))| - \mathrm{cc}(J'(m, 2))$.

For the graph $J'(m, 2)$, in an optimal positive zero forcing process there is initially one white vertex for each clique in the min-max clique covering. In fact, we can assign each vertex $\{i\}$ to be white, then within each maximal clique there is exactly one white vertex.

Similarly, the Johnson graph $J(m, 2)$ also has a min-max clique covering with simple intersection.

Lemma 10. *For any $m > 3$, $J(m, 2)$ has the following properties:*
(1) $V(J(m, 2)) = \binom{m}{2}$, (2) $\mathrm{cc}(J(m, 2)) = m$, (3) $Z_+(J(m, 2)) = \binom{m}{2} - m + 2$, and (4) $Z_+(J(m, 2)) = |V(J(m, 2))| - \mathrm{cc}(J(m, 2)) + 2$.

In this case we need that $m > 3$ since the vertices of graph $J(3, 2)$ are $\{1, 2\}, \{1, 3\}, \{2, 3\}$ and these form a complete graph on three vertices. This is an exceptional maximal clique in $J(m, 2)$; every other maximal clique consists of all sets that contain a fixed element. In fact, any maximal clique in the compressed cliques graph with more than three vertices will be the set of all vertices whose corresponding sets all contain a common element.

9   Forbidden Subgraphs of Compressed Cliques Graphs

In this section we illustrate some forbidden subgraphs of the compressed cliques graph. A *claw* is a graph with four vertices and three edges in which one vertex is adjacent to the other three. A graph is called *claw-free* if no set of four vertices induce a subgraph that is a claw. Claw-free graphs have been widely studied [10].

Lemma 11. *If G is a graph that has a min-max clique covering with simple intersection, then both G and the compressed cliques graph $\mathcal{C}(G)$ are claw-free.*

Note the graph on the left in Fig. 2 is special in the sense that it alone cannot be a compressed cliques graph, but rather if a compressed cliques graph contains the subgraph on the left as an induced subgraph, then it must contain other vertices and edges. For example, consider the graph T_3 on the right in Fig. 2. Observe that the graph T_3 has edge clique cover number three, and this min-max clique cover has simple intersection. Furthermore, $\mathcal{C}(T_3)$ is itself T_3, and includes the graph on the left in Fig. 2 as an induced subgraph.

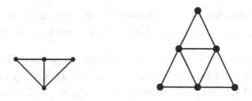

Fig. 2. A non-compressed cliques graph (left), and a self-compressed graph T_3 (right).

Before we come to our next claim regarding compressed cliques graphs, we define some particular subgraphs. We call a cycle $u_1 u_2 \cdots u_t u_1$ *suspended* when exactly one of $u_1, u_2, \ldots, u_t$ has degree larger than two in G, and all of the remaining vertices from this set have degree two in G.

Lemma 12. *If G is a graph with a min-max clique cover satisfying simple intersection, then $\mathcal{C}(G)$ does not contain a suspended cycle.*

10   Vertex-Clique Graphs

For any graph G with vertex set V and edge set E, we construct a new graph H, obtained from G, that has a min-max covering with simple intersection. For each vertex $v \in V$, we construct a clique $K_{d(v)}$ where $d(v)$ is the degree of v in G. For each edge $\{u, v\} \in E$, we add an edge between a vertex in $K_{d(u)}$ and a vertex in $K_{d(v)}$ such that each vertex in $K_{d(u)}$ (or $K_{d(v)}$) has at most one neighbour outside of $K_{d(u)}$ (or $K_{d(v)}$). This new graph is called the *vertex-clique graph of* G. By definition a vertex-clique graph cannot be a complete graph. The next lemma provides some interesting properties of a vertex-clique graph. Recall that a *line graph* of a given graph X, denoted by $L(X)$, is obtained by associating a vertex with each edge of X and connecting two vertices with an edge if and only if the corresponding edges of X have a vertex in common.

Lemma 13. *Let H be a vertex-clique graph of G. Then (1) H is a line graph; (2) H has a min-max clique covering that satisfies simple intersection; and (3) $\mathcal{C}(H) = H$.*

For a graph G, we can form a simplified graph called the *reduced graph of* G that has the same positive zero forcing number as the original graph. First we define an induced path $u_1 u_2 \cdots u_t$ in a graph to be a *suspended path* if the vertices $u_2, u_3, \ldots, u_{t-1}$ all have degree two. To form the reduced graph of G, first recursively delete all vertices with degree one. Once all the vertices of degree one have been removed, contract any induced suspended paths of length at least two to an edge. The graph that remains after performing both of these operations is called the *reduced graph of G*, and is denoted by $\mathcal{R}(G)$. It is clear that these two operations do not effect the size of a positive zero forcing set.

Lemma 14. *Let G be a graph, then $Z_+(G) = Z_+(\mathcal{R}(G))$.*

Applying this reduction to a vertex-clique graph produces a new graph for which we can determine bounds on the positive zero forcing number.

Theorem 6. *Let G be a connected vertex-clique graph and $\mathcal{R}(G)$ be the reduced graph of G. $\mathcal{R}(G)$ has the following properties: (1) If $\mathcal{R}(G)$ has only one vertex, then $Z_+(G) = 1$; (2) If $\mathcal{R}(G)$ is a 3-cycle, then $Z_+(G) = 2$; and (3) If $\mathcal{R}(G)$ has more than three vertices, let k be the number of edges in $\mathcal{R}(G)$ that are themselves maximal cliques in $\mathcal{R}(G)$, then $Z_+(G) \leq k$. Further, this upper bound can be tight for some graphs G.*

Finally, we give an example that shows the upper bound in the theorem above cannot be improved in general. Let H be the vertex-clique graph of complete bipartite graph $K_{2,3}$. Then $\mathcal{R}(H) = K_2 \square K_3$, where $\square$ denotes the Cartesian product of graphs. Note that the number of edges in $K_2 \square K_3$ that are themselves maximal cliques is three, and $Z_+(H) = 3$. In general, if H is the vertex clique graph of $K_{2,n}$, then $\mathcal{R}(H) = K_2 \square K_n$. In this case, $Z_+(\mathcal{R}(H)) = n$ and the number of edges in $\mathcal{R}(H)$ that form maximal cliques is precisely n.

11   Relations Between $Z_+(G)$ and $\mathrm{cc}(G)$

Lemma 15. [3] *For any graph G, $|V(G)| - \mathrm{cc}(G) \leq Z_+(G)$.*

Our first result gives a large family of graphs for which the positive zero forcing number is equal to the difference between the number of vertices and the edge clique covering number.

Theorem 7. *If G has a min-max clique cover with simple intersection and the compressed cliques graph has no induced cycles, other than K_3, then $Z_+(G) = |V(G)| - \mathrm{cc}(G)$.*

If G consists of a series of cliques $C_1, \ldots, C_k$ in which only consecutive cliques intersect, then we call G a *path of cliques*. In this case we have equality in the inequality of Lemma 15.

Corollary 5. *Let G be a graph that is a path of cliques, then $|V(G)| - \mathrm{cc}(G) = Z_+(G)$.*

The next family of graphs that we consider are a generalization of the *musical graph* M_n defined in [12]. For $n \geq 3$, the graph M_n has $2n$ vertices, $5n$ edges, and is isomorphic to the Cayley graph $Cay(\mathbb{Z}_{2n}, \{\pm 1, \pm(n-1), n\})$.

The graph M_n has a min-max clique covering with n cliques that satisfies simple intersection. The positive zero forcing number for M_n can then be easily calculated.

Lemma 16. *If $n \geq 3$, then $Z_+(M_n) = n + 2$.*

The musical graph is an example of a graph for which $Z_+(G) \leq |V(G)| - |\text{cc}(G)| + 2$. We can generalize this to a family of related graphs. Let G be a graph for which $\{C_1, C_2, \ldots, C_\ell\}$ is a min-max clique covering. If each of $C_i \cap C_{i+1}$ for $i = 1, \ldots, \ell - 1$, as well as $C_1 \cap C_\ell$, are non-empty, while all other intersections of these cliques are empty, then G is called a *cycle of cliques*. These graphs are a generalization of the graphs defined as $C_t(K_n)$ [13].

Theorem 8. *If G is a cycle of cliques, then $Z_+(G) \leq |V(G)| - \text{cc}(G) + 2$.*

There is a subfamily of the cycles of cliques for which the positive zero forcing number is equal to the number of vertices minus the edge clique covering number of the graph.

Theorem 9. *Let G be a cycle of cliques labeled $\{C_1, C_2, \ldots, C_\ell\}$. If there are two cliques C_i and C_j with $C_{i,i}$ and $C_{j,j}$ non-empty, then $Z_+(G) = |V(G)| - \text{cc}(G)$.*

Our next example is a family of graphs for which there is a large gap between the positive zero forcing number and the number of vertices minus the edge clique covering number of the graph.

Define a graph $X(n; \ell_1, \ldots, \ell_k)$ that has vertices $x_1, \ldots, x_n$ which induce a clique. In addition, this graph also contains disjoint cycles $C_1, \ldots, C_k$, in which C_i has length ℓ_i, and each cycle contains exactly two vertices from the set $\{x_1 \ldots, x_n\}$. In this case, the number of vertices in this graph is $n + \sum_{i=1}^{k}(\ell_i - 2)$ and $\text{cc}(X) = 1 + \sum_{i=1}^{k}(\ell_i - 1)$. Moreover, the only such clique cover is a min-max clique cover that has simple intersection.

Theorem 10. *The graph $X = X(n; \ell_1, \ldots, \ell_k)$ defined above has a min-max clique covering with simple intersection such that*

$$Z_+(X(n; \ell_1, \ldots, \ell_k)) = n - 1 = |V(X)| - \text{cc}(X) + k.$$

12   Concluding Remarks

In this paper, we defined a simple intersecting clique covering of a graph. This means that any vertex of the graph is contained in at most two cliques in the covering. Clearly, this can be generalized. We will say a clique covering of a graph has *s-wise intersection* if any vertex is contains in at most s cliques in the clique covering. Then we could generalize the Johnson graph $J(n, s)$, to the graph $J'(n, s)$. The vertices of this graph will be subsets of size at most s from $\{1, \ldots, n\}$ and two vertices will be adjacent if and only if their sets are intersecting. Then every graph will have a clique covering that has s-wise intersection for s sufficiently large. Then for any graph G, the compressed cliques graph of G is an induced subgraph of $J'(n, s)$. Moreover, if we can determine the positive zero forcing number for the compressed cliques graph, then we can determine the positive zero forcing number for the original graph G.

Recenly, Yang [16] proved that computing the positive zero forcing number of a graph is NP-complete. From Theorem 5, we know that for graphs that have an edge clique covering satisfying simple intersection, we can determine the positive zero forcing number of the graph from the positive zero forcing number of the compressed cliques graph. From [5,11], we know that finding $cc(G)$ is fixed-paramerter tractable. Using this result and Theorem 5, we could show that if G has an edge clique covering satisfying simple intersection, then the problem of computing $Z_+(G)$ is fixed-paramerter tractable, parametered by the edge clique cover number of G. This could be extended to more general graphs.

References

1. AIM Minimum Rank-Special Graphs Work Group: Zero forcing sets and the minimum rank of graphs. Linear Algebra Appl. **428**(7), 1628–1648 (2008)
2. Barioli, F., Barrett, W., Fallat, S., Hall, H.T., Hogben, L., Shader, B., van den Driessche, P., van der Holst, H.: Parameters related to tree-width, zero forcing, and maximum nullity of a graph. J. Graph Theor. **72**, 146–177 (2013)
3. Barioli, F., Barrett, W., Fallat, S., Hall, H.T., Hogben, L., Shader, B., van den Driessche, P., van der Holst, H.: Zero forcing parameters and minimum rank problems. Linear Algebra Appl. **433**(2), 401–411 (2010)
4. Burgarth, D., Giovannetti, V.: Full control by locally induced relaxation. Phys. Rev. Lett. **99**(10), 100–501 (2007)
5. Cygan, M., Pilipczuk, M., Pilipczuk, M.: Known algorithms for edge clique cover are probably optimal. In: Proceedings of the 24th ACM-SIAM Symposium on Discrete Algorithms (SODA), pp. 1044–1053 (2013)
6. Doob, M.: Spectral graph theory. In: Gross, J.L., Yellen, J. (eds.) Handbook of Graph Theory. CRC Press, Boca Raton (2004)
7. Ekstrand, J., Erickson, C., Hall, H.T., Hay, D., Hogben, L., Johnson, R., Kingsley, N., Osborne, S., Peters, T., Roat, J., et al.: Positive semidefinite zero forcing. Linear Algebra Appl. **439**, 1862–1874 (2013)
8. Ekstrand, J., Erickson, C., Hay, D., Hogben, L., Roat, J.: Note on positive semidefinite maximum nullity and positive semidefinite zero forcing number of partial 2-trees. Electron. J. Linear Algebra **23**, 79–97 (2012)
9. Fallat, S., Meagher, K., Yang, B.: On the complexity of the positive semidefinite zero forcing number. Linear Algebra Appl. **491**, 101–122 (2016)
10. Faudree, R., Flandrin, E., Ryjacek, Z.: Claw-free graphs - a survey. Discrete Math. **164**, 87–147 (1997)
11. Gramm, J., Guo, J., Huffner, F., Niedermeier, R.: Data reduction and exact algorithms for clique cover. ACM J. Exp. Algorithmics **13** (2009)
12. Knuth, D.E.: The Art of Computer Programming. Introduction to Combinatorial Algorithms and Boolean Functions, vol. 4. Addison-Wesley Professional, New York (2008)
13. Peters, T.A.: Positive semidefinite maximum nullity and zero forcing number. Ph.D. thesis, Iowa State University (2012)
14. Severini, S.: Nondiscriminatory propagation on trees. J. Phys. A **41**(48) (2008)
15. Yang, B.: Fast-mixed searching and related problems on graphs. Theor. Comput. Sci. **507**(7), 100–113 (2013)
16. Yang, B.: Lower bounds for positive semidefinite zero forcing and their applications. Accepted J. Comb. Optim. (2015). http://dx.doi.org/10.1007/s10878-015-9936-0

Improved Algorithms for Several Parameterized Problems Based on Random Methods

Qilong Feng[✉], Xiong Jiang, and Jianxin Wang

School of Information Science and Engineering,
Central South University, Changsha 410083, People's Republic of China
csufeng@mail.csu.edu.cn

Abstract. In this paper, we apply random methods to solve several NP-hard problems. For the Weighted P_3-Packing problem, by randomly partitioning the vertices in given graph, a randomized parameterized algorithm of running time $O^*(32^k)$ is given. For the Weighted Load Coloring problem, a randomized parameterized algorithm of running time $O^*(11.32^k)$ is presented. For the Claw-free Edge Deletion problem on diamond-free graphs, a parameterized algorithm of running time $O^*(2.895^k)$ is given.

1 Introduction

Random methods have been used to solve many NP-hard problems in the field of parameterized computation [6,7,17,18]. For a given set U and an integer k, many problems are to find a subset S of size k in U with certain properties. Generally, the random partition on U is to divide the elements in U into two subsets U_1 and U_2 such that with certain probability, for any subsets S_1 and S_2 of S, where $|S_1| + |S_2| = k$, S_1 and S_2 are contained in U_1 and U_2, respectively. It is possible that U can be partitioned into more than two parts. This method is called random-partition method. For many other problems, especially for the problems that can be solved using branching methods, assume that in each branching step of a given instance, r elements should be chosen from d candidate elements to be contained in objective solution. By applying branching method, a branching recurrence $T(k) \leq \binom{d}{r} T(k-r)$ can be obtained, where k is the size of objective solution, and $T(k)$ is the running time of the algorithm solving the problem. It is easy to see that the running time of the algorithm is bounded by $O^*(\binom{d}{r}^{k/r})$. However, for the problem with r elements chosen from d candidate elements in each step, if choose an element randomly from the given d elements and put the chosen element into objective solution, then with probability r/d, the element can be correctly handled. Therefore, the given problem can be solved randomly in time $O^*((d/r)^k)$ with certain probability. It is easy to see that $(d/r)^k \leq \binom{d}{r}^{k/r}$.

This work is supported by the National Natural Science Foundation of China under Grants (61232001, 61472449, 61572190, 61420106009).

D. Zhu and S. Bereg (Eds.): FAW 2016, LNCS 9711, pp. 65–74, 2016.
DOI: 10.1007/978-3-319-39817-4_7

This method is called random-choosing method. In this paper, we apply random-partition and random-choosing methods to several parameterized problems and get corresponding randomized algorithms respectively.

Parameterized Weighted P_3-Packing Problem. Packing problems are important class of NP-hard problems, which have lots of applications in scheduling [2] and code optimization [13]. Given a graph G and a subgraph H of G, the subgraph packing problem is to find the maximum number of vertex disjoint subgraphs such that each subgraph is isomorphic to H. If H is a K_2, the subgraph packing problem becomes the famous maximum matching problem, which can be solved in polynomial time [10]. The subgraph packing problem is NP-hard when the number of vertices in H is not less than three [16].

Many results were presented for the subgraph packing problem from approximation algorithm perspective [12,14]. For a given subgraph H, a parameterized algorithm of running time $O^*(2^{O(|H|k\log k+k|H|\log|H|)})$ for the H-packing problem was given in [9]. In this paper, we study a special subgraph packing problem, called Weighted P_3-Packing, where a P_3 is a simple path with four vertices and three edges. For a given graph $G = (V, E)$, a P_3-Packing $\mathcal{P}$ of G is a collection of vertex-disjoint P_3s. The definition of the problem is as follows.

Parameterized Weighted P_3-Packing problem: Given a graph $G = (V, E)$ and an integer k, where each edge is associated with a positive weight, find a P_3-Packing of size k in G with maximum weight, or report that no such packing exists.

For the unweighted P_3-Packing problem, a parameterized algorithm of running time $O^*(4.18^{4k})$ is given in [3]. Since the Weighted P_3-Packing problem can be reduced to the Weighted 4-Set Packing problem, all the algorithms for the Weighted 4-Set Packing problem can be applied to solve the Weighted P_3-Packing problem. The Weighted d-Set Packing problem ($d \geq 4$) has been extensively studied in the literature (see, for example [5,19]). In particular, Zehavi [19] presented a parameterized algorithm for Weighted d-Set Packing problem with running time $O^*((0.563 \cdot 2.851^d)^k)$, implying that the Weighted P_3-Packing problem can be solved in time $O^*(37.2^k)$, which is the current best result for the problem. For the unweighted 4-Set Packing, Björklund [4] presented a randomized parameterized algorithm of running time $O^*(1.642^{4k})$ based on algebraic technique. However, it is unknown whether the methods in [4] can be applied to the weighted case.

In this paper, we apply random-partition method to solve the Weighted P_3-Packing problem. We give that the Weighted P_3-Packing problem on tripartite graphs can be solved in polynomial time, and a parameterized algorithm of running time $O^*(32^k)$ for the Weighted P_3-Packing problem is given.

Parameterized Weighted Load Coloring Problem. The other problem applied random-partition method is Load Coloring problem, which has applications in computer networks. For a given graph $G = (V, E)$, a coloring on vertex set V is a mapping function f from the vertices in V to color set, i.e., $f : V \to C$.

We assume that the color set C just contains red and blue colors. In a colored graph $G = (V, E)$, for an edge $[u, v]$, if both u and v are colored red, then $[u, v]$ is called a *red edge*. Similarly, if both u and v are colored blue, then $[u, v]$ is called a *blue edge*. The Parameterized Load Coloring problem is defined as follows.

Parameterized Load Coloring: Given a graph $G = (V, E)$ and an integer k, find a coloring of G such that at least k edges in G are colored red and at least k edges in G are colored blue, or report that no such coloring exists.

The Load Coloring problem is proved to be NP-hard in [11]. Gutin and Jones [11] gave a parameterized algorithm of running time $O^*(4^k)$. In this paper, we consider a generalized version of the Parameterized Load Coloring problem, the Parameterized Weighted Load Coloring problem. Based on the random-partition method, the vertices in the given instance can be partitioned into two parts, which results in a parameterized algorithm of running time $O^*(11.32^k)$.

Parameterized Claw-Free Edge Deletion Problem on Diamond-Free Graphs. Given a graph G and an integer k, the Parameterized Claw-free Edge Deletion problem is to delete at most k edges such that the remaining graph has no induced claw ($K_{1,3}$). A trivial algorithm of running time $O^*(3^k)$ can be obtained for the Parameterized Claw-free Edge Deletion problem. Whether this trivial result can be improved is still open. Diamond structure plays important role in the study of claw-free problems. Cygan et al. [8] proved that the claw and diamond free edge deletion problem admits polynomial kernel. In this paper, we focus on parameterized algorithm of the Parameterized Claw-free Edge Deletion problem on diamond-free graphs. By applying random-choosing technique, a parameterized algorithm of running time $O^*(2.895^k)$ is given.

2 Algorithms for the Parameterized Weighted P_3-Packing Problem

For a given P_3 $l = (a, b, c, d)$, a, d are called the end-points of l, and b, c are called the internal-points of l. For P_3-Packing P, let $V(P)$ denote the set of vertices contained in the P_3s of P. For a graph G, a matching M of G is a set of edges in G such that no two edges in M have common endpoint. For a subset $E' \subseteq E$, let $V(E')$ denote the set of vertices contained in the edges of E'. For an edge $[u, v]$ in G, let $wt[u, v]$ denote the weight of edge $[u, v]$.

2.1 Weighted P_3-Packing on Tripartite Graphs

We first study a constrained Weighted P_3-Packing problem on tripartite graphs. Given a tripartite graph $H = (X \cup Y \cup Z, E_H)$, and a P_3-Packing P in H, for each P_3 l of P, if the end-points of l are from X, one internal-point of l is from Y, and the other one is from Z, then P is called a special P_3-Packing of H. We first give the definition of the Weighted P_3-Packing problem on tripartite graphs.

Constrained Weighted P_3-Packing on Tripartite Graphs: Given a tripartite graph $H = (X \cup Y \cup Z, E_H)$ and an integer k, where each edge is associated with a positive weight, find a special P_3-Packing of size k in H with maximum weight, or report that no such packing exists.

For any instance $(H = (X \cup Y \cup Z, E_H), k)$ of the Constrained Weighted P_3-Packing on Tripartite Graphs problem, if any vertex u in $Y \cup Z$ has no neighbor in X, then u can be deleted. Thus, in the following, we assume that each vertex in $Y \cup Z$ must have at least one neighbor in X. Based on tripartite graph H, a weighted auxiliary graph H' can be constructed in the following way.

For a tripartite graph $H = (X \cup Y \cup Z, E_H)$, let $|X|, |Y|$ and $|Z|$ denote the number of vertices in X, Y and Z, respectively. The auxiliary graph H' contains the vertex set $X \cup Y \cup Z \cup Y' \cup Z'$, where Y' and Z' are the copies of Y and Z, respectively. All the edges in H are put into H'. For each vertex u' in Y', let u be the corresponding vertex in Y. If $[u, v]$ $(v \in X)$ is an edge in H, then add edge $[u', v]$ into H' with same weight as $[u, v]$, and delete $[u, v]$ from H'. For each vertex w' in Z', let w be the corresponding vertex in Z. If $[w, v]$ $(v \in X)$ is an edge in H, then add edge $[w', v]$ into H' with same weight as $[w, v]$, and delete edge $[w, v]$ from H'. Let a be the sum weights of the edges in $\{[u, v] | u \in Y' \cup Z', v \in X, [u, v]$ is an edge in $H'\}$. Moreover, for each edge $[u, v]$ with $u \in Y, v \in Z$, assign weight $2a + 1 + wt[u, v]$. Let b be the sum weights of edges in $\{[u, v] | u \in Y, v \in Z, [u, v]$ is an edge in $H'\}$. For each vertex u in Y and its corresponding vertex u' in Y', add edge $[u, u']$ into H' with weight $2b + 1$. Similarly, for each vertex v in Z and its corresponding vertex v' in Z', add edge $[v, v']$ into H' with weight $2b + 1$. Let $\bar{E} = \{[u, u'] | u \in Y, u' \in Y', [u, u']$ is an edge in $H'\} \cup \{[v, v'] | v \in Z, v' \in Z', [v, v']$ is an edge in $H'\}$, and $\tilde{E} = \{[v, w] | v \in Y, w \in Z, [v, w]$ is an edge in $H'\}$.

Lemma 1. *For a tripartite graph $H = (X \cup Y \cup Z, E_H)$, an integer k, and any real number $d \geq 0$, the tripartite graph H has a special P_3-Packing of size k with weight d if and only if the auxiliary graph H' has a matching M of size $|Y| + |Z| + k$ with weight $(|Y| + |Z| - 2k)(2b + 1) + k(2a + 1) + d$. Moreover, the special P_3-Packing of H can be constructed from M in polynomial time.*

Based on Lemma 1, we can get the following result for the Constrained Weighted P_3-Packing on Tripartite Graphs problem.

Theorem 1. *The Constrained Weighted P_3-Packing on Tripartite Graphs problem can be solved in time $O(nm + n^2 \log n)$, where n and m are the number of vertices and edges in the input graph, respectively.*

2.2 Randomized Algorithm for Parameterized Weighted P_3-Packing

Given an instance $(G = (V, E), k)$ of the Parameterized Weighted P_3-Packing problem, assume that $P = \{l_1, l_2, \cdots, l_k\}$ is a P_3-Packing of size k with maximum weight in G. It is easy to see that $V(P)$ contains $4k$ vertices, consisting of $2k$ end-points and $2k$ internal-points. In order to solve the Parameterized Weighted P_3-Packing problem by the algorithm solving Constrained Weighted

P_3-Packing on Tripartite Graphs problem, we need to partition the vertices in G into three parts V_1, V_2 and V_3 such that for each l_i ($1 \leq i \leq k$), the two end-points of l_i is in V_1, one internal-point of l_i is V_2, and the other one is in V_3. For any vertex v in G, we do the following random process: v is put into V_1 with probability $1/2$; v is put into V_2 with probability $1/4$; v is put into V_3 with probability $1/4$. The probability that the $2k$ end-points of $V(P)$ are put into V_1, k internal-points of $V(P)$ are in V_2, and the other k internal-points are in V_3, is $(1/2)^{2k}(1/4)^k(1/4)^k2^k = (1/32)^k$ (Note that for a P_3 (u, v, w, r) in $\mathcal{P}$, in a successful partition, any vertex from v and w can be in V_2, and the other in V_3). A tripartite graph H can be constructed in the following way. Let $V_1 \cup V_2 \cup V_3$ be the vertex set of H. First delete all the edges in G whose two endpoints are both in V_1, V_2, or V_3, and for any vertex u in $V_2 \cup V_3$, if u has no neighbor in V_1, then delete u from V_2 or V_3. Let $H = (V_1 \cup V_2 \cup V_3, E_H)$ be the tripartite graph constructed by the above process. Then, an instance (H, k) of the Constrained Weighted P_3-Packing on Tripartite Graphs problem can be obtained. The specific process solving the Parameterized Weighted P_3-Packing problem is given in Fig. 1.

Algorithm R-P3P(G, k)
Input: a weighted graph G, and parameter k
Output: a P_3-Packing of size k with maximum weight in G, or report no such packing exists.
1. $\mathcal{Q} = \emptyset$;
2. **loop** $c \cdot 32^k$ times
2.1   randomly partition the vertices of G into three sets V_1, V_2 and V_3;
2.2   construct a weighted tripartite graph $H = (V_1 \cup V_2 \cup V_3, E_H)$ from G by removing the edges with both endpoints contained in V_1, V_2, or V_3, and deleting the vertices in $V_2 \cup V_3$ with no neighbor in V_1;
2.3   construct a special P_3-Packing P of size k with maximum weight in H;
2.4   **if** $P \neq \emptyset$ and the weight of P_3-Packing in $\mathcal{Q}$ is less than the weight of P **then** replace the P_3-Packing in $\mathcal{Q}$ with P;
3. **if** $\mathcal{Q} \neq \emptyset$ **then** return $\mathcal{Q}$;
4. return("no such packing exists.").

Fig. 1. Randomized algorithm for Parameterized Weighted P_3-Packing problem

Theorem 2. *The Parameterized Weighted P_3-Packing problem can be solved randomly in time $O^*(32^k)$.*

3   Randomized Algorithm for Weighted Load Coloring

In this section, we consider the Parameterized Weighted Load-Coloring problem, where each edge in the given graph has a positive weight. For a given weighted graph $G = (V, E)$ and a coloring f in G, let E_{fr} and E_{fb} be the set of red edges and blue edges under coloring f, respectively. Assume that the number of edges in E_{fr} and E_{fb} are at least k. The k edges in E_{fr} with maximum sum weight can be found, whose sum weight is denoted by W_r. Similarly, the k edges in E_{fb} with maximum sum weight can also be found, whose sum weight is denoted by W_b. For coloring f, let W_f denote the weight of f, and define W_f to be the value $W_r + W_b$. The Parameterized Weighted Load Coloring problem can be defined as follows.

Parameterized Weighted Load Coloring: Given a weighted graph $G = (V, E)$ and an integer k, find a maximum weighted coloring f of G such that at least k edges in G are colored red, at least k edges in G are colored blue, or report that no such coloring exists.

The general idea solving the Parameterized Weighted Load Coloring problem is as follows: randomly partition the vertices of G into two parts V_1 and V_2; color all the vertices in V_1 red, and color all the vertices in V_2 blue. The specific process solving the Parameterized Weighted Load Coloring problem is given in Fig. 2.

Theorem 3. *For an instance (G, k) of the Parameterized Weighted Load Coloring problem, if G can be colored with k red edges and k blue edges, then with probability larger than $1 - (1/e)^c$, a maximum weighted coloring of G can be returned by Algorithm R-WLC in time $O^*(11.32^k)$ such that at least k edges are colored red and at least k edges are colored blue, where c is a constant.*

Proof. If (G, k) is a no-instance, then no matter how the vertices in G are partitioned in step 2.1, the conditions in step 2.4 are not satisfied, which can be handled correctly by step 3, i.e., "no such coloring exists." is returned.

Assume that f is the maximum weighted coloring of G, i.e., k red edges with maximum sum weight W_r and k blue edges with maximum sum weight W_b are contained in the colored graph. Let E_1 be the set of k edges to get weight W_r, and let E_2 be the set of k edges to get weight W_b. Let V_r be the set of vertices contained in E_1, and let V_b be the set of vertices contained in E_2. If the vertices in V_r and V_b are partitioned correctly in step 2.1 (Without loss of generality, assume that V_r is partitioned into V_1, and V_b is partitioned into V_2.), by coloring the vertices in V_1 and V_2 by red and blue, respectively, at least k red edges can be found in $G[V_1]$, and at least k blue edges can be found in $G[V_2]$. Thus, a maximum weighted coloring can be returned in step 2.4.

We now discuss the probability that Algorithm R-WLC fails to find the maximum weighted coloring. Assume that there does not exist isolated vertex in $G[V_r]$ and $G[V_b]$. Assume that in subgraph $G[V_r]$, the number of connected components containing i vertices is r_i, where $2 \leq i \leq k$, and in subgraph $G[V_b]$, the number of connected components containing j vertices is t_j, where $2 \leq j \leq k$. For a connected component X in $G[V_r]$ or $G[V_b]$, if X contains i vertices, then the number of edges in X is at least $i - 1$. It is easy to get that $\sum_{i=2}^{k}(i - 1)r_i \leq k$, $\sum_{j=2}^{k}(j - 1)t_j \leq k$. The number of vertices in $G[V_r]$ and $G[V_b]$ are $\sum_{i=2}^{k} ir_i$ and $\sum_{j=2}^{k} jt_j$, respectively.

Algorithm R-WLC(G, k)
Input: a weighted graph G, and parameter k
Output: a maximum weighted coloring in G with at least k red edges and k blue edges, or report no such coloring exists.

1. $F = \emptyset$;
2. **loop** $c \cdot 11.32^k$ times
   2.1 randomly partition the vertices of G into two disjoint sets V_1, V_2;
   2.2 delete all the edges with one endpoint in V_1 and the other one in V_2;
   2.3 color the vertices in V_1 and V_2 with red and blue, respectively, denoted the coloring by f;
   2.4 **if** there are k red edges in $G[V_1]$ and k blue edges in $G[V_2]$ **then**
       if F is empty **then**
       add f into F;
       else if the weight of the coloring in F is smaller than the weight of f
       then replace the coloring in F with f;
3. **if** F is not empty **then** return the coloring in F; **else** return ("no such coloring exists.").

Fig. 2. Randomized algorithm for Parameterized Weighted Load Coloring

Without loss of generality, assume that $r_2 \le t_2$. In step 2.1, each vertex is put into V_1 or V_2 with probability $1/2$. Thus, the probability that the vertices in V_r and V_b are correctly partitioned is $2^{r_2}(1/2)^{\sum_{i=2}^{k} i r_i}(1/2)^{\sum_{j=2}^{k} j t_j} = (1/2)^{r_2 + \sum_{i=3}^{k} i r_i + \sum_{j=2}^{k} j t_j} \le (1/2)^{\sum_{i=2}^{k}(i-1)r_i + \sum_{j=2}^{k}(j-1)t_j + \sum_{i=3}^{k} r_i + \sum_{j=2}^{k} t_j}$. Based on the inequalities, $\sum_{i=2}^{k}(i-1)r_i \le k$, $\sum_{j=2}^{k}(j-1)t_j \le k$, we can get that the maximum value of $\sum_{i=3}^{k} r_i$ is $k/2$ and the maximum value of $\sum_{j=2}^{k} t_j$ is k. Therefore, $(1/2)^{\sum_{i=2}^{k}(i-1)r_i + \sum_{j=2}^{k}(j-1)t_j + \sum_{i=3}^{k} r_i + \sum_{j=2}^{k} t_j} \le (1/2)^{2k + k/2 + k} = (1/2)^{3.5k}$. Consequently, in each loop of step 2, the probability that the vertices in V_r and V_b are not correctly partitioned is $1 - (1/2)^{3.5k}$. Then, the probability that none of the $c \cdot 11.32^k$ loops can correctly divide the vertices in V_r and V_b is $(1 - (1/2)^{3.5k})^{c \cdot 11.32^k} \le (1/e)^c$. Therefore, Algorithm R-WLC can correctly partition the vertices in V_r and V_b with probability larger than $1 - (1/e)^c$.

For the running time, step 2.1 can be done in $O(n)$ time, where n is the number of vertices in G. Steps 2.3, 2.4 run in time $O(n + m)$, where m is the number of edges in G. Therefore, the running time of algorithm R-WLC is $O(11.32^k k(m + n)) = O^*(11.32^k)$. □

3.1 Algorithm for the Parameterized Claw-Free Edge Deletion Problem on Diamond-Free Graphs

For a given graph $G = (V, E)$, a subgraph C with four vertices $\{v, u_1, u_2, u_3\}$ and three edges $\{[v, u_1], [v, u_2], [v, u_3]\}$ is called a claw of G, and a subgraph D with four vertices $\{u, v, w_1, w_2\}$ and five edges $\{[u, v], [u, w_1], [u, w_2], [v, w_1], [v, w_2]\}$ is called a diamond of G. For claw C, v is called central-vertex, and u_1, u_2, u_3

are called fringe-vertices. A graph $G = (V, E)$ is a Claw-free Graph if G has no induced subgraph that is isomorphic to claw. For a subset $E' \subseteq E$ of edges, let $G[E']$ denote the subgraph induced by the vertices contained in $V(E')$.

For the Parameterized Claw-free Edge Deletion problem, diamond structure is one of the major obstacle to get algorithm better than $O^*(3^k)$. In this paper, we apply random-choosing method to deal with the Parameterized Claw-free Edge Deletion problem on diamond-free graphs.

For any vertex v with degree larger than two in G and for any vertex u in $N(v)$, if there is no edge from u to any vertex in $N(v)$, then it is called that v has a dangled-edge $[v, u]$. Assume that there exists a vertex w in $N(v)\backslash\{u\}$ with $[u, w]$ in G. Since G is diamond-free graph, w is the unique vertex in $N(v)\backslash\{u\}$ connected to u. It is called that v has a dangled-triangle (v, u, w). For a given instance (G, k) of the Parameterized Claw-free Edge Deletion problem on diamond-free graphs, assume that $D \subseteq E$ is of size k and graph $G[E\backslash D]$ is a claw-free graph. In the following, we give the strategy to deal with the claws in G based on the number of dangled-edges and dangled-triangles of claws.

(1) for any vertex v with r $(r \geq 4)$ dangled-edges, let H_v be the set of dangled-edges of v. Randomly choose one edge e from H_v, and put it into D.

   Since v has r $(r \geq 4)$ dangled-edges, at least $r/2$ edges from $H(v)$ must be in D. Therefore, by randomly choosing one edge to put into D, with probability at least $1/2$, this case can be rightly handled.

(2) for any vertex v with r $(r \geq 3)$ dangled-triangles, let T_v be the set of dangled-triangles of v. Randomly choose one triangle (v, u, w) from T_v, and put edges $[v, u], [v, w]$ into D.

   Since v has r $(r \geq 3)$ dangled-triangles, at least $\lfloor r/2 \rfloor$ triangles from T_v must be deleted. Therefore, by randomly choosing one triangle from T_v to delete, with probability at least $1/3$, this case can be rightly handled. Since at least two edges are put into D each time, the number of executions of this case is bounded by $k/2$.

(3) for any vertex v with two dangled-triangles and one dangled-edge, assume that the two dangled-triangles are $(v, u_1, w_1), (v, u_2, w_2)$, and the dangled-edge is $[v, x]$.

   Instead of solving this case randomly, we deal with it using branching method. In order to destroy the claw structure, we have the following three cases. (1) edges $[v, u_1], [v, w_1]$ are deleted; (2) edges $[v, u_2], [v, w_2]$ are deleted; (3) edge $[v, x]$ is deleted. The branching factor is $(2, 2, 1)$.

(4) for any vertex v with one dangled-triangle and two dangled-edges, assume that the dangled-triangle is (v, u_1, w_1), and the dangled-edges are $[v, x], [v, y]$. We still deal with this case using branching method. In order to destroy the claw structure, we have the following three cases. (1) edges $[v, u_1], [v, w_1]$ are deleted; (2) edge $[v, x]$ is deleted; (3) edge $[v, y]$ is deleted. The branching factor is $(2, 1, 1)$.

   For any claw C with v as central vertex, and a dangled edge $[v, u]$ of v, if there is no other claw except C containing $[v, u]$, then $[v, u]$ is called a unique-claw edge, otherwise, $[v, u]$ is called a multiple-claw edge.

(5) for any claw C with v as central vertex and $[v, u], [v, x], [v, y]$ as unique-claw edges, arbitrarily choose any edge from $\{[v, u], [v, x], [v, y]\}$ and delete. Under this case, any edge from $\{[v, u], [v, x], [v, y]\}$ can be deleted to destroy claw C.

(6) for any claw C with v as central vertex, $[v, u], [v, x]$ as unique-claw edges, and $[v, y]$ as a multiple-claw edge, edge $[v, y]$ can be deleted directly. Under this case, since $[v, u], [v, x]$ are unique-claw edges, edge $[v, y]$ can be deleted to destroy claw C.

(7) for any claw C with v as central vertex, $[v, u]$ as unique-claw edge, and $[v, x], [v, y]$ as multiple-claw edges, randomly choose one edge from $\{[v, x], [v, y]\}$ and delete. Under this case, at least one of $\{[v, x], [v, y]\}$ must be deleted to destroy claw C. By randomly choosing one edge from $\{[v, x], [v, y]\}$, with probability $1/2$, this case can be rightly handled.

(8) for any claw C with v as central vertex, and $[v, u], [v, x], [v, y]$ as multiple-claw edges, randomly choose one edge from $\{[v, u], [v, x], [v, y]\}$ and delete. Under this case, at least one edge from $\{[v, u], [v, x], [v, y]\}$ can be deleted to destroy claw C. By randomly choosing one edge from $\{[v, u], [v, x], [v, y]\}$, with probability $1/3$, this case can be rightly handled. Without loss of generality, assume that edge $[v, u]$ is deleted. Then, at least two edges will become unique-claw edges. Therefore, the number of executions of this case is bounded by $k/3$.

The algorithm solving the Parameterized Claw-free Edge Deletion problem on diamond-free graphs is based on the above cases. When case (2) is executed $k/2$ times, case (4) is executed $k/6$ times, and case (8) is executed $k/3$ times, we can get the worst running time, i.e., the running time of the algorithm solving the Parameterized Claw-free Edge Deletion problem on diamond-free graphs is bounded by $3^{k/3}3^{k/2}2.42^{k/6} = 2.895^k$. Summarizing the above discussion, we can get the following result.

Theorem 4. *The Parameterized Claw-free Edge Deletion problem on diamond-free graphs can be solved randomly in time $O^*(2.895^k)$.*

References

1. Ahuja, N., Baltz, A., Doerr, B., Privtivy, A., Srivastav, A.: On the minimum load coloring problem. J. Discrete Algorithms 5(3), 533–545 (2007)
2. Bar-Yehuda, R., Halldórsson, M., Naor, J., Shachnai, H., Shapira, I.: Scheduling split intervals. In: Proceedings of the 13th Annual ACM-SIAM Symposium on Discrete Algorithms, pp. 732–741 (2002)
3. Binkele-Raible, D., Fernau, H.: Packing paths: recycling saves time. Discrete Appl. Math. **161**, 1686–1689 (2013)
4. Björklund, A., Husfeldt, T., Kaski, P., Koivisto, M.: Narrow sieves for parameterized paths and packings (2010). CoRR abs/1007.1161
5. Chen, J., Feng, Q., Liu, Y., Lu, S., Wang, J.: Improved deterministic algorithms for weighted matching and packing problems. Theoret. Comput. Sci. **412**(23), 2503–2512 (2011)

6. Chen, J., Lu, S.: Improved parameterized set splitting algorithms: a probabilistic approach. Algorithmica **54**(4), 472–489 (2008)
7. Chen, J., Lu, S., Sze, S.H., Zhang, F.: Improved algorithms for path, matching, and packing problems. In: Proceedings of the 17th Annual ACM-SIAM Symposium on Discrete Algorithms, pp. 298–307 (2007)
8. Cygan, M., Pilipczuk, M., Pilipczuk, M., Leeuwen, E., Wrochna, E.: Polynomial kernelization for removing induced claws and diamonds (2015). arXiv preprint arXiv:1503.00704
9. Fellows, M., Heggernes, P., Rosamond, F.A., Sloper, C., Telle, J.A.: Finding k disjoint triangles in an arbitrary graph. In: Hromkovič, J., Nagl, M., Westfechtel, B. (eds.) WG 2004. LNCS, vol. 3353, pp. 235–244. Springer, Heidelberg (2004)
10. Gabow, H.: Data structures for weighted matching and nearest common ancestoers. In: Proceedings of the 1st Annual ACM-SIAM Symposium on Discrete Algorithms, pp. 434–443 (1990)
11. Gutin, G., Jones, M.: Parameterized algorithms for load coloring problem. Inf. Process. Lett. **114**, 446–449 (2014)
12. Hassin, R., Rubinstein, S.: An approximation algorithm for maximum triangle packing. Discrete Appl. Math. **154**, 971–979 (2006)
13. Hell, P., Kirkpatrick, D.: On the complexity of a generalized matching problem. In: Proceedings of the 10th Annual ACM Symposium on Theory of Computing, pp. 240–245 (1978)
14. Hurkens, C., Schrijver, A.: On the size of systems of sets every t of which have an SDR, with application to worst case ratio of heuristics for packing problems. SIAM J. Discrete Math. **2**, 68–72 (1989)
15. Kann, V.: Maximum bounded H-matching is MAX-SNP-complete. Inf. Process. Lett. **49**, 309–318 (1994)
16. Kirkpatrick, D.G., Hell, P.: On the complexity of general graph factor problems. SIAM J. Comput. **12**, 601–609 (1983)
17. Marx, D., Razgon, I.: Fixed-parameter tractability of multicut parameterized by the size of the cutset. In: Proceedings of the 43rd Annual ACM Symposium on Theory of Computing, pp. 469–478 (2011)
18. Marx, D.: Randomized techniques for parameterized algorithms. In: Thilikos, D.M., Woeginger, G.J. (eds.) IPEC 2012. LNCS, vol. 7535, p. 2. Springer, Heidelberg (2012)
19. Zehavi, M.: Deterministic parameterized algorithms for matching and packing problems (2013). arXiv:1311.0484v2

Parameterized Algorithms for Maximum Edge Biclique and Related Problems

Qilong Feng$^{(\boxtimes)}$, Zeyang Zhou, and Jianxin Wang

School of Information Science and Engineering,
Central South University, Changsha 410083, People's Republic of China
csufeng@mail.csu.edu.cn

Abstract. Maximum Edge Biclique and related problems have wide applications in management science, bioinformatics, etc. In this paper, we study the parameterized algorithms for the Parameterized Edge Biclique problem, the Parameterized Edge Biclique Packing problem, and the Parameterized Biclique Edge Deletion problem. For the Parameterized Edge Biclique problem, the current best result is of running time $O^*(2^k)$, and we give a parameterized algorithm of running time $O^*(k^{\lceil \sqrt{k} \rceil})$. For the Parameterized Edge Biclique Packing problem, based on randomized divide-and-conquer technique, a parameterized algorithm of running time $O^*(4^{(2k-1)t)}k^{\lceil \sqrt{k} \rceil})$ is given. We study the Parameterized Biclique Edge Deletion problem on bipartite graphs and general graphs, and give parameterized algorithms of running time $O^*(2^k)$ and $O^*(3^k)$, respectively.

1 Introduction

Maximum edge biclique problem is to find a biclique subgraph with maximum number of edges for a given bipartite graph G, which has wide applications in management science [11], machine learning [9], and bioinformatics [2, 4, 5, 13–15].

The weighted and unweighted maximum edge biclique problems are both NP-hard [6, 10]. Tan [12] studied the inapproximability of the maximum weighted edge biclique problem. Feige and Kogan [7] proved that the unweighted maximum edge biclique problem is hard to approximate in time $O(n^\epsilon)$ where $\epsilon > 0$. Dawande et al. [6] gave an approximation algorithm with expected ratio 2 for sufficiently dense random bipartite graphs. Tanay et al. [14] gave an exact algorithm for the weighted maximum edge biclique problem with running time $O(n2^d)$, where n is number of vertices in the graph, and d is the maximum degree in the graph. In this paper, we study the following parameterized problem.

Parameterized Edge Biclique: Given a bipartite graph $G = (L \cup R, E)$ and an integer k, find a biclique subgraph of at least k edges of G, or report that no such subgraph exists.

This work is supported by the National Natural Science Foundation of China under Grants (61232001, 61472449, 61572414, 61420106009).

D. Zhu and S. Bereg (Eds.): FAW 2016, LNCS 9711, pp. 75–83, 2016.
DOI: 10.1007/978-3-319-39817-4_8

The algorithm in [14] implies a parameterized algorithm of running time $O^*(2^k)$ for the Parameterized Edge Biclique problem. In this paper, we give new structure properties for the problem, which result in a parameterized algorithm of running time $O^*(k^{\lceil\sqrt{k}\rceil})$.

Acuña et al. [1] studied a related model of the maximum edge biclique problem, called edge biclique packing problem, where for a given bipartite graph G and a positive integer k, find a set of t vertex-disjoint bicliques such that the number of edges inside the bicliques is maximized. As pointed out in [1], the problem has backgrounds in bioinformatics and consumer products bundling. For the data analysis in bioinformatics, the relationship between individuals and conditions can be modeled by a bipartite graph. Let one side be set of genes, the other side be set of conditions, and the edge between a gene and a condition means that the gene satisfies the condition. The objective is to find a set of clusters of genes with common conditions. The edge biclique packing problem was also applied in metabolic networks. For the consumer product bundling problem, the relationship between products and clients can also be modeled by a bipartite graph $G = (L \cup R, E)$, where vertices in L represent products and vertices in R represent clients and the edge between a client and a product means that the client consumes the product. The objective is to find t product bundlings that maximize the supplied demand. In this paper, we study the following parameterized problem.

Parameterized Edge Biclique Packing: Given a bipartite graph $G = (L \cup R, E)$ and two integers t, k, find t vertex-disjoint bicliques of G, each of which contains at least k edges of G, or report that no such packing exists.

Based on randomized divide-and-conquer technique, a parameterized algorithm of running time $O^*(4^{(2k-1)t}k^{\lceil\sqrt{k}\rceil})$ is given.

The biclique deletion problem is to delete some elements of a given graph to make remaining graph a biclique. Hochbaum [8] studied several weighted biclique deletion problems, based on edges or vertices are deleted, the given graph is bipartite graph or general graph, and the two parts of the remaining graph is independent set or not. Hochbaum [8] proved the vertex deletion version on bipartite graphs and general graphs without the independent set requirement can be solved in polynomial time. The author also proved that other versions of the problem are NP-hard and gave a 2-approximation algorithm for the edge deletion on bipartite graph. In this paper, we study the following problems.

Parameterized Bipartite Biclique Edge Deletion: Given a bipartite graph $G = (L \cup R, E)$ and an integer k, delete a subset $E' \subseteq E$ of at most k edges of G such that the remaining graph is a biclique, or report that no such subset exists.

Parameterized General Biclique Edge Deletion: Given a graph $G = (V, E)$ and an integer k, delete a subset $E' \subseteq E$ of at most k edges of G such that the remaining graph is a biclique, or report that no such subset exists.

For the Parameterized Bipartite Biclique Edge Deletion problem, a parameterized algorithm of running time $O^*(2^k)$ is given. We also study the Parameterized General Biclique Edge Deletion problem under the constraint that the left and right parts of the objective biclique form independent set respectively, a detailed structure analysis of the problem is presented, which results in a parameterized algorithm of running time $O^*(3^k)$.

2   Algorithm for Parameterized Edge Biclique Problem

A biclique is a complete bipartite graph that each pair of vertices in different side has an edge. For a given bipartite graph $G = (L \cup R, E)$ and a biclique $B = (L' \cup R', E')$ of G, there are $|L'| + |R'|$ vertices and $|L'| * |R'|$ edges in B. For two vertices u, v in G, let $[u, v]$ denote the edge between u and v. For a vertex v, let $N(v)$ denote the set of neighbors of v, i.e., $N(v) = \{u | [u, v] \in E\}$. For a vertex v in G, let $deg(v)$ denote the degree of v.

Lemma 1. *Given a biclique $B = (L' \cup R', E')$ and two subsets $L_1 \subseteq L'$, $R_1 \subseteq R'$, where L_1 and R_1 are not empty, the subgraph induced by $L_1 \cup R_1$ is also a biclique.*

Lemma 2. *Given an instance (G, k) of the Parameterized Edge Biclique problem, if there exists a vertex v in G with degree at least k, then (G, k) is a yes-instance.*

For Lemma 2, if v has degree at least k, then the subgraph induced by $\{v\} \cup N(v)$ is a biclique with at least k edges.

Lemma 3. *If a given instance (G, k) of the Parameterized Edge Biclique problem is a yes-instance, then there exists a biclique B of at least k edges with left part L' and right part R' such that one of $\{L', R'\}$ contains at most $\lceil \sqrt{k} \rceil$ vertices.*

It is easy to see that if both L' and R' contain more than $\lceil \sqrt{k} \rceil$ vertices, then by choosing one part with at most $\lceil \sqrt{k} \rceil$ vertices, a biclique with at least k edges can be obtained. The algorithm solving the Parameterized Edge Biclique problem is given in Fig. 1.

Theorem 1. *The Parameterized Edge Biclique problem can be solved in time $O^*(k^{\lceil \sqrt{k} \rceil})$.*

Proof. Given an instance (G, k) of the Parameterized Edge Biclique problem, if (G, k) is a no-instance, then no matter how the vertices in G and their neighbors are enumerated, no biclique with at least k edges can be returned, which will be handled correctly in step 3.

Now assume that (G, k) is a yes-instance of the Parameterized Edge Biclique problem. If there exists a vertex v of degree at least k, then by Lemma 2, a biclique with at least k edges can be found based on v and its neighbors, which is rightly handled by steps 1-1.2. By Lemma 3, we can find a biclique C with

---

Algorithm APEB(G, k)
Input: a bipartite graph G, and parameter k
Output: a biclique with at least k edges, or report no such biclique exists.
1.   **if** there exists a vertex v with degree at least k **then**
1.1   let C be any subset of $N(v)$ of size k;
1.2   return the subgraph induced by $\{v\} \cup C$;
2.   **for** each vertex v in G **do**
2.1   **for** $i = 1$ to $\lceil \sqrt{k} \rceil$ **do**
2.2   **for** each subset S of size i of $N(v)$ **do**
2.3   if the vertices in S and the vertices in $\bigcap_{w \in S} N(w)$ induce a biclique
            with at least k edges **then**
2.4   return the biclique induced by $S \cup (\bigcap_{w \in S} N(w))$;
3.   return("no such biclique exists.").

Fig. 1. Algorithm for the Parameterized Edge Biclique problem

at least k edges of G with left part L' and right part R', such that at least one of L' and R' contains at most $\lceil \sqrt{k} \rceil$ vertices. Without loss of generality, assume that L' contains at most $\lceil \sqrt{k} \rceil$ vertices. For any vertex v in R', L' must be a subset of $N(v)$. Therefore, when v is chosen in step 2, all the subsets of $N(v)$ are enumerated, which contains L'. Then, C can be constructed by L' and $\bigcap_{w \in L'} N(w)$, which will be returned in step 2.4.

Steps 1-1.2 can be done in $O(n^2)$ time, and all the enumerations in step 2 can be done in $O(nk^{2.5}k^{\lceil \sqrt{k} \rceil})$. Therefore, the Parameterized Edge Biclique problem can be solved in time $O^*(k^{\lceil \sqrt{k} \rceil})$. $\square$

Based on "and-composition" method in [3], it is easy to get the following result.

Lemma 4. *The Parameterized Edge Biclique problem does not admit a polynomial kernel unless the polynomial hierarchy collapses to the third level.*

3   Algorithm for Parameterized Edge Biclique Packing Problem

Assume that (G, t, k) is an instance of the Parameterized Edge Biclique Packing problem.

Lemma 5. *If B is a biclique of k edges in G, then B contains at most $k + 1$ vertices.*

Proof. Let L' and R' be the left and right parts of B respectively. Then, $|L'| \cdot |R'| = k$. The number of vertices in B is $|L'| + |R'|$, which can get maximum value $k + 1$. $\square$

Given a biclique B with left part L' and right part R', for a non-empty subset $L'' \subset L'$ and a non-empty subset $R'' \subset R'$, the biclique induced by the vertices in $L'' \cup R''$ is called a sub-biclique of B. For a set of bicliques $Q = \{C_1, C_2, \cdots, C_t\}$, if no two bicliques in Q have common vertex, then Q is called a *t-packing*; if each biclique in Q has at least k edges, then Q is called a *proper t-packing*.

Lemma 6. *For a biclique B of G, if each sub-biclique of B contains at most $k - 1$ edges, then the number of edges in B is less than $2k$.*

Proof. Assume that $B = (L' \cup R', E')$. For each sub-biclique C of B, the maximum number of edges in C is bounded by $max\{(|L'| - 1)|R'|, |L'|(|R'| - 1)\}$, and we can get that

$$max\{(|L'| - 1) * |R'|, |L'| * (|R'| - 1)\} \leq k - 1$$
$$\begin{cases} (|L'| - 1) * |R'| \leq k - 1 \\ |L'| * (|R'| - 1) \leq k - 1 \end{cases}$$
$$2(|L'| * |R'|) - (|L'| + |R'|) \leq 2k - 2$$
$$2(|L'| * |R'|) - (|L'| * |R'| + 1) \leq 2k - 2$$
$$|L'| * |R'| \leq 2k - 1$$

$\square$

Lemma 7. *For a biclique B with at least $2k$ edges of G, there must exist a sub-biclique C of B such that the number of vertices in C is at most $2k$, and the number of edges in C is at least k and less than $2k$.*

Proof. Let C be a minimal sub-biclique with at least k edges of B such that the number of edges contained in any sub-biclique of C is less than k. By Lemma 6, the number of edges in C is less than $2k$. By Lemma 5, the number of vertices in C is bounded by $2k$. $\square$

The algorithm solving the Parameterized Edge Biclique Packing problem is given in Fig. 2.

Theorem 2. *For an instance (G, t, k) of the Parameterized Edge Biclique Packing problem, if G contains proper t-packing, then with probability larger than $1 - (1/e)^c$, Algorithm PEBP can return a proper t-packing in time $O^*(4^{(2k-1)t}k^{\lceil\sqrt{k}\rceil})$, where c is a constant.*

Proof. If the given instance (G, t, k) is a no-instance, no proper t-packing can be found in step 3.7, and an empty set will be returned in step 4.

Assume that (G, t, k) is a yes-instance of the Parameterized Edge Biclique Packing problem. If $t = 1$, then it is the Parameterized Edge Biclique problem, which can be solved in $O^*(k^{\lceil\sqrt{k}\rceil})$ time. Now suppose $t > 1$. By Lemma 7, there exists a proper t-packing Q_t with at most $2kt$ vertices. With probability at least $\binom{t}{t/2}/2^{2tk}$, in step 3.1, at most kt vertices of $t/2$ bicliques are partitioned into V_1, and the remaining vertices of the other $t/2$ bicliques are in V_2. Clearly, G_1

Algorithm PEBP(G, t, k)
Input: a bipartite graph G, and two integers t, k
Output: t vertex-disjoint bicliques, each of which contains at least k edges, or
report no proper t-packing exists
1.   **if** $t = 1$ **then**
1.1   **if** there exists a biclique C with at least k edges **then**
         return C; **else** return $\emptyset$;
2. $Q = \emptyset$;
3.   loop $c.2^{(2k-1)t}$ times **do**
3.1   randomly partition the vertices in G into two parts V_1 and V_2;
3.2   let G_1 be the graph induced by the vertices in V_1;
3.3   let G_2 be the graph induced by the vertices in V_2;
3.4 $Q_1 =$ PEBP$(G_1, \lceil t/2 \rceil, k)$;
3.5 $Q_2 =$ PEBP$(G_2, t - \lceil t/2 \rceil, k)$;
3.6   **if** $Q_1 \neq \emptyset$ and $Q_2 \neq \emptyset$ **then**
3.7 $Q = Q_1 \cup Q_2$;
4.   return Q.

Fig. 2. Algorithm for the Parameterized Edge Biclique Packing problem

and G_2 both contain a proper $t/2$-packing, and they are vertex-disjoint. Thus, by the inductive hypothesis, step 3.4 returns a proper $t/2$-packing and step 3.5 also returns a proper $t/2$-packing. Therefore, a proper t-packing can be obtained in step 3.7.

As explained earlier, the vertices in Q_t can be correctly partitioned into V_1, V_2 with probability at least $\binom{t}{t/2}/2^{2tk}$. Thus, the probability that step 3.1 does not partition the vertices in Q_t correctly is $1 - \binom{t}{t/2}/2^{2tk}$. Therefore, the probability that none of the $c \cdot 2^{(2k-1)t}$ executions of loops in step 3 correctly partitions the vertices in Q_t is $(1 - \binom{t}{t/2}/2^{2tk})^{c \cdot 2^{(2k-1)t}} < (1/e)^c$. Consequently, Algorithm PEBP can correctly construct a proper t-packing with probability larger than $1 - (1/e)^c$.

We now analyze the running time of the Algorithm PEBP. Let $T(t)$ be the running time of algorithm PEBP(G, t, k). Then, $T(1) = O(nk^{2.5}k^{\lceil \sqrt{k} \rceil})$. When $t > 1$, we can get the following recurrence relation: $T(t) = c \cdot 2^{(2k-1)t}(nk^{2.5}k^{\lceil \sqrt{k} \rceil} + T(\lceil t/2 \rceil) + T(t - \lceil t/2 \rceil)$, where c is a constant. It is easy to get that $T(t) \leq O(nk^{2.5}4^{(2k-1)t)}k^{\lceil \sqrt{k} \rceil})$. $\square$

4   Algorithm for Parameterized Bipartite Biclique Edge Deletion Problem

Assume that (G, k) is an instance of the Parameterized Bipartite Biclique Edge Deletion problem. We first deal with the vertices with degree at most k.

Lemma 8. *For a vertex v of G with degree at most k, a biclique containing v with maximum number of edges can be found in time $O(n^2k2^k)$, where n is the number of vertices in G.*

Since v has degree at most k, by enumerating all the possible vertices in $N(v)$, the biclique containing v with maximum number of edges can be obtained. The algorithm solving the Parameterized Bipartite Biclique Edge Deletion problem is given in Fig. 3.

Algorithm PBED(G, k, Q)
Input: a bipartite graph $G = (V_1 \cup V_2, E)$, an integer k, and a set Q of the deleted edges in G
Output: a set of k edges whose deletion makes the remaining graph a biclique if it exists
1. **if** $(|Q| > k)$ or $(|Q| = k)$ but the remaining graph by deleting the edges in Q is not a biclique **then** stop;
2. **if** $|Q| \leq k$ and the remaining graph by deleting the edges in Q is a biclique **then** return Q;
3. **if** there exists a vertex v with degree at most k **then**
3.1   find a biclique B containing v with maximum number of edges;
3.2   PBED$(G - (E - B), k - |E - B|, Q \cup (E - B))$;
3.3   PBED$(G \backslash \{v\}, k - deg(v), Q \cup \{[v, u] | u \in N(v)\})$.

Fig. 3. Algorithm for the Parameterized Bipartite Biclique Edge Deletion problem

Theorem 3. *For an instance (G, k) of the Parameterized Biclique Edge Deletion problem, by calling algorithm $PBED(G, k, \emptyset)$, the Parameterized Bipartite Biclique Edge Deletion problem can be solved in time $O^*(2^k)$.*

Proof. If (G, k) is a no-instance of the problem, then no Q can be returned in step 2, which is handled by step 1.

Assume that (G, k) is a yes-instance of the Parameterized Biclique Edge Deletion problem. For a vertex u of G, if the degree of u is at least $k + 1$, then u must be contained in the final biclique. Therefore, for any vertex v with degree at most k, it has two cases: either v is the final biclique, or the edges incident to v are deleted. If v is contained in final biclique, then by Lemma 8, a biclique B containing v with maximum number of edges can be found. If the number of edges in $E - B$ is less than $k - |Q|$, then $Q \cup (E - B)$ can be returned as a valid solution. If v is not contained in the final biclique, then all the edges incident to v must be deleted, which is handled by step 3.3.

By Lemma 8, step 3.1 can be done in $O(n^2k2^k)$. Since at most k edges can be deleted, step 3.3 can be executed k times. Therefore, the total running time of algorithm is bounded by $O^*(2^k)$. $\qquad\square$

5   Algorithm for Parameterized General Biclique Edge Deletion Problem

In this section, we focus on the Parameterized General Biclique Edge Deletion problem under the constraint that the left and right parts of the biclique form independent set respectively. Assume that a biclique B with left part L' and right part R' can be obtained by deleting at most k edges of G. Let GIBD(G, k, Q) be the algorithm solving the Parameterized General Biclique Edge Deletion problem, where G is a general graph, Q is the set of edges to be deleted. We first deal with the vertices with degree at most k.

For a vertex v with degree at most k, we deal with vertex v by the following cases.

(1) vertex v is contained in the final biclique.

Partition the vertices in $N(v)$ into two parts $N_1(v), N_2(v)$ such that the vertices in $N_1(v)$ is in the final biclique with v, and the vertices in $N_2(v)$ is not in the final biclique with v. Then, a biclique $B = (L' \cup R', E')$ with maximum number of edges can be found in time $O(n^2 k 2^k)$ such that the vertices in $N_1(v)$ are contained in R'. If the number of edges in $E - B$ is less than $k - |Q|$, then $Q \cup (E - B)$ can be returned as a valid solution.

(2) vertex v is not contained in the final biclique.

Under this case, we can get that GIBD$(G\backslash\{v\}, k - deg(v), Q \cup \{[v, u] | u \in N(v)\})$.

We now assume that graph G contains no vertex with degree at most k. For a vertex v of G, if the degree of v is at least $k + 1$, then v must be contained in the final biclique. For three vertices x, y, z in G, if there exist edges $[x, y], [x, z], [z, y]$, then x, y, z form a triangle, denoted by (x, y, z).

Lemma 9. *For any triangle (x, y, z) of G, in order to transform G into a biclique, at least one edge from $\{[x, y], [x, z], [z, y]\}$ is deleted.*

Proof. Assume that B is a biclique obtained from G by deleting edges such that the left and right parts of the biclique form independent set respectively. Let L' and R' be the left and right parts of B. Assume that no edge from $\{[x, y], [x, z], [z, y]\}$ is deleted. For vertices x, y, z, two vertices of $\{x, y, z\}$ are in one part of B, and one vertex is in the other part of B. Without loss of generality, assume that x, y are in L', and z is in R'. Since no edge from $\{[x, y], [x, z], [z, y]\}$ is deleted, edge $[x, y]$ is in $G[L']$, contradicting the fact that B is a biclique with each part being an independent set. □

Since all the vertices in G have degree at least $k + 1$ and all the vertices in G must be contained in final biclique, triangle is the only obstruction for the getting the final biclique. By Lemma 9, in order to get a biclique from G, we have to destroy all the triangles in G. For each triangle in G, we have three branchings to delete the edges in triangle.

Summarizing above discussion, we can get the following result for the General Biclique Edge Deletion problem.

Theorem 4. *The Parameterized General Biclique Edge Deletion problem can be solved in time $O^*(3^k)$.*

References

1. Acuna, V., Ferreira, C., Freire, A., Moreno, E.: Solving the maximum edge biclique packing problem on unbalanced bipartite graphs. Discrete Appl. Math. **164**, 2–12 (2014)
2. Ben-Dor, A., Chor, B., Karp, R., Yakhini, Z., Discovering local structure in gene expression data: the order-preserving submatrix problem. In: Proceedings of 6th Annual International Conference on Computational Biology (RECOMB), pp. 49–57 (2002)
3. Bodlaender, H., Downey, R., Fellows, M., Hermeliny, D.: On problems without polynomial kernels. J. Comput. Syst. Sci. **75**(8), 423–434 (2009)
4. Bu, S.: The summarization of hierarchical data with exceptions, Master Thesis, Department of Computer Science, University of British Columbia (2004)
5. Cheng, Y., Church, G.: Biclustering of expression data. In: Proceedings of 8th International Conference on Intelligent Systems for Molecular Biology (ISMB), pp. 93–103 (2000)
6. Dawande, M., Keskinocak, P., Tayur, S.: On the Biclique Problem in Bipartite Graphs, GSIA working paper, 1996–04 (1996)
7. Feige, U., Kogan, S.: Hardness of approximation of the balanced complete bipartite subgraph problem, Manuscript (2004)
8. Hochbaum, D.: Approximating clique and biclique problems. J. Algorithms **29**, 174–200 (1998)
9. Mishra, N., Ron, D., Swaminathan, R.: On finding large conjunctive clusters. In: Schölkopf, B., Warmuth, M.K. (eds.) COLT/Kernel 2003. LNCS (LNAI), vol. 2777, pp. 448–462. Springer, Heidelberg (2003)
10. Peeters, R.: The maximum edge biclique problem is NP-complete. Discrete Appl. Math. **131**(3), 651–654 (2003)
11. Swaminathan, J., Tayur, S.: Managing broader product lines through delayed differentiation using vanilla boxes. Manage. Sci. **44**, 161–172 (1998)
12. Tan, J.: Inapproximability of maximum weighted edge biclique and its applications. In: Agrawal, M., Du, D.-Z., Duan, Z., Li, A. (eds.) TAMC 2008. LNCS, vol. 4978, pp. 282–293. Springer, Heidelberg (2008)
13. Tan, J., Chua, K., Zhang, L., Zhu, S.: Algorithmic and complexity issues of three clustering methods in microarray data analysis. Algorithmica **48**(2), 203–219 (2007)
14. Tanay, A., Sharan, R., Shamir, R.: Discovering statistically significant biclusters in gene expression data. Bioinformatics **18**, 136–144 (2002)
15. Zhang, L., Zhu, S.: New clustering method for microarray data analysis. In: Proceedings of 1st IEEE Computer Society Bioinformatics Conference (CSB), pp. 268–275 (2002)

The Scheduling Strategy of Virtual Machine Migration Based on the Gray Forecasting Model

He Hong[(⊠)] and Cao Boyan

School of Mechanical, Electrical and Information Engineering,
Shandong University, Weihai 264209, China
hehong@sdu.edu.cn

Abstract. In stage of Infrastructures Providers (IP) provides cloud service for Service Providers (SP), in order to maximize the profits of IP, saving energy and reducing consumption is taken into consideration. As a new application mode of virtualization technology, virtual machine migration is of great practical meaning. We present a forecast model based on gray and credibility ant colony scheduling algorithm for virtual machine migration scheduling policy. The model can estimate the future utilization of a period of virtual machine CPU node. In determining whether the virtual machine should be moved out, the dual-threshold mechanism is set up to avoid frequent migration shocks caused by the transient oscillation of CPU resource utilization. Positioning probability is defined to improve the convergence speed when target node is selected. The experiments show that the algorithm can effectively avoid the frequent migration of virtual machine, which is a result of the shock caused by the change in CPU utilization, reduce energy consumption, and improve IP gains.

Keywords: Virtual machine · Online migration · Gray forecasting model · Dual-threshold · Positioning probability

1 Introduction

As a new computing model, the cloud computing can provide a flexible and on-demand storage and computing resources to users via computer network. Virtualization technology as enabler and important technical support of Cloud computing, is a way to be able to represent abstract methods of computer resources [1]. The simulation, gathering, sharing and isolation of resources can be achieved with the aids of virtualization technology. Furthermore, virtualization technology can also take advantage of the virtual machine to provide the necessary environment [2] for the reliable operation of a variety of applications and rapid deployment. Core feature of cloud computing is on-demand service, which makes cloud computing task allocation and resource scheduling to become technical problems. Because of the interest conflicts between the ordinary User, Infrastructures Providers and Service Providers, current studies only focus on one of them and how to make one of them benefit. In other words, this is actually the three communities of interest. Thus to make cloud computing resource management, you must put the interests of the three as a whole, not only maximize

© Springer International Publishing Switzerland 2016
D. Zhu and S. Bereg (Eds.): FAW 2016, LNCS 9711, pp. 84–91, 2016.
DOI: 10.1007/978-3-319-39817-4_9

cloud service providers and infrastructure providers profits, but also improve the general user satisfaction.

In the stage of service provider (SP) provides cloud services for ordinary users, the user satisfaction and enhance revenue are taken into consideration. In the stage of SP purchase virtual resources from the infrastructure provider (IP), a virtual machine providing model is created. And the dynamic double subpopulation particle swarm optimization is introduced, the particle velocity and position of algorithm are redefined based on virtual machine providing model. In order to improve the convergence speed of double subpopulation particle swarm optimization, the particle velocity update weights is dynamically adjusted based on the fitness value of particles in an iterative process changes. PSO algorithm is easy to fall into local optimum, the immune algorithm is introduced to enhance the diversity of particles, making the algorithm can adjust the global factor dynamically. The improved PSO algorithm can not only find more solution at the beginning of search, but also capable of rapid convergence in the latter so as to achieve the optimal solution. The fusion of ant colony algorithm and genetic algorithm scheduling policy is introduced when SP provide cloud services to ordinary users. First, genetic operators is used in globally quick search, the initial value of ant colony algorithm's pheromone is the result of global genetic algorithm, then the exact solution of task scheduling is achieved by using ACO operator, full use the dual advantages of the ant colony algorithm and genetic algorithms on solving NP problems. Experiments show that, while in two stages of purchase virtual resources and provide cloud resources to the general users, the two algorithms can not only improve SP profits, but also improve customer satisfaction.

In stage of IP provides cloud service for SP, in order to maximize the profits of IP, the energy consumption is taken into consideration. This paper presents a predictive model based on gray and credible ant colony scheduling algorithm for virtual machine migration scheduling policy. CPU resource utilization is an important index for live migration of virtual machines, and when there is a mutation in the arrival of CPU utilization, if there is no effective scheduling policy, the virtual machine migration occurs unnecessary, thus wasting system overhead. Gray prediction model in the second part can estimate the future utilization of a period of virtual machine CPU node. If a load on the host at the current time the CPU utilization is greater than the larger threshold (CPU utilization is less than a small threshold value), and the next three consecutive load prediction values are greater than the threshold (smaller than the threshold), the virtual machine will perform the migration. Experiments show that the algorithm can effectively avoid the frequent migration of virtual machine as a result of the shock caused by the change in CPU utilization, reduce energy consumption, and improve IP gains.

2 Gray Forecasting Model

We forecast future load values by the gray forecasting model before migrating virtual machines to target nodes. The gray forecast model is established by light and incompletability information. It is used to make predictions. The gray system theory research and solve the problem of how to analyze, modeling, forecast, make policy and control the gray system. Gray forecasting is the forecasting of a gray system. Some forecasting methods commonly used at present (such as regression analysis) need a larger sample. A smaller sample will cause greater error and make the target failure. The model given here needs less modeling information. It is of high precision and convenient operation. So it has a wide range of application in various forecasting fields being an effective tool to deal with the small sample forecasting problem.

In the following, the methods of building a gray forecast based on model are described with a time series of data by data analyzing and processing.

2.1 Data Preprocessing

For example, may wish to set up the original data sequence

$$x^{(0)} = \left\{ x^{(0)}(1), x^{(0)}(2), \ldots, x^{(0)}(N) \right\} = \{6, 3, 8, 10, 7\}$$

Accumulation of raw data:

$$x^{(1)}(1) = x^{(0)}(1) = 6,$$
$$x^{(1)}(2) = x^{(0)}(1) + x^{(0)}(2) = 6 + 3 = 9,$$
$$x^{(1)}(3) = x^{(0)}(1) + x^{(0)}(2) + x^{(0)}(3) = 6 + 3 + 8 = 17,$$
$$\ldots$$

A new data series gotten is that $x^{(1)} = \{6, 9, 17, 27, 34\}$.

The formula above can be summarized as $x^{(1)}(i) = \left\{ \sum_{j=1}^{i} x^{(0)}(j) | i = 1, 2, \ldots, N \right\}$.

We call the data series represented by this formula a primary accumulation generation of raw data column (a primary accumulation generation in short).

Following we define: $\Delta x^{(1)}(i) = x^{(1)}(i) - x^{(1)}(i - 1) = x^{(0)}(i)$, in which $i = 1, 2, \ldots, N, x^{(0)}(0) = 0$.

2.2 The Principle of Modeling

Given observation data series $x^{(0)} = \{x^{(0)}(1), x^{(0)}(2), \ldots, x^{(0)}(N)\}$, after a primary accumulation generation we got: $x^{(1)} = \{x^{(1)}(1), x^{(1)}(2), \ldots, x^{(1)}(N)\}$. Suppose that $x^{(1)}$ satisfy the first order ordinary differential equation: $\frac{dx^{(1)}}{dt} + ax^{(1)} = u$, where u is a constant, and a is called the development gray number. This equation satisfied the following initial condition:

$$\text{When } t = t_0, x^{(1)} = x^{(1)}(t_0) \tag{1}$$

Then the solution of this equation is:

$$x^{(1)}(t) = \left[x^{(1)}(t_0) - \frac{u}{a}\right]e^{-a(t-t_0)} + \frac{u}{a} \tag{2}$$

For discrete values of equal interval sampling (take note of $t_0 = 1$), the solution is

$$x^{(1)}(k+1) = \left[x^{(1)}(1) - \frac{u}{a}\right]e^{-ak} + \frac{u}{a} \tag{3}$$

Because $x^{(1)}(1)$ is used as an initial value, $x^{(1)}(2), x^{(1)}(3), \ldots, x^{(1)}(N)$ are brought into the Eq. (1). Suppose difference is used to replace differential coefficient, because of the equal interval sampling $\Delta t = (t+1) - t = 1$, therefore we get $\frac{\Delta x^{(1)}(2)}{\Delta t} = x^{(0)}(2)$. Similarly $\frac{\Delta x^{(1)}(3)}{\Delta t} = x^{(0)}(3), \ldots, \frac{\Delta x^{(1)}(N)}{\Delta t} = x^{(0)}(N)$.

By the formula (1), we get the following:

$$\begin{bmatrix} x^{(0)}(2) \\ x^{(0)}(3) \\ \vdots \\ x^{(0)}(N) \end{bmatrix} = \begin{bmatrix} -\frac{1}{2}[x^{(1)}(2) + x^{(1)}(1)] & 1 \\ -\frac{1}{2}[x^{(1)}(3) + x^{(1)}(2)] & 1 \\ \vdots & \vdots \\ -\frac{1}{2}[x^{(1)}(N) + x^{(1)}(N-1)] & 1 \end{bmatrix} \begin{bmatrix} a \\ u \end{bmatrix} \tag{4}$$

By matrix (4), the estimate values of a and u can be got. They are a' and b'. Bringing a' and b' into formula (3), the forecast value can be obtained. By recording a sequence value of the load: $k_{(1)}, k_{(2)}, \ldots, k_{(t)}$, the gray forecasting model can find out load forecasting values at $t+1$, $t+2$ and $t+3$. If all of these three load forecasting values are greater than the upper limit threshold value or all of them are less than the lower threshold value, the virtual machine migration will be triggered.

3 The Selection and Positioning Strategy of Virtual Machine to Be Migrated

When the virtual machine to be migrated is selected, it is considered that the virtual machine to be moved out will release most of resources. At the same time, migration costs are small. Generally, memory resources and CPU resources are synthetically considered at first, then it is decided which virtual machine will be moved out.

In a virtual machine, Let w = utilization ratio of CPU/utilization ratio of memory. The w is greater, the utilization ratio of memory is lower and utilization ratio of CPU is higher in the virtual machine. So it will release more resources after the migration. Moreover the amount of data to be transmitted is small [3]. Let x = utilization ratio of CPU * utilization ratio of memory. The x is greater, the utilization ratio of CPU and memory are higher. Though the amount of data to be transmitted is greater when migrating the virtual machine being of maximum x, the most resources will be released.

Suppose H_{t+1} indicates the set of continuous three forecasting values of the utilization ratio of CPU after the moment t, H_{max} shows the upper limit threshold value of the utilization ratio of CPU when the host is triggered to start a migration, and H_{min}

shows the lower limit threshold value. M_{thre} indicates the threshold value of utilization ratio of memory. The current memory occupancy of the physical node is M.

The dynamic migration strategy of a virtual machine can be described as follows:

While $H_{t+1} < H_{min}$, all virtual machines on the physical node are moved out and close the physical node to save energy;

While $H_{min} < H_{t+1} < H_{max}$, no migration because of load balancing on the physical node;

While $H_{t+1} > H_{max}$ and $M < M_{thre}$, the virtual machine being of the biggest w will be moved out;

While $H_{t+1} > H_{max}$ and $M > M_{thre}$, the virtual machine being of the biggest x will be moved out.

When a virtual machine is moved out, the CPU utilization may become too high and the memory utilization is deficiency in the target physical node, or the CPU utilization is deficiency and the memory utilization may become too high [4]. We have defined w and x. When the target physical node is selection, the virtual machine being of a bigger w should be moved into a target node which has a smaller w value. Otherwise, if w value is smaller, the virtual machine should be moved into a target node which being of a bigger w value. After the values match of w, m optimum values are selected. Let $p = x / \sum x$, in it x is current available resources on someone target node among these m optimum solutions, and $\sum x$ shows total available resources of these m optimum solutions, so that $\sum_{i=1}^{m} p_i = 1$. In case of there are five optimum solutions: $S = \{S_1, S_2, S_3, S_4, S_5\}$, by computing, their proportion available resources are 0.1, 0.3, 0.2, 0.2, 0.2, then $S_1, \ldots, S_5$ can be intercalated into five interval $S_1 : (0, 0.1]$, $S_2 : (0.1, 0.4]$, $S_3 : (0.4, 0.6]$, $S_4 : (0.6, 0.8]$, $S_5 : (0.8, 1]$.

When the migration target node is to be selected, first, the random function is used to generate a random number between $(0, 1]$, then select an appropriate interval in $S_1, \ldots, S_5$ according to the value of the random number. The target node represented by the selected interval is the ultimate goal of the virtual machine migration. Based on the positioning probability management, target nodes with more available resources are being of greater probability of receiving a virtual machine, the probability of being selected as a target node where the physical nodes with high resource utilization will become low. Thus load balancing for each physical node is realized in the data center.

4   Experiment and Analysis

CloudSim simulation software is used in the experiment to verify the efficiency of the algorithm. CloudSim Toolkit supports the system components in a virtualized environment, such as data center, hosts, virtual machines, scheduling and resources allocation strategies [5]. The CloudSim-3.0, Windows XP SP3, jdk6.5, MyEclipse ver8.5 and Ant1.8.1 are also used in the experiment. The hardware environment is Pentium Dual Core Processor with main frequency 2.6 GHz. With the CloudSim-3.0, we deploy the same 1000 physical nodes and meanwhile provide 3000 virtual machines of the same performance. The number of virtual machines migration and degree of system load balance are calculated every 10 s. After 20 times of sampling, find the average values as the migration times and the degree of system load balance under a certain threshold value. In the experiments, different threshold values are used in order that we can observe the number of virtual machines migration, shown in Fig. 1.

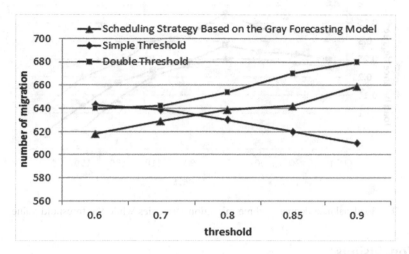

Fig. 1. The number of migration of the three algorithms under different thresholds

The number of migration caused by scheduling strategy based on the gray forecasting model is obviously smaller than those caused by Double Threshold (DT) algorithm. This shows that the scheduling strategy based on the gray forecasting model can make the migration more efficient. With the increase of the threshold setting, the number of migration caused by both the scheduling strategy based on the gray forecasting model and DT algorithm showed a clear upward trend, but the number of migration caused by the Simple Threshold (ST) algorithm showed a clear downward trend. This is because that when the dual-threshold is set up, threshold enhancement, low threshold is also to become high. Many virtual machines of low CPU and memory utilization rate in the physical machine will be moved out so that the number of migration will increase. When the threshold is 0.7, the numbers of migration caused by the three scheduling algorithms are most similar, shown in Fig. 2.

Figure 2 shows three load balance curves of the three algorithms while the threshold value is 0.7. It is found that there is a rise in the load balance in the early stage, but with the passage of the time, the load balance of the three scheduling strategies is gradually becoming smaller. The load balance of ST algorithm with a single limit threshold is greater than that of DT algorithm with dual-threshold and the dynamic migrating strategy based on gray forecasting model. But the scheduling strategy based on gray forecasting model has better load balancing effect. It can improve utilization ratio of the system resources.

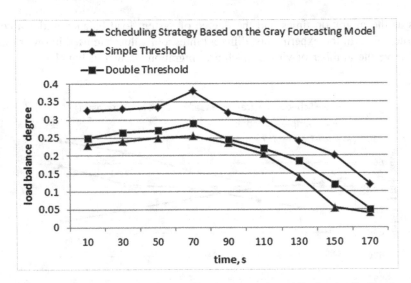

Fig. 2. The load balance degree of three migration strategies while the threshold value is 0.7

5 Conclusions

A kind of scheduling strategy of virtual machine migration based on the gray fore-casting model is proposed in this paper. The model is of high precision and high efficiency. In addition, the dual-threshold value is applied in the virtual machine migration strategy. It effectively eliminates the frequent migration caused by the CPU utilization ratio shocks. The idea of positioning probability is introduced while the target node is being selected, to solve the clustering effect and greatly enhance the migration success rate. In order to realize the goal of saving energy, [6] the scheduling strategy will be gradually improved in the following research.

Acknowledgements. Supported by Natural Science Foundation of Shandong Province (ZR2013FM031); Supported by State Key Lab of High Performance Server and Storage (2014HSSA03).

References

1. Pallis, G.: Cloud computing: the new frontier of internet computing. IEEE Internet Comput. **14**(5), 65–73 (2010)
2. Fei, M.A., Feng, L.I.U., Zhuyi, L.I.: Rapid real time migration method for virtual machine in cloud computing environment. J. Beijing Univ. Posts. Telecommun. **35**(001), 103–106 (2012)
3. Quan, C., Qianni, D.: Cloud computing and its key technologies. Comput. Appl. **29**(9), 2562–2567 (2009)
4. Tarighi, M., Motamedi, S.A., Sharifian, S.: A new model for virtual machine migration in virtualized cluster server based on fuzzy decision making. J. Telecommun. **1**(1), 40–51 (2010)

5. Karagiannis, T., Roido, A., Aloutsos, M., et al.: Transport layer identification of P2P traffic. In: Proceedings of the 2004 ACM SIGCOMM Internet Measurement Conference, Taormina, Italy (2004)
6. Kim, N., Cho, J., Seo, E.: Energy-credit scheduler: an energy-aware virtual machine scheduler for cloud systems. Future Gener. Comput. Syst. **32**(2), 128–137 (2014)

A Polynomial Time Algorithm for Finding a Spanning Tree with Maximum Number of Internal Vertices on Interval Graphs

Xingfu Li[1(✉)], Haodi Feng[2], Haitao Jiang[2], and Binhai Zhu[3]

[1] School of Software, Shanxi Agriculture University,
Taigu, Jinzhong, Shanxi 030801, China
xingfulisdu@qq.com

[2] School of Computer Science and Technology, Shandong University, Jinan,
Shandong 250100, China
{fenghaodi,htjiang}@sdu.edu.cn

[3] Department of Computer Science, Montana State University, Bozeman, MT, USA
bhz@montana.edu

Abstract. This paper studies the Maximum Internal Spanning Tree problem which is to find a spanning tree with the maximum number of internal vertices on a graph. We prove that the problem can be solved in polynomial time on interval graphs. The idea is based on the observation that the number of internal vertices in a maximum internal spanning tree is at most one less than the number of edges in a maximum path cover on any graph. On an interval graph, we present an $O(n^2)$-algorithm to find a spanning tree in which the number of internal vertices is exactly one less than the number of edges in a maximum path cover of the graph, where n is the number of vertices in the interval graph.

Keywords: Polynomial · Algorithm · Maximum internal spanning tree · Interval graph

1 Introduction

The *Maximum Internal Spanning Tree* problem, MIST briefly, is motivated by the design of cost-efficient communication networks [14]. It asks to find a spanning tree of a graph such that the number of its internal vertices is maximized. Since a Hamilton path (if exists) of a graph is also a spanning tree of that graph with its internal vertices maximized in number, and finding a Hamilton path in a graph is NP-Hard classically [7], MIST is NP-hard trivially. The complement problem of MIST is the so called *Minimum Leaves Spanning Tree* problem, MLST briefly. MLST asks to find a spanning tree of a graph such that the number of its leaves is minimized. One application of MLST appears in water resources engineering [2].

MLST is NP-hard and cannot be approximated to within any constant performance ratio [10], if $P \neq NP$. As algorithmic approaches, Flandrin et al. in

© Springer International Publishing Switzerland 2016
D. Zhu and S. Bereg (Eds.): FAW 2016, LNCS 9711, pp. 92–101, 2016.
DOI: 10.1007/978-3-319-39817-4_10

[4] and Kyaw in [9] have respectively studied the conditions of whether a graph
has a spanning tree with a bounded number of leaves.

Unlike MLST, MIST admits approximation algorithms with a constant performance ratio. Prieto et al. [11] first presented a 2-approximation using local
search in 2003. Later, by a slight modification of depth-first search, Salamon
et al. [14] improved Prieto's 2-approximation algorithm to running in linear-time.
Besides, they proposed a $\frac{3}{2}$-approximation algorithm on claw-free graphs and a
$\frac{6}{5}$-approximation algorithm on cubic graphs [14]. Salamon even showed that his
2-approximation algorithm in [14] can achieve a performance ratio $\frac{r+1}{3}$ on r-
regular graphs [16]. Furthermore, by local optimization, Salamon [15] devised an
$O(n^4)$-time and $\frac{7}{4}$-approximation algorithm on graphs without leaves. Through
a different analysis, Knauer et al. [8] showed that Salamon's algorithm in [15]
can actually take $O(n^3)$ time to achieve a performance ratio $\frac{5}{3}$ even on general
undirected simple graphs. In 2014, Li, Wang and Chen [21] presented a 1.5-
approximation algorithm by a local search method for the maximum internal
spanning tree problem on general graphs. Meanwhile, Li and Zhu [22] gave a
1.5-approximation algorithm using a greedy method for the maximum internal
spanning tree problem on general graphs.

Salamon et al. [15] also studied the vertex-weighted cases of MIST which asks
for a maximum weighted spanning tree of a vertex weighted graph. They gave an
$O(n^4)$-time and $(2\Delta - 3)$-approximation algorithm for weighted MIST on graphs
without leaves, where Δ is the maximum degree of the graph. They also gave
an $O(n^4)$-time and 2-approximation algorithm for weighted MIST on claw-free
graphs without leaves. Later, Knauer et al. [8] presented a $(3+\epsilon)$-approximation
algorithm for weighted MIST on undirected simple graphs.

Fixed parameter algorithms of MIST have also been extensively studied in
the recent years. Prieto and Sloper [11] designed the first FPT-algorithm with
running time $O^*(2^{4k\log k})$ in 2003. Coben et al. [3] improved this algorithm to
achieve a time complexity $O^*(49.4^k)$. Then an FPT-algorithm for MIST with
time complexity $O^*(8^k)$ was proposed by Fomin et al. [6], who also gave an
FPT-algorithm for its directed version with time complexity $O^*(16^{k+o(k)})$ [5].
For directed graphs, a randomized FPT algorithm proposed by M. Zehavi is by
now the fastest one which runs in $O^*(2^{(2-\frac{\Delta+1}{\Delta(\Delta-1)})k})$ time [18], where Δ is the
vertex degree bound of a graph. On cubic graphs in which each vertex has degree
three, Binkele-Raible et al. [2] proposed an $O^*(2.1364^k)$ time algorithm.

For the kernalization of MIST, Prieto and Sloper first presented an $O(k^3)$-
vertex kernel [11,12]. Later, they improved it to $O(k^2)$ [13]. Recently, Fomin
et al. [6] gave a $3k$-vertex kernel for this problem, which is the best by now.

As for the exact exponential algorithms to solve MIST, Binkele-Raible et al.
[2] proposed a dynamic programming algorithm with time complexity $O^*(2^n)$.
On graphs with maximum vertex degree bounded by 3 especially, they devised
a branching algorithm of $O(1.8612^n)$ time and polynomial space.

In this paper, we pay attention to the maximum internal spanning tree problem on interval graphs. Interval graphs have received a lot of attention due
to their applicability to DNA physical mapping problems [23], and find many

applications in several fields and disciplines such as genetics, molecular biology, scheduling, VLSI circuit design, archaeology and psychology [24]. We prove that there is a polynomial algorithm to find a maximum internal spanning tree on interval graphs. We present an algorithm with time complexity $O(n^2)$, where n is the number of vertices in an interval graph.

This paper is organized as follows. Section 2 presents basic notations and properties of interval graphs. In Sect. 3, we present our polynomial algorithm and its analysis. Section 4 concludes this paper by looking forward to some future work on the maximum internal spanning tree problem.

2   Preliminaries

In this paper, all graphs in which we are going to find spanning trees are undirected, simple and connected. Each path or cycle in a graph is always simple. The first and the last vertices of a path are the *endpoints* of that path, while the others except the endpoints of a path are the *inner* vertices of that path. The *length* of a path or cycle is the number of edges in it. A *connected component* of a graph is a subgraph in which any two vertices are connected to each other by paths, and which is connected to no additional vertices in the supergraph. A connected component of a graph is referred to as a *path (cycle) component* if it is a path (cycle) of that graph. A vertex in a graph is a *leaf* if its degree is 1, and *internal* otherwise.

A spanning subgraph of G is a *path cover* if every connected component of it is a path. A path cover of G is *maximum* if its edges are maximized in number over all path covers of G.

A *maximum internal spanning tree* of G is a spanning tree of G whose internal vertices are maximized in number over all spanning trees of G. The *Maximum Internal Spanning Tree* problem, MIST namely, is given by an undirected simple graph, and asks to find a maximum internal spanning tree for that graph.

A graph G is called an *interval graph* if its vertices can be put in a one-to-one correspondence with a family F of intervals on the real line such that two vertices are adjacent in G if and only if the corresponding intervals intersect. F is called an *intersection model* for G. A *right-end ordering* of an interval graph G can be obtained by sorting the intervals of the intersection model of G on their right ends in time $O(|V(G)| + |E(G)|)$ [20]. An ordering of the vertices according to this numbering is found to be quite useful in solving some graph-theoretic problems on interval graphs [20]. Throughout this paper, an interval graph is represented by its right-end ordering graph. Figure 1 shows an interval graph and its corresponding right-end ordering graph.

Lemma 1 (Ramadingam and Rangan [20]). *In a right-end ordering graph of an interval graph, for every three indices i, j, k, if $i < j < k$ and there is an edge between v_i and v_k, then there must be an edge between v_j and v_k.*

Let G be an interval graph. Let π be a right-end ordering of G. The path $P = v_{\pi_0}, v_{\pi_1}...v_{\pi_s}$ is *typical* if for every $1 \leq i \leq s$, $\pi_0 < \pi_i$. Arikati and Rangan

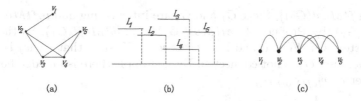

Fig. 1. (a) An interval graph G; (b) An intersection model F of G; (c) The right-end ordering of G

[19] proved that, in any interval graph, every path can be transformed into a typical one with its vertices unchanged. That is,

Lemma 2 (Arikati and Rangan [19]). *Let p be a path of an interval graph G, then there is a typical path q of G such that $V(p) = V(q)$.*

By Lemma 2, throughout this paper, every path is typical whenever it is mentioned.

Let G_1 be a subgraph of an interval graph G. Note that the interval graph G is represented by its right-end ordering. The left-most vertex of G_1 is $v_{leftMost(G_1)} = \{v_j | j = \min_{v_i \in V(G_1)}(i)\}$, and the right-most vertex of G_1 is $v_{rightMost(G_1)} = \{v_j | j = \max_{v_i \in V(G_1)}(i)\}$. Particularly, the left-most vertex of a path is the left endpoint of the path, since the path is typical.

Two subgraphs G_1 and G_2 of an interval graph are said to be *intersecting* if there is a vertex v_i of G_1(or G_2) such that $leftMost(G_2) < i < rightMost(G_2)$(or $leftMost(G_1) < i < rightMost(G_1)$).

Every graph considered in this paper is connected, so every path in a right-end ordering graph must intersect with another path. Otherwise, the path must be isolated with the other parts of the right-end ordering graph, then the graph is not connected. So we have

Property 1. For every path p on a right-end ordering graph, there must be another path $q \neq p$ such that p and q are intersecting.

For every two intersecting subgraphs, we have the following property.

Property 2. Let G_1 and G_2 be two intersecting subgraphs of an interval graph. If $leftMost(G_1) < leftMost(G_2)$, then $leftMost(G_2) < rightMost(G_1)$.

Proof. Assume that $leftMost(G_2) > rightMost(G_1)$, then G_1 and G_2 would not be intersecting. This comes to a contradiction. □

Lemma 3. *Let G_1, G_2 be two connected subgraphs of an interval graph G, If G_1 and G_2 are intersecting and $leftMost(G_1) < leftMost(G_2)$, then there exist another vertex $w \in V(G_1)$, such that w and $v_{leftMost(G_2)}$ are connected by an edge of G.*

Proof. We partition the vertex set of G_1 into two sets V_1 and V_2, where $V_1 = \{v_s | v_s \text{ is in } V(G_1) \text{ and } s < leftMost(G_2)\}$ and $V_2 = \{v_s | v_s \text{ is in } V(G_1)$

and $s > leftMost(G_2)$}. Since G_1 and G_2 are intersecting and $leftMost(G_1) < leftMost(G_2)$, by Property 2, $leftMost(G_2) < rightMost(G_1)$. So there must be one vertex u in V_1 and another vertex w in V_2 such that (u, w) is an edge of G_1, otherwise, G_1 is not connected. By Lemma 1, there is an edge between w and the vertex $v_{leftMost(G_2)}$. $\qquad\square$

3   The Polynomial Algorithm

In this section, we will present a polynomial algorithm for the maximum internal spanning tree problem on interval graphs. Given an interval graph G, the scheme of our algorithm is first to find a maximum path cover of G using the algorithm from [19] and finally connect every path together to form a spanning tree of G. Due to Arikati and Rangan [19], a maximum path cover of an interval graph can be found in *linear* time. The key point of our algorithm is to decide how to connect these paths together in polynomial time so that we can obtain a maximum internal spanning tree of an interval graph.

In the following, we fist prove an upper bound of the number of internal vertices of a spanning tree. Then, we give a method to connect the paths together to form a spanning tree so that the number of internal vertices of the spanning tree is just equal to the proven upper bound. Based on these, we can show that our algorithm can find a maximum internal spanning tree on an interval graph in polynomial time.

3.1   The Upper Bound of Internal Vertices in a Spanning Tree

Lemma 4. *Let T^* be a maximum internal spanning tree of a graph. If T^* has at least two vertices, then there is a path cover of T^* which has less path components than the number of leaves of T^*.*

Proof. Let x be the number of leaves in T^*. The proof is by induction on x. If $x = 2$, T^* is a path component, the lemma holds true trivially. Then the inductive assumption is, if a tree has at most $x - 1$ leaves, it must have a path cover with less path components than the leaves in that tree. Later, we show that if T^* has x (> 2) leaves, it must have a path cover with at most $x - 1$ path components.

Since $x > 2$, a path, say P with at least one edge, can be identified in T^* with both of its endpoints as leaves of T^*. We then delete those edges incident to the vertices of P except those in P. This gives rise to a spanning forest of T^*. Let $T_1, ..., T_j, T_{j+1}, ..., T_k$ be all the trees in the forest except P, where T_i for $1 \leq i \leq j$ has only one vertex while the others do not. Note that the vertex in T_i for $1 \leq i \leq j$ is also a leaf of T^*. Namely, one path can cover T_i for $1 \leq i \leq j$. Moreover, T_i for $j + 1 \leq i \leq k$ has at most $x - 1$ leaves because it has at most one leaf which does not act as a leaf in T^*. Let $T_{j+1}, T_{j+2}, ..., T_k$ have $x_{j+1}, x_{j+2}, ..., x_k$ leaves respectively. By the inductive assumption, T_i for $j + 1 \leq i \leq k$ must have a path cover with at most $x_i - 1$ path components. Hence T^* has a path cover with at most $1 + j + \sum_{j+1 \leq i \leq k}(x_i - 1) \leq 1 + j + \sum_{j+1 \leq i \leq k}(x_i) - (k - j) \leq 1 + j + (x - (2 + j) + (k - j)) - (k - j) = x - 1$ path components. $\square$

Lemma 5. *The number of internal vertices of a maximum internal spanning tree is less than the number of edges of a maximum path cover in a graph.*

Proof. Let T^* be a maximum internal spanning tree of a graph G. Let P^* be a maximum path cover of G. By Lemma 4, the number of leaves in T^* is larger than the number of path components in P^*. So the number of internal vertices in T^* is less than the number of edges in P^*. □

3.2 Algorithm

Our algorithm starts with a maximum path cover P^* of an interval graph G, which can be done in linear time [19]. Now we are going to connect all path components of P^* together to form a spanning tree of G so that the internal vertices of the spanning tree is exactly one less than the number of edges of P^*, just meeting with the upper bound stated in Lemma 5.

The idea to connect the path components is as following. Every time we maintain a *currently-optimal* tree T_c and a set of paths P_c. Initially, we arbitrarily choose a path component, say p_0, from P^* and let $T_c \leftarrow p_0$, $P_c \leftarrow P^* \setminus \{p_0\}$. We want to connect all path components in P_c to T_c so as to get a spanning tree. T_c grows by connecting path components chosen from P_c one by one. Every time the path component chosen from P_c should be intersecting with T_c. The connection rules are as follows.

Let p be a path component which is in P_c and intersecting with T_c. Now we are going to connect p to T_c. There are two cases.

Case 1: $leftMost(p) < leftMost(T_c)$. By Lemma 3, there exists a vertex $w \in V(p)$ such that $(w, v_{leftMost(T_c)})$ is an edge of G. We connect p with T_c by adding the edge between w and $v_{leftMost(T_c)}$.

Case 2: $leftMost(p) > leftMost(T_c)$. By Lemma 3, there exists a vertex $w \in V(T_c)$ such that $(w, v_{leftMost(p)})$ is an edge of G. We connect p with T_c by adding the edge between w and $v_{leftMost(p)}$.

We summarize the polynomial algorithm as Algorithm 1.

3.3 Analysis

In this section, we prove that the Algorithm 1 can find a maximum internal spanning tree for an interval graph in polynomial time.

By Property 1, every path component in P^* can be connected to T_c to form a spanning tree when the algorithm finishes, otherwise, the interval graph would not be connected. In the following, we will count the number of internal vertices in the output tree of the algorithm.

Lemma 6. *At any time, the degree of the left-most vertex of T_c which is $v_{leftMost(T_c)}$ is always one.*

Proof. We proof this lemma by induction. Initially, T_c is a simple path. So the left-most vertex of T_c is naturally a leaf of T_c. That is, the degree of the left-most vertex is one. Assume that the degree of the left-most vertex of T_c is one

---

Algorithm 1. Finding a maximum internal spanning tree on an interval graph

---

Input:
    An interval graph G which has already been right-end ordered.
Output:
    A maximum internal spanning tree of G
1: Find a maximum path-cover P^* of G, which can be done in linear time [19].
2: $T_c \leftarrow \{p|\ p$ is a path component of $P^*\}$, $P_c \leftarrow P^* \setminus \{p\}$
3: **while** P_c is not empty **do**
4:     Choose a path component q from P_c, where q is intersecting with T_c.
5:     **if** $leftMost(q) < leftMost(T_c)$ **then**
6:         By **Case 1**, choose a vertex $w \in V(q)$ which is adjacent to $v_{leftMost(T_c)}$. Let T_c be the resultant new tree by adding the edge between w and $v_{leftMost(T_c)}$.
7:     **end if**
8:     **if** $leftMost(q) > leftMost(T_c)$ **then**
9:         By **Case 2**, choose a vertex $w \in V(T_c)$ which is adjacent to $v_{leftMost(q)}$. Let T_c be the resultant new tree by adding the edge between w and $v_{leftMost(q)}$.
10:    **end if**
11:    $P_c \leftarrow P_c \setminus \{q\}$
12: **end while**
13: **return** T_c.

---

before connecting some path component $p \in P_c$. Now we will prove that after connecting the path p to T_c, the degree of the left-most vertex in the resultant newly tree T'_c is still one. If $leftMost(p) < leftMost(T_c)$, then by **Case 1**, after connecting p to T_c, the left-most vertex in the resultant newly tree T'_c turns to be the left-most vertex of the path p. So the degree of the left-most vertex in the resultant newly tree T'_c is still one. If $leftMost(p) > leftMost(T_c)$, then by **Case 2**, after connecting p to T_c, the left-most vertex in the resultant newly tree T'_c remains to be the left-most vertex of the tree T_c. By the induction, the degree of the left-most vertex of T_c is one. So the lemma holds. □

Lemma 7. *After a path component $p \in P_c$ is connected to the currently-optimal tree T_c, the internal vertices of the currently-optimal tree increase exactly by $|E(p)|$.*

Proof. If $leftMost(p) < leftMost(T_c)$, then by **Case 1**, we use the left-most vertex of T_c to connect p and T_c. By Lemma 6, the left-most vertex of T_c has degree one. So after this connection, the left-most vertex of T_c will become an internal vertex. Moreover, the internal vertices of p will turn to be internal in the newly resultant tree, and the number of internal vertices of p is $|E(p)| - 1$. So after connecting p to T_c, the internal vertices of the currently-optimal tree increase exactly by $|E(p)|$.

If $leftMost(p) > leftMost(T_c)$, then by **Case 2**, we use the left-most vertex of p to accomplish this connection. So the endpoint of p turns to be an internal vertex of the resultant tree. Moreover, the internal vertices of p are still be internal in the newly resultant tree, where the number of internal vertices of p

is $|E(p)| - 1$. So after connecting p to T_c, The internal vertices of the currently-optimal tree increases exactly by $|E(p)|$. □

Lemma 8. *Let P^* be a maximum path cover of an interval graph G. Then the Algorithm 1 returns a spanning tree with the number of internal vertices equal to $|E(P^*)| - 1$.*

Proof. Consider the change of the number of internal vertices in T_c. Initially, T_c is a simple path of P^*. Let p_0 be such a chosen path. So before the *while* loop in the Algorithm 1, the number of internal vertices in T_c is equal to $|E(p_0)| - 1$. By Lemma 7, after each connection, the increment of the number of internal vertices in T_c is equal to the number of edges of the path component which is connected to T_c. So when the algorithm finishes, we can obtain a spanning tree with $|E(P^*)| - 1$ internal vertices. □

Theorem 1. *The Algorithm 1 can find a maximum internal spanning tree for an interval graph with time complexity $O(n^2)$, where n is the number of vertices in the interval graph.*

Proof. By Lemmas 5 and 8, the Algorithm 1 returns a spanning tree with the number of internal vertices equal to the upper bound of the number of internal vertices in a spanning tree. So the Algorithm 1 can find a maximum internal spanning tree on an interval graph.

Let G be an interval graph. The time complexity to find a maximum path cover in G is $O(|V(G)| + |E(G)|)$ [19]. So the step 1 in the Algorithm 1 takes $O(|V(G)| + |E(G)|)$ time. Now we will calculate the time complexity of the *while-loop* in the Algorithm 1. It takes $O(|V(G)|)$ time to find the left-most vertex and the right-most vertex of a subgraph. Once we know the left-most vertex and the right-most vertex of a subgraph, we just spend $O(1)$ time to decide whether two subgraphs are intersecting. Let q be a path component which is intersecting with a currently-optimal tree T_c. If $leftMost(q) < leftMost(T_c)$, then we will spend $O(|V(q)|) \leq O(|V(G)|)$ time to find a vertex $w \in V(q)$ such that w and $v_{leftMost(T_c)}$ are connected by an edge in G. If $leftMost(q) > leftMost(T_c)$, then we will spend $O(|V(T_c)|) \leq O(|V(G)|)$ time to find a vertex $w \in V(T_c)$ such that w and $v_{leftMost(q)}$ are connected by an edge in G. So the time complexity of the *while-loop* in the Algorithm 1 is $O(|P^*| * (3|V(G)|)) \leq O(|V(G)|^2)$, where P^* is a maximum path cover of G. So the time complexity of the Algorithm 1 is $O(n^2)$ where n is the number of vertices of an interval graph. □

4   Conclusions and Discussions

We have present a polynomial algorithm for the maximum internal spanning tree problem on interval graphs. We claim that the maximum internal spanning tree problem on circular-arc graphs can also be solved in polynomial time, since these two graph classes are very similar. Moreover, the maximum internal spanning tree problem on other special graph classes can also be studied, including their algorithms and complexity.

Acknowledgments. This research is supported by the Doctoral Science Foundation of Shanxi Agriculture University under the grant of 2015YJ19.

References

1. Akiyama, T., Nishizeki, T., Saito, N.: NP-completeness of the Hamiltonian cycle problem for bipartite graphs. J. Inf. Process. **3**(2), 73–76 (1980)
2. Binkele-Raible, D., Fernau, H., Gaspers, S., Liedloff, M.: Exact and parameterized algorithms for max internal spanning tree. Algorithmica **65**, 95–128 (2013)
3. Coben, N., Fomin, F.V., Gutin, G., Kim, E.J., Saurabh, S., Yeo, A.: Algorithm for finding k-vertex out-trees and its application to k-internal out-branching problem. JCSS **76**(7), 650–662 (2010)
4. Flandrin, E., Kaiser, T., Kuzel, R., Li, H., Ryjacek, Z.: Neighborhood unions and extremal spanning trees. Discrete Math. **308**(12), 2343–2350 (2008)
5. Fomin, F.V., Lokshtanov, D., Grandoni, F., Saurabh, S.: Sharp seperation and applications to exact and parameterized algorithms. Algorithmica **63**(3), 692–706 (2012)
6. Fomin, F.V., Gaspers, S., Saurabh, S., Thomassé, S.: A linear vertex kernel for maximum internal spanning tree. JCSS **79**, 1–6 (2013)
7. Garey, M.R., Johnson, D.S.: Computers and Intractability: A Guide to the Theory of NP-Completeness. Freeman, W. H (1979)
8. Knauer, M., Spoerhase, J.: Better approximation algorithms for the maximum internal spanning tree problem. In: Dehne, F., Gavrilova, M., Sack, J.-R., Tóth, C.D. (eds.) WADS 2009. LNCS, vol. 5664, pp. 459–470. Springer, Heidelberg (2009)
9. Kyaw, A.: Spanning trees with at most 3 leaves in $K_{1,4}$-free graphs. Discrete Math. **309**(20), 6146–6148 (2009)
10. Lu, H., Ravi, R.: The power of local optimization approximation algorithms for maximum-leaf spanning tree. Technical report, Department of Computer Science, Brown University (1996)
11. Prieto, E., Sloper, C.: Either/Or: using VERTEX COVER structure in designing FPT-Algorithms — the case of k-INTERNAL SPANNING TREE. In: Dehne, F., Sack, J.-R., Smid, M. (eds.) WADS 2003. LNCS, vol. 2748, pp. 474–483. Springer, Heidelberg (2003)
12. Prieto, E.: Systematic kernelization in FPT algorithm design. Ph.D. thesis, The University of Newcastle, Australia (2005)
13. Prieto, E., Sloper, C.: Reducing to independent set structure the case of k-internal spanning tree. Nord. J. Comput. **12**(3), 308–318 (2005)
14. Salamon, G., Wiener, G.: On finding spanning trees with few leaves. Inf. Process. Lett. **105**(5), 164–169 (2008)
15. Salamon, G.: Approximating the maximum internal spanning tree problem. Theor. Comput. Sci. **410**(50), 5273–5284 (2009)
16. Salamon, G.: Degree-Based Spanning Tree Optimization. Ph.D. thesis, Budapest University of Technology and Ecnomics, Hungary (2009)
17. Shiloach, Y.: Another look at the degree constrained subgraph problem. Inf. Process. Lett. **12**(2), 89–92 (1981)
18. Zehavi, M.: Algorithms for k-internal out-branching. In: Gutin, G., Szeider, S. (eds.) IPEC 2013. LNCS, vol. 8246, pp. 361–373. Springer, Heidelberg (2013)
19. Arikati, S.R., Rangan, C.: Linear algorithm for optimal path cover problem on interval graphs. Inform. Process. Lett. **35**, 149–153 (1990)

20. Ramalingam, G., Rangan, C.: A unified approach to domination problems on interval graphs. Inform. Proc. Lett. **27**, 271–274 (1988)
21. Li, W., Chen, J., Wang, J.: Deeper local search for better approximation on maximum internal spanning trees. In: Schulz, A.S., Wagner, D. (eds.) ESA 2014. LNCS, vol. 8737, pp. 642–653. Springer, Heidelberg (2014)
22. Li, X., Zhu, D.: Approximating the maximum internal spanning tree problem via a maximum path-cycle cover. In: Ahn, H.-K., Shin, C.-S. (eds.) ISAAC 2014. LNCS, vol. 8889, pp. 467–478. Springer, Heidelberg (2014)
23. Goldberg, P.W., Golumbic, M.C., Kaplan, H., Shamir, R.: Four strikes against physical mapping of DNA. J. Comp. Biol. **2**, 139–152 (1995)
24. Golumbic, M.C.: Algorithmic Graph Theory and Perfect Graphs (Annals of Discrete Mathematics, vol. 57). North-Holland Publishing Co., Amsterdam (2004)

Fractional Edge Cover Number of Model RB

Tian Liu(✉)

Key Laboratory of High Confidence Software Technologies, Ministry of Education, Peking University, Beijing 100871, China
lt@pku.edu.cn

Abstract. Model RB is a random constraint satisfaction problem with a growing domain size, which exhibits exact phase transition phenomena. Many hard instances with planted solutions can be generated via Model RB, to be used as benchmarks for algorithmic competitions and researches. In the past, some structural parameters of constraint hypergraphs are analyzed to show hardness of Model RB, such as hinge width, decycling number, treewidth, and hypertree width. In this paper, one more structural parameter of constraint hypergraphs of Model RB, namely the fractional edge cover number, is analyzed. We show upper and lower bounds on the fractional edge cover number of Model RB. In particular, the fractional edge cover number of Model RB is shown to be asymptotically linear in the number of variables, like hinge width, decycling number, treewidth and hypertree width. These results together provide further evidences on the hardness of Model RB.

Keywords: Model RB · Fractional edge cover · Hardness

1 Introduction

Constraint satisfaction problems (CSPs) can model many real world problems, such as n-queens, Latin squares, etc. A CSP instance is consist of a constraint hypergraph on a set of variables and many constraints on the hyperedges. Each constraint gives the compatible assignments of values to the variables in a hyperedge. The task is to decide if the instance is satisfiable, that is, if there is an assignment of values to all the variables, such that it is compatible with all the constraints. The Boolean satisfaction problem 3-SAT is a special case of CSPs, where each variable only takes two different values and each constraint is a conjunction of three variables or their negations.

Since 3-SAT is already an $\mathcal{NP}$-complete problem, CSPs are hard to solve in general. Many structural decomposition methods are developed to find tractable classes of CSPs, such as tree decomposition [13], hypertree decomposition [1], and fractional edge cover [11]. For constraint hypergraphs with a structural parameter w and input size $||I||$, usually we can solve these instances in time

Partially supported by Natural Science Foundation of China (Grant Nos. 61370052 and 61370156).

D. Zhu and S. Bereg (Eds.): FAW 2016, LNCS 9711, pp. 102–110, 2016.
DOI: 10.1007/978-3-319-39817-4_11

$O\left(\|I\|^{f(w)}\right)$, where $f(w)$ is a function of w, such as a low degree polynomial of w, or linear in w. When the structural parameter w is constantly bounded, we get a tractable class of CSPs. At the moment, the most powerful structural decomposition method is fractional hypertree decomposition [17].

On the other hand, random instances of CSPs can be generated by randomly setting constraint hypergraphs, and then randomly setting compatible assignments for each constraint. A parameter called density, which is the ratio between the number of constraints and the number of variables, can be used to control the number of constraints. With a small density and thus a small number of constraints, the instances are likely to be satisfiable. With a large density and thus a large number of constraints, the instances are unlikely to be satisfiable. When the number of variables goes to infinity, such a change of satisfiability may happen suddenly around a critical value of density. Such phenomena are called the satisfiability phase transition of random CSPs. Moreover, the hardest instances of CSPs are located around the satisfiability thresholds [2,3,5,6,19,20]. However, a rigorous link between the hardness of random instances and the satisfiability phase transition is still unknown.

The most common random CSPs are random 3-SAT in theoretical computer science and Model A,B,C,D in artificial intelligence. For random 3-SAT, the exact satisfiability threshold is still unknown, although some upper and lower bounds are shown in the past [4,18]. Moreover, if planted solutions are used to generated satisfiable random instances with known solutions, the instances usually become much easier to solve. For Model A,B,C,D, when the number of variables goes to infinity, the satisfiability thresholds will go to zero, thus they are useless in generating large hard instances [10]. A random CSP model, called Model RB, is defined by Xu and Li [23], which has an increasing domain size and exhibits exact phase transition phenomena. Many hard instances with planted solutions can be generated via Model RB [24,26], to be successfully used as benchmarks in various algorithmic competitions and in many research papers, such as the annual CSP solver competitions, the annual Pseudo-Boolean (0-1 Integer Programming) solver competitions, the annual MAX-SAT solver competitions, and the annual SAT solver competitions, etc. [24].

In the past, the hardness of model RB is theoretically shown by exponential lower bounds on the length of resolution [25], and some analysis on the evolution of its solution space [27,28], besides many experimental results [26]. Some structural parameters of constraint hypergraphs are also analyzed to show hardness of Model RB with respective to the structural decomposition methods, such as hinge width [15], decycling number [12], treewidth [22], and hypertree width [16]. In this paper, one more structural parameter of constraint hypergraphs of Model RB, namely the fractional edge cover number [11], is analyzed. We show upper and lower bounds on the fractional edge cover number of Model RB. In particular, the fractional edge cover number of Model RB is shown to be asymptotically linear in the number of variables, like hinge width, decycling number, treewidth and hypertree width. These results together provide further evidences on the hardness of Model RB.

This paper is structured as follows. After introducing definitions and some facts on Model RB and fractional edge cover respectively, as well as a version of Chernoff bound in Sect. 2, lower and upper bounds on fractional edge cover number of Model RB are shown in Sect. 3. In Sect. 4, we conclude the paper with remarks on our results and open problems.

2   Preliminaries

In this section, we give definitions and facts on Model RB and fractional edge cover respectively, as well as a version of the Chernoff bound.

2.1   Model RB

An instance I of *constraint satisfaction problem* (CSP) is a triple (V, D, C). V is a finite set of variables. D is a finite set of values, called domain. C is a set of constraints. For each constraint, there is a subset of variables, called the scope of this constraint. For a constraint scope with k variables, a subset of D^k is given as compatible assignment of values to the variables in this scope. A solution of I is an assignment of values to all variables which is compatible to all constraint. If there is at least one solution, I is called satisfiable, otherwise unsatisfiable. Given an instance, we are asked to decide if it is satisfiable, and in some cases to find a solution if it exists.

A hypergraph is just a set system, which is consist of some subsets (called hyperedges) of a finite set (called vertex set). A hypergraph H with vertex set V and hyperedge set E is denoted by $H = (V, E)$. The constraint hypergraph of a CSP is consist of the scopes of the constraints in this CSP.

Model RB is a random CSP defined as follows [23].

1. Given n variables, each variable takes values in $\{1, 2, ..., d\}$, where $d = n^\alpha$ and $\alpha > 0$ is a constant;
2. Select with repetition $m = rn \ln n$ random constraints, where r is a constant. For each constraint, select without repetition k of n variables, where $k \geq 2$ is an integer constant;
3. Select uniformly at random without repetition $(1 - p)d^k$ compatible assignments for each constraint, where $0 < p < 1$ is a constant.

For an instance I of Model RB, the input size $||I||$ is about $O\left(m(k \log n + D^k)\right)$, that is, for each of the m constraints, we list the k variables involved and at most D^k compatible assignments.

It is known that in Model RB there exists satisfiability thresholds $r_{cr} = -\frac{\alpha}{\ln(1-p)}$ [23].

Let $H_{n,r,k}^{RB}$ denote a random constraint hypergraph in Model RB.

2.2 Fractional Edge Cover

Given a hypergraph $H = (V, E)$, if there is a mapping

$$\psi : E \to [0, \infty),$$

such that

$$\sum_{e \in E, v \in e} \psi(e) \geq 1, \text{ for every } v \in V,$$

then ψ is called a *fractional edge cover* of H [11].

For a fractional edge cover ψ, The weight of hyperedge e under ψ is $\psi(e)$. The weight of ψ is $\sum_{e \in E} \psi(e)$. The optimal fractional edge cover ψ^* of H is a fractional edge cover with the minimum weight over all possible fractional edge covers of H. The *fractional edge cover number* of H, denoted by $\rho^*(H)$, is the weight of an optimal fractional edge cover ψ^* of H [11], that is,

$$\rho^*(H) = \min_{\psi} \sum_{e \in E} \psi(e) = \sum_{e \in E} \psi^*(e).$$

It is known that for a CSP instance I, the number of solutions of I is at most $||I||^{\rho^*(H_I)}$, where $||I||$ is the size of I and H_I is the constraint hypergraph of I. It is known that the solutions of I can be enumerated in time $||I||^{\rho^*(H_I)+O(1)}$ [11].

2.3 Chernoff Bound

We say that a random event Q happens *with high probability* if the probability of this event $\Pr(Q)$ goes to 1 asymptotically. We will use the following version of Chernoff Bound [21].

Lemma 1 (Chernoff Bound). *Given a random variable X, X follows a binomial distribution, i.e., $X \sim B(n, \frac{\mu}{n})$. If $0 < \epsilon < 1$, then*

$$\Pr(X \leq (1 - \epsilon)\mu) \leq e^{-\mu\epsilon^2/2}.$$

3 Fractional Edge Cover Number of Model RB

In this section, we will give lower and upper bounds on fractional edge cover number of Model RB.

Suppose that $\rho^*(H^{RB}_{n,r,k})$ is the fractional edge cover number of Model RB, where n is the total number of vertices in the constraint hypergraph, k is the number of vertices contained in each hyperedge, and $rn \ln n$ is the maximum number of hyperedges. We can first give a lower bounds on $H^{RB}_{n,r,k}$ as follows.

Theorem 1. $\rho^*(H^{RB}_{n,r,k}) \geq \frac{n}{k}.$

Proof. Let ψ be an arbitrary fractional edge cover of $H_{n,r,k}^{RB}$. Then by definition of fractional edge cover,

$$\sum_{e \in E, v \in e} \psi(e) \geq 1, \text{ for all } v \in V.$$

Now we summarize all these n inequalities over all vertices $v \in V$,

$$\sum_{v \in V} \left(\sum_{e \in E, v \in e} \psi(e) \right) \geq n.$$

Since every hyperedge contains exactly k vertices, the weight $\psi(e)$ of every hyperedge e will appear exactly k times in the left hand side. Thus,

$$\sum_{v \in V} \left(\sum_{e \in E, v \in e} \psi(e) \right) = k \cdot \sum_{e \in E} \psi(e).$$

From the above two inequalities, we have

$$k \cdot \sum_{e \in E} \psi(e) \geq n,$$

or equivalently,

$$\sum_{e \in E} \psi(e) \geq \frac{n}{k}.$$

Since ψ is an arbitrary fractional edge cover, the last inequality also holds for the optimal fractional edge cover ψ^*. Therefore,

$$\rho^*(H_{n,r,k}^{RB}) = \sum_{e \in E} \psi^*(e) \geq \frac{n}{k}.$$

We have finished the proof of this theorem.   □

After we get a lower bound on $\rho^*(H_{n,r,k}^{RB})$ by a counting argument as above, we will give a matching upper bound on $\rho^*(H_{n,r,k}^{RB})$, by an explicit construction of a fractional edge cover for $H_{n,r,k}^{RB}$. To this end, we need a lower bound on the minimum degree of vertices in $H_{n,r,k}^{RB}$ by Chernoff bound as follows.

Suppose that the degree of a vertex v is denoted by $deg(v)$, and the minimum degree of $H_{n,r,k}^{RB}$ is denoted by $\delta(H_{n,r,k}^{RB})$.

Lemma 2. $\delta(H_{n,r,k}^{RB}) \geq (1 - \frac{2}{\sqrt{kr}}) \cdot kr \ln n$ *with high probability.*

Proof. Let v be an arbitrary vertex in $H_{n,r,k}^{RB}$. By definition of $H_{n,r,k}^{RB}$, we repeat for $rn \ln n$ times to randomly select hyperedges, and each time independently select k different vertices from all n vertices to form an hyperedge. For an arbitrary hyperedge e, the vertex v is contained in e with probability $\frac{k}{n}$.

For $i = 1, ..., rn \ln n$, let X_i be a random variable with probability $\frac{k}{n}$ to be 1, and with probability $1 - \frac{k}{n}$ to be 0, respectively. All these X_i's are independent and identically distributed 0–1 variables. Then,

$$deg(v) = \sum_{i=1}^{rn \ln n} X_i.$$

The random variable $deg(v)$ has a binomial distribution $B\left(rn \ln n, \frac{k}{n}\right)$. The expectation μ of $deg(v)$ is

$$\mu = (rn \ln n)\frac{k}{n} = kr \ln n.$$

By the Chernoff bound, for any $0 < \delta < 1$,

$$\Pr\left(deg(v) \le (1 - \delta) \cdot kr \ln n\right) \le e^{-(kr \ln n)\delta^2/2}.$$

Let $\delta = \frac{2}{\sqrt{kr}}$. If $kr \le 4$, $(1 - \delta) \cdot kr \ln n \le 0$, this will lead to a trivial case. Otherwise for $kr > 4$, we have $0 < \delta < 1$. Then

$$\Pr\left(deg(v) \le \left(1 - \frac{2}{\sqrt{kr}}\right) \cdot kr \ln n\right) \le e^{-2 \ln n} = \frac{1}{n^2}.$$

By the Union bound,

$$\Pr\left(\exists v, deg(v) \le \left(1 - \frac{2}{\sqrt{kr}}\right) \cdot kr \ln n\right) \le n \cdot \frac{1}{n^2} = \frac{1}{n}.$$

Thus,

$$\lim_{n \to \infty} \Pr\left(\delta(H_{n,r,k}^{RB}) \le \left(1 - \frac{2}{\sqrt{kr}}\right) \cdot kr \ln n\right) = 0.$$

We have finished the proof of this lemma. □

Once we have a lower bound δ_L on the minimum degree of variables in Model RB, we can construct a fractional edge cover of Model RB by putting weight $\frac{1}{\delta_L}$ on each hyperedge. Since each variable is contained in at least δ_L hyperedges, the sum of weight of hyperedges containing this variable is at least 1, which makes sure that it is a fractional edge cover. In this way, we can get a matching upper bound of $\rho^*(H_{n,r,k}^{RB})$ as follows.

Theorem 2. $\rho^*(H_{n,r,k}^{RB}) \le \frac{n}{k} \cdot \left(1 - \frac{2}{\sqrt{kr}}\right)^{-1}$, *with high probability.*

Proof. For a random constraint hypergrah $H_{n,r,k}^{RB} = (V, E)$ of Model RB, we define a mapping $\psi_0 : E \to [0, \infty)$ as follows.

$$\psi_0(e) = \left(1 - \frac{2}{\sqrt{kr}}\right)^{-1} \cdot \frac{1}{kr \ln n}, \quad \text{for all } e \in E.$$

Recall that with high probability,

$$\delta(H_{n,r,k}^{RB}) \leq \left(1 - \frac{2}{\sqrt{kr}}\right) \cdot kr \ln n.$$

Thus with high probability, for each variable v,

$$\sum_{e \in E, v \in e} \psi_0(e) = deg(v) \cdot \left(1 - \frac{2}{\sqrt{kr}}\right)^{-1} \cdot \frac{1}{kr \ln n} \geq 1.$$

Therefore, ψ_0 is a fractional edge cover of $H_{n,r,k}^{RB}$ with high probability. The weight of ψ_0 is

$$\sum_{e \in E} \psi_0(e) \leq (rn \ln n) \cdot \left(1 - \frac{2}{\sqrt{kr}}\right)^{-1} \cdot \frac{1}{kr \ln n} = \frac{n}{k} \cdot \left(1 - \frac{2}{\sqrt{kr}}\right)^{-1}.$$

Thus fractional edge cover number of $H_{n,r,k}^{RB}$ is no larger than $\frac{n}{k} \cdot \left(1 - \frac{2}{\sqrt{kr}}\right)^{-1}$ with high probability. We have finished the proof.   □

Note that the upper bound is only within a constant ratio $\left(1 - \frac{2}{\sqrt{kr}}\right)^{-1}$ to the lower bound. For fixed r, the more larger k is, the more tighter upper bounds we get.

4   Conclusions

In this paper, we show linear lower and upper bounds on fractional edge cover number of Model RB. Since the structural decomposition method based on fractional edge cover runs in time exponential in fractional edge cover number, these results provide evidence for hardness of Model RB.

The fractional edge cover number and hypertree width are incomparable in the sense that, there are hypergraphs with bounded hypertree width and unbounded fractional edge cover number, and vice versa [11]. The most powerful structural parameter is fractional hypertree width, which supersedes both fractional edge cover number and hypertree width [17]. To show lower and upper bounds on fractional hypertree width for Model RB is an open problem.

Among these structural parameters, perhaps tree width is the best understood on the classical random graphs. It is known that there is a threshold where the tree width suddenly jumps from constant to linear in the number of vertices [7–9,12–14]. Whether there are similar phenomena for the fractional edge cover number and the fractional hypertree width on classical random graphs is also unknown.

Acknowledgments. We thank Ms. Yu Song for drafting an earlier version of this paper.

References

1. Adler, I., Gottlob, G., Grohe, M.: Hypertree width and related hypergraph invariants. Eur. J. Comb. **28**, 2167–2181 (2007)
2. Cheeseman, P., Kanefsky, R., Taylor, W.: Where the really hard problems are. In: Proceedings of IJCAI 1991, pp. 163–169 (1991)
3. Cook, S.A., Mitchell, D.G.: Finding hard instances of the satisfiability problem: a survey. DIMACS Ser. **35**, 1–17 (1997)
4. Dubois, O., Boufkhad, Y., Mandler, J.: Typical random 3-SAT formulae and the satisfiability threshold. In: Proceedings of SODA 2000, pp. 126–127 (2000)
5. Fan, Y., Shen, J.: On the phase transitions of random k-constraint satisfaction problems. Artif. Intell. **175**, 914–927 (2011)
6. Fan, Y., Shen, J., Xu, K.: A general model and thresholds for random constraint satisfaction problems. Artif. Intell. **193**, 1–17 (2012)
7. Gao, Y.: Phase transition of tractability in constraint satisfaction and Bayesian network inference. In: Proceedings of UAI, pp. 265–271 (2003)
8. Gao, Y.: On the threshold of having a linear treewidth in random graphs. In: Chen, D.Z., Lee, D.T. (eds.) COCOON 2006. LNCS, vol. 4112, pp. 226–234. Springer, Heidelberg (2006)
9. Gao, Y.: Treewidth of Erdos-Renyi random graphs, random intersection graphs, and scale-free random graphs. Discrete Appl. Math. **160**(4–5), 566–578 (2012)
10. Gao, Y., Culberson, J.: Consistency and random constraint satisfaction problems. J. Artif. Intell. Res. **28**, 517–557 (2007)
11. Grohe, M., Marx, D.: Constraint solving via fractional edge covers. ACM Trans. Alg. 11(1), Article 4 (2014)
12. Jiang, W., Liu, T., Ren, T., Xu, K.: Two hardness results on feedback vertex sets. In: Atallah, M., Li, X.-Y., Zhu, B. (eds.) FAW-AAIM 2011. LNCS, vol. 6681, pp. 233–243. Springer, Heidelberg (2011)
13. Kloks, T.: Treewidth: Computations and Approximations, pp. 18–55. Springer, Berlin (1994)
14. Lee, C., Lee, J., Oum, S.: Rank-width of random graphs. J. Graph. Theor. **70**(3), 339–347 (2012)
15. Liu, T., Lin, X., Wang, C., Su, K., Xu, K.: Large hinge width on sparse random hypergraphs. In: Proceedings of IJCAI, pp. 611–616 (2011)
16. Liu, T., Wang, C., Xu, K.: Large hypertree width for sparse random hypergraphs. J. Comb. Optim. **29**, 531–540 (2015)
17. Marx, D.: Tractable hypergraph properties for constraint satisfaction and conjunctive queries. J. ACM **60**(6), 42 (2013)
18. Mezard, M., Parisi, G., Zecchina, R.: Analytic and algorithmic solution of random satisfiability problems. Science **297**(5582), 812–815 (2002)
19. Mitchell, D.G., Selman, B., Levesque, H.J.: Hard and easy distributions of sat problems. In: Proceedings of AAAI 1992, pp. 459–465 (1992)
20. Selman, B., Mitchell, D.G., Levesque, H.J.: Generating hard satisfiability problems. Artif. Intell. **81**, 17–29 (1996)
21. Rucinski, A., Janson, S., Luczak, T.: Random Graphs. Wiley, New York (2000)
22. Wang, C., Liu, T., Cui, P., Xu, K.: A note on treewidth in random graphs. In: Wang, W., Zhu, X., Du, D.-Z. (eds.) COCOA 2011. LNCS, vol. 6831, pp. 491–499. Springer, Heidelberg (2011)
23. Xu, K., Li, W.: Exact phase transitions in random constraint satisfaction problems. J. Artif. Intell. Res. **12**, 93–103 (2000)

24. Xu, K.: BHOSLIB: Benchmarks with Hidden Optimum Solutions for Graph Problems. http://www.nlsde.buaa.edu.cn/kexu/benchmarks/graph-benchmarks.htm
25. Xu, K., Li, W.: Many hard examples in exact phase transitions. Theor. Comput. Sci. **355**, 291–302 (2006)
26. Xu, K., Boussemart, F., Hemery, F., Lecoutre, C.: Random constraint satisfaction: easy generation of hard (satisfiable) instances. Artif. Intell. **171**, 514–534 (2007)
27. Xu, W., Zhang, P., Liu, T., Gong, F.: The solution space structure of random constraint satisfaction problems with growing domains. J. Stat. Mech. Theor. Exp. **2015**, P12006 (2015)
28. Zhao, C., Zhang, P., Zheng, Z., Xu, K.: Analytical and belief-propagation studies of random constraint satisfaction problems with growing domains. Phys. Rev. E **85**, 016106 (2012)

Breaking Cycle Structure to Improve Lower Bound for Max-SAT

Yan-Li Liu[1,2], Chu-Min Li[1,3(✉)], Kun He[1(✉)], and Yi Fan[1]

[1] Department of Computer Science and Technology,
Huazhong University of Science and Technology, Wuhan, China
yanli12008@163.com, chu-min.li@u-picardie.fr, brooklet60@gmail.com
[2] Department of Science, Wuhan University of Science and Technology,
Wuhan, China
[3] MIS, University of Picardie Jules Verne, Amiens, France

Abstract. Many practical optimization problems can be translated to Max-SAT and solved using a Branch-and-Bound (BnB) Max-SAT solver. The performance of a BnB Max-SAT solver heavily depends on the quality of the lower bound. Lower bounds in state-of-the-art BnB Max-SAT solvers are based on detecting inconsistent subsets of clauses and then on applying Max-SAT resolution to transform each inconsistent subset of clauses into an equivalent set containing an empty clause and a number of compensation clauses. In this paper, we focus on the transformation of the inconsistent subsets of clauses containing one unit clause and a number of binary clauses. We show that Max-SAT resolution generates a lot of ternary compensation clauses when transforming such an inconsistent set, deteriorating the quality of the lower bound, and propose a new inference rule, called cycle breaking rule, to transform the inconsistent set. We prove the correctness of the rule and implement it in a new BnB Max-SAT solver called Brmaxsat. Experimental results showed that cycle breaking rule is very effective, especially on Max-2SAT.

Keywords: NP-complete · Max-SAT · Branch and bound · Lower bound

1 Introduction

The Maximum Satisfiability Problem (Max-SAT) on a conjunctive formula is to find an assignment of variables such that the number of unsatisfied clauses is minimized. The decision version of Max-SAT is NP-complete [1], even for Max-2SAT problem where each clause has at most two literals.

Given a *propositional variables* set $V = \{x_1, x_2, x_3, \ldots, x_n\}$, x_i may take values 0 (*false*) or 1 (*true*). A *literal* l_i is a variable x_i or its negation $\bar{x}_i$. A *clause* $c = l_1 \vee l_2 \vee \ldots \vee l_k$ is a disjunction of literals. If a clause has one literal, it is called

This work is supported by National Natural Science Foundation (61472147, 51306133).

unit clause. A *conjunctive normal form* (CNF) is a conjunction of clauses denoted as F, which is usually represented as a clauses set $F = \{c_1, c_2, c_3, \ldots, c_m\}$. The literal x_i is satisfied when propositional variable x_i takes *true*, and the literal $\bar{x}_i$ is satisfied when variable x_i takes *false*. A clause c is satisfied if at least one literal of the clause is satisfied. The *empty clause*, also called *conflict clause*, is the case that all its literals are unsatisfied, denoted as $\square$. An *assignment* is a mapping from V' to $\{0, 1\}$, $V' \subseteq V$. The assignment is complete if $V' = V$; otherwise it is partial. The space of all possible assignments of a CNF can be represented as a 2^n (n is the number of propositional variables) size of binary search tree, where internal nodes represent partial assignments and leaf nodes represent complete assignments. Max-SAT is called Max-kSAT when all the clauses have k literals per clause. The subtraction of two clause sets, denoted as $F_2 - F_1$, is the set containing those clauses that are in F_2 but not in F_1.

In recent years, we have seen considerable progress on exact Max-SAT solvers. Some exact Max-SAT solvers based on SAT algorithm are more efficient on some crafted instances and most industrial instances. These Max-SAT solvers use learning clauses [2], blocking variable [3], mixed integer programming [4] which are successful in SAT solvers. some researchers proposed new reductions to transform Max-SAT problem to Max-CSP [5] and worked well. Max-SAT algorithms based branch and bound (BnB) have shown their efficiency, especially on random and crafted instances. The most competitive exact BnB Max-SAT solvers are based on the Davis-Putnam-Loveland procedure(DPLL) [6], including akmaxsat [7], wmaxsatz+ [8], maxsatz2013f [9], ahmaxsat-ls-1.55 [10], ISAC+2014 [11]. Main research work for BnB Max-SAT algorithms is to reduce upper bound (UB) and improve lower bound (LB). At present, lots of algorithms utilize a local search algorithm to optimize the initial UB [7,9,10,12].

More work focuses on how to improve LB. LB includes two parts: the number of empty clauses derived by current partial assignment and the number of non-empty clauses (underestimation of LB) which will become unsatisfied when partial assignment is extend to complete assignment. The former is fixed at each node of the search tree, so the latter has a big impact on the efficiency of BnB solvers. Many computation methods [13–18] for the underestimation of LB are based on Max-SAT resolution [19]. These methods can be divided into two classes: The first class is inference rules for some special inconsistent sets which have at least one unsatisfied clause under arbitrary assignment. For example: IC Rule [13] is suited for inconsistent set$\{l, \bar{l}\}$; Star Rule [14] is for $\{l_1, \ldots, l_k, \bar{l}_1, \bar{l}_2, \ldots, \bar{l}_k\}$. These inference rules improve LB faster but they can't be applied on all inconsistent sets. The second class is general Max-SAT resolution which is applied on each inconsistent set instead of some special cases. However, Max-SAT resolution creates much more compensation clauses.

At each node of search tree, there are enormous cycle structures where only two paths lead to the empty clause in implication graph. If directly applying Max-SAT resolution or inference rules on this case, lots of new ternary clauses or k-ary ($k \geqslant 3$) clauses containing at least three literals occur and lead to more complicated problem. Different from the traditional reasoning methods, cycle breaking rule breaks the cycle structure by adding a unit clause which has the same literal with the first joint clause of cycle structure. New binary clauses created by cycle breaking rule have a big effect on unit propagation. Experimental results show it improves LB efficiently.

2 A Basic Max-SAT Solver

We first introduce a basic BnB Max-SAT solver, maxSat in Algorithm 1. maxSat explores the binary search tree in a depth-first manner. In the pseudo-code of maxSat, F is the input Max-SAT instance. UB is an upper bound of the number of unsatisfied clauses found by the best complete assignment so far. The initial UB is set by a local search solver such as *ubcsat* [12] or *CCLS* [20], and it will be updated when the algorithm reaches the leaf node of search tree. *simplify-Formula(F)* is a procedure that simplifies F by applying some sound inference rules, such as Rule 1 [21] and Rule 2 [22]. Rules $1 \sim 2$ simplify F and promise the same number of unsatisfied clauses under any assignment of variables.

Rule 1. *If* $F_1 = \{l_1 \vee l_2 \ldots \vee l_k, \bar{l}_1 \vee l_2 \ldots \vee l_k\} \cup F'$, *then* $F_2 = \{l_2 \vee \ldots \vee l_k\} \cup F'$ *is equivalent to* F_1.

Rule 2. *If* $F_1 = \{l, \bar{l}\} \cup F'$, *then* $F_2 = \{\Box\} \cup F'$ *is equivalent to* F_1.

Algorithm 1. maxSat (F, UB)

Input: a CNF formula F and an upper bound UB
Output: minimal number of unsatisfied clauses, i.e. the final UB and the corresponding assignment
1. $F \leftarrow simplifyFormula(F)$;
2. **if** $F = \emptyset$ or F only contains empty clauses **then**
3.   **return** $\#emptyClauses(F)$;
4. **end if**
5. $LB \leftarrow \#emptyClauses(F) + underestimation(F, UB)$;
6. **if** $LB \geqslant UB$ **then**
7.   **return** UB;
8. **end if**
9. $x \leftarrow selectVariable(F)$;
10. $UB \leftarrow \min(UB, maxSat(F_x, UB))$;
11. **return** $UB \leftarrow \min(UB, maxSat(F_{\bar{x}}, UB))$;

$\#emptyClauses(F)$ is a function that returns the number of unsatisfied clauses in formula F under the current assignment. $underestimation(F, UB)$ is a function that returns the minimum number of non-empty clauses in F that will become unsatisfied if the current partial assignment is extended to a complete assignment. In practice, the underestimated LB is the number of disjoint inconsistent sets with unit propagation. That will be explained in detail in Sects. 3.1

and 3.2. *selectVariable(F)* returns a chosen branch variable by heuristics. Generally, the algorithm chooses a variable which affects the size of search tree greatly. $F_x(F_{\bar{x}})$ is a procedure of applying one-literal rule on F when branch literal is satisfied. Specifically, when branch literal $x_i(\bar{x}_i)$ is satisfied, one-literal rule is to delete all the satisfied clauses containing the literal $x_i(\bar{x}_i)$ and remove all the occurrences of the literal $\bar{x}_i(x_i)$ from those clauses which form is defined as $\bar{x}_i \vee x_m \ldots \vee x_r(x_i \vee x_m \ldots \vee x_r)$.

At every node, maxSat compares UB and LB. Obviously, if $LB \geqslant UB$, a better solution can't be found from this sub-tree. Algorithm maxSat prunes the sub-tree below the current node and backtracks to the parent node. If $LB < UB$, the algorithm tries to find a possible better solution by picking one unassigned variable as new branch node. The final UB obtained after maxSat has searched the entire search tree space is the optimal solution.

2.1   Underestimation of Lower Bound

Good quality underestimation of lower bound is very crucial for BnB Max-SAT solvers. The main methods for underestimated LB include unit propagation, inference rules and failed literal detecting [17,18]. All of them are to find the disjoint inconsistent sets.

Definition 1 (Inconsistent Set and Disjoint Inconsistent Set). *A clause set φ is called an inconsistent set if φ contains at least one unsatisfied clause under any complete assignment. Two inconsistent sets φ_i and φ_j are disjoint if $\varphi_i \cap \varphi_j = \emptyset$.*

Example 1. $\varphi = \{x_1, \bar{x}_1 \vee x_2, \bar{x}_2\}$, there exists at least one unsatisfied clause under any assignment of x_1, x_2:(0,0)(0,1)(1,0)(1,1). So φ is an inconsistent set.

Definition 2 (Minimal Inconsistent Set). *An inconsistent set φ is minimal if it becomes satisfiable after removing any one clause from φ.*

The number of disjoint inconsistent sets is small than or equal to the minimum number of clauses which will become unsatisfied if the current assignment is extended to a complete assignment. So the number of disjoint inconsistent sets is used as the underestimated LB. Unit propagation is the main way for efficiently detecting an inconsistent set. Unit propagation is to repeatedly apply the one-literal rule until reaching an empty clause or a saturation state where no any unit clause can be propagated. Example 2 illustrates the resolution process of an inconsistent set. Each inconsistent set matches a directed implication graph. A directed implication graph $G = (K, E)$ where K is a node set and E is an edge set. Node k_r ($k_r \in K$) is the satisfied literal x_i in clause; there is a directed edge e from node k_r to node k_m if literal $x_i = 1$ causes that $x_j = 1$.

Example 2. Let $F = \{x_1, \bar{x}_1 \vee x_2, \bar{x}_1 \vee x_3, \bar{x}_1 \vee x_7, \bar{x}_2 \vee \bar{x}_3 \vee x_4, \bar{x}_4 \vee x_5, \bar{x}_5 \vee x_6, \bar{x}_6\}$. Figure 1 shows the propagating process for unit clause x_1. If $x_1 = 1$, x_2 must be *true* in order to satisfy clause $\bar{x}_1 \vee x_2$; repeatedly apply one-literal rule until unit clause $\bar{x}_6$ becomes empty. Obviously, $F - \{\bar{x}_1 \vee x_7\}$ is a minimal inconsistent set.

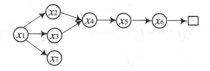

Fig. 1. Finding an inconsistent set by unit propagation

2.2 Inference Rules

Inference rules based on Max-SAT resolution [19] are proposed to transform an inconsistent set of clauses with special structure into an empty clause and a number of compensation clauses. The transformation should preserve the number of unsatisfied clauses. In fact, the equivalence of two Max-SAT instances is defined as follows:

Definition 3 (Max-SAT Problem Equivalence). *Two conjunctive formulas F_1 and F_2 are equivalent for Max-SAT if and only if F_1 and F_2 have the same number of unsatisfied clauses under any complete assignment.*

Li et al. [16, 17] proposed new inference rules Rules 3 and 4 for two inconsistent sets with special structures in 2007.

Definition 4 (Chain Structure). *An implication graph contains only one path where all nodes link one by one, is called chain structure.*

Figure 2 shows the chain structure. From the view of inconsistent set, a chain structure as a minimal inconsistent set contains two unit clauses and a number of binary clauses, and each variable in binary clauses occurs exactly negatively one time and positively one time, except the two variables in unit clauses which occurs one time in binary clauses with the sign opposite to the unit clauses.

Fig. 2. Implication graph of chain structure

Rule 3 (Chain Rule). *If $F_1 = \{l_1, \bar{l}_1 \vee l_2, \bar{l}_2 \vee l_3, \ldots, \bar{l}_{k-1} \vee l_k, \bar{l}_k\} \cup F'$, then $F_2 = \{\Box, l_1 \vee \bar{l}_2, l_2 \vee \bar{l}_3, \ldots, l_{k-1} \vee \bar{l}_k\} \cup F'$ is equivalent to F_1.*

Figure 3 is for a special structure containing a cycle. In Fig. 3, unit propagation starts with clause l_1 and deduced conflict clause $\bar{l}_k \vee \bar{l}_{k+1}$.

Rule 4 (Cycle Rule). *If $F_1 = \{l_1, \bar{l}_1 \vee l_2, \ldots, \bar{l}_{k-1} \vee l_k, \bar{l}_{k-1} \vee l_{k+1}, \bar{l}_k \vee \bar{l}_{k+1}\} \cup F'$, then $F_2 = \{\Box, l_1 \vee \bar{l}_2, l_2 \vee \bar{l}_3, \ldots, l_{k-1} \vee \bar{l}_k \vee \bar{l}_{k+1}, \bar{l}_{k-1} \vee l_k \vee \bar{l}_{k+1}\} \cup F'$ is equivalent to F_1.*

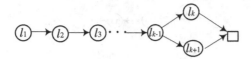

Fig. 3. Implication graph containing a cycle

It has been proved that F_1 is equivalent to F_2 for Max-SAT problem [16]. When chain inconsistent sets or inconsistent sets containing a cycle are detected, algorithm replaces F_1 with F_2. Inference rules are more efficient than unit propagation for computing LB because new clauses in F_2 can make up more inconsistent sets. Example 3 shows how inference rules improve the underestimated LB.

Example 3. Let $\varphi = \{x_1, \bar{x}_1 \lor x_2, \bar{x}_1 \lor x_3, \bar{x}_2 \lor \bar{x}_3, x_4, \bar{x}_4 \lor x_1, \bar{x}_2 \lor \bar{x}_4, \bar{x}_3 \lor \bar{x}_4\}$. According to the occurrence order of unit clause, unit propagation detects one inconsistent set $\varphi_1 = \{x_1, \bar{x}_1 \lor x_2, \bar{x}_1 \lor x_3, \bar{x}_2 \lor \bar{x}_3\}$. The remaining clauses are satisfied once φ_1 is removed, so LB is 1. However cycle inference rule is applied on φ_1, new clauses $\bar{x}_1 \lor x_2 \lor x_3$ and $x_1 \lor \bar{x}_2 \lor \bar{x}_3$ are added to φ. Another inconsistent set $\varphi_2 = \{x_4, \bar{x}_4 \lor x_1, \bar{x}_2 \lor \bar{x}_4, \bar{x}_3 \lor \bar{x}_4, \bar{x}_1 \lor x_2 \lor x_3\}$ is found. So LB is 2.

3   Breaking Cycle Structure

Inference rules are good for improving LB, but they can't be directly applied on arbitrary structure, especially complicated structure. There are lots of general cycle structures during the computing underestimation of LB, so we wish to find general cycle rule to improve the LB.

3.1   Generalized Cycle Structure

Definition 5 (Simple Cycle Structure). *A directed implication graph containing only two paths of length greater than zero that begins at the same node A and ends at another same node B, is called a simple cycle structure.*

Figure 4 shows the simplest cycle implication graph. Figure 5 shows a general cycle implication graph($i \geqslant 2$, $2i > k \geqslant i + 1$).

A simple cycle structure as a minimal inconsistent set containing one unit clause and a number of binary clauses, in which each variable exactly occurs one

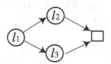

Fig. 4. The simplest cycle

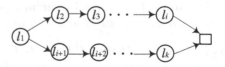

Fig. 5. The length k of a cycle

time negatively and one time positively in binary clauses, except the variable in the unit clause which occurs two times in binary clauses with the sign opposite to the unit clause.

Definition 6 (Length of a Cycle). *In a directed cycle implication graph, the number of all nodes in two paths of the cycle except the starting node is the length of the cycle.*

By definition of length of a cycle, the length of the simplest cycle in Fig. 4 is 3. The length of the cycle in Fig. 5 is k. Actually, the length of a simple cycle is the number of the binary clauses in a cycle inconsistent set.

We adopt the integer programming to prove the equivalence of two formulas. Assume $F = \{c_1, c_2, c_3, \ldots, c_m\}$ is a formula over a variable set $\{x_1, x_2, x_3, \ldots, x_n\}$ and $c_i = x_{i_1} \vee \ldots \vee x_{i_k} \vee \bar{x}_{i_{k+1}} \vee \ldots \vee \bar{x}_{i_{k+r}}$ $(1 \leqslant i \leqslant m)$. Note that all positive literals in c_i are stored in the front of the negative ones. Each propositional variable maps an integer variable taking values 0 or 1. A positive literal x_i in clause is transformed to $1-x_i$ of integer expression and a negative literal $\bar{x}_i$ in clause is transformed to integer expression x_i. The disjunction operation maps the multiplication of integers. So, define the integer transformation of clause c_i as

$$\varepsilon_i = (1 - x_{i_1}) \ldots (1 - x_{i_k}) x_{i_{k+1}} \ldots x_{i_{k+r}}$$

ε_i is a integer variable taking 0 or 1. Obviously, ε_i is 0 iff at least one of x_{i_j} $(1 \leqslant j \leqslant k)$ is 1 or at least one of x_{i_s} $(k + 1 \leqslant s \leqslant k + m)$ is 0. So, when $\varepsilon_i = 0$, clause c_i is satisfied. Otherwise $\varepsilon_i = 1$ iff c_i isn't satisfied. For formula F, the number of unsatisfied clauses can be defined as

$$\xi(F) = \sum_{i=1}^{m} \varepsilon_i$$

Lemma 1. $F_1 = \{\bar{l}_1 \vee l_2, \bar{l}_1 \vee l_3, \bar{l}_2 \vee \bar{l}_3\}$, then $F_2 = \{\bar{l}_1, \bar{l}_1 \vee l_2 \vee l_3, l_1 \vee \bar{l}_2 \vee \bar{l}_3\}$ is equivalent to F_1.

Proof.
$$\xi(F_2) = l_1 + l_1(1 - l_2)(1 - l_3) + (1 - l_1)l_2l_3$$
$$= l_1 + l_1 - l_1l_2 - l_1l_3 + l_1l_2l_3 + l_2l_3 - l_1l_2l_3$$
$$= l_1 - l_1l_2 + l_1 - l_1l_3 + l_2l_3 = \xi(F_1)$$

Lemma 2. $F_1 = \{\bar{l}_1 \vee l_2, \bar{l}_2 \vee l_3\}$, then $F_2 = \{\bar{l}_1 \vee l_3, \bar{l}_2 \vee l_1 \vee l_3, l_2 \vee \bar{l}_1 \vee \bar{l}_3\}$ is equivalent to F_1.

Proof.
$$\xi(F_2) = l_1(1 - l_3) + l_2(1 - l_1)(1 - l_3) + (1 - l_2)l_1l_3$$
$$= l_1 - l_1l_3 + l_2 - l_1l_2 - l_2l_3 + l_1l_2l_3 + l_1l_3 - l_1l_2l_3$$
$$= l_1 - l_1l_2 + l_2 - l_2l_3 = \xi(F_1)$$

Lemma 2 is that Max-SAT resolution is limited to two binary clauses. Theorem 1 means that inference rule is difficult to deal with cycle structure because new lots of ternary clauses will deteriorate the quality of the lower bound.

Theorem 1. *If an inconsistent set φ_1 is a simple cycle structure and the length of the cycle is $n\,(n \geqslant 3)$, φ_2 is equivalent to φ_1 and there are $2\,(n-2)$ ternary clauses and one empty clause in φ_2.*

Proof. (a) If $n = 3$, $\varphi_1 = \{l_1, \bar{l}_1 \vee l_2, \bar{l}_1 \vee l_3, \bar{l}_2 \vee \bar{l}_3\}$, as showed in Fig. 4. Apply Lemma 1, get $\{l_1, \bar{l}_1, \bar{l}_1 \vee l_2 \vee l_3, l_1 \vee \bar{l}_2 \vee \bar{l}_3\} = \{\Box, \bar{l}_1 \vee l_2 \vee l_3, l_1 \vee \bar{l}_2 \vee \bar{l}_3\} = \varphi_2$. The length of cycle in Fig. 4 is 3, so Theorem 1 is sound.
(b) If $n = k - 1$, Theorem 1 is sound. When $n = k$ as showed in Fig. 5.
$\varphi_1 = \{l_1, \bar{l}_1 \vee l_2, \bar{l}_2 \vee l_3, \dots, \bar{l}_{i-1} \vee l_i, \bar{l}_1 \vee l_{i+1}, \bar{l}_{i+1} \vee l_{i+2}, \dots, \bar{l}_{k-1} \vee l_k, \bar{l}_i \vee \bar{l}_k\}$
Repeatedly apply Lemma 2 on path from node l_2 to l_i,
$\varphi_1 \Leftrightarrow \{l_1, \bar{l}_2 \vee l_1 \vee l_3, l_2 \vee \bar{l}_1 \vee \bar{l}_3, \bar{l}_1 \vee l_3, \bar{l}_3 \vee l_4, \dots, \bar{l}_{i-1} \vee l_i, \bar{l}_1 \vee l_{i+1}, \dots, \bar{l}_i \vee \bar{l}_k\}$
$\Leftrightarrow \{l_1, \bar{l}_2 \vee l_1 \vee l_3, \dots, \bar{l}_{i-1} \vee l_1 \vee l_i, l_{i-1} \vee \bar{l}_1 \vee \bar{l}_i, \bar{l}_1 \vee l_i, \bar{l}_1 \vee l_{i+1}, \dots, \bar{l}_i \vee \bar{l}_{k+1}\}$
Repeatedly apply Lemma 2 on path from node l_{i+1} to l_k,
$\varphi_1 \Leftrightarrow \{l_1, \bar{l}_2 \vee l_1 \vee l_3, l_2 \vee \bar{l}_1 \vee \bar{l}_3, \dots, \bar{l}_{i-1} \vee l_1 \vee l_i, l_{i-1} \vee \bar{l}_1 \vee \bar{l}_i, \bar{l}_1 \vee l_i, \bar{l}_{i+1} \vee l_1 \vee l_{i+2}, l_{i+1} \vee \bar{l}_1 \vee \bar{l}_{i+2}, \dots, \bar{l}_{k-1} \vee l_1 \vee l_k, l_{k-1} \vee \bar{l}_1 \vee \bar{l}_k, \bar{l}_1 \vee l_k, \bar{l}_i \vee \bar{l}_k\}$
Apply the Lemma 1 on the $\{\bar{l}_1 \vee l_i, \bar{l}_1 \vee l_k, \bar{l}_i \vee \bar{l}_k\}$
we get $\{\bar{l}_1, \bar{l}_1 \vee l_i \vee l_k, l_1 \vee \bar{l}_i \vee \bar{l}_k\}$
$\varphi_2 = \{\Box, \bar{l}_2 \vee l_1 \vee l_3, l_2 \vee \bar{l}_1 \vee \bar{l}_3, \dots, \bar{l}_{i-1} \vee l_1 \vee l_i, l_{i-1} \vee \bar{l}_1 \vee \bar{l}_i, \bar{l}_{i+1} \vee l_1 \vee l_{i+2}, l_{i+1} \vee \bar{l}_1 \vee \bar{l}_{i+2}, \dots, \bar{l}_{k-1} \vee l_1 \vee l_k, l_{k-1} \vee \bar{l}_1 \vee \bar{l}_k, \bar{l}_1 \vee l_i \vee l_k, l_1 \vee \bar{l}_i \vee \bar{l}_k\}$
So, the Theorem 1 is sound for the length of k cycle structure.

A more common structure of inconsistent set is *generalized cycle structure* that can be regarded as a simple cycle structure by adding a chain. As Fig. 6 $(0 \leqslant i < j, k \geqslant 2)$ shown, the generalized cycle structure can be equivalently transformed into a chain and a simple cycle structure by Lemma 3.

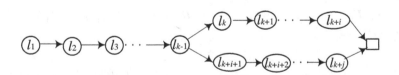

Fig. 6. Implication graph of generalized cycle structure

Lemma 3. *If $F_1 = \{l_1, \bar{l}_1 \vee l_2\}$, then $F_2 = \{l_1 \vee \bar{l}_2, l_2\}$ is equivalent to F_1.*

Proof. $\xi(F_1) = (1 - l_1) + l_1(1 - l_2) = 1 - l_1 l_2 = (1 - l_1)l_2 + (1 - l_2) = \xi(F_2)$.

Definition 7 (First Joint Node). *In a directed implication graph containing a cycle structure, node A is the first joint node (FJD) of node $\Box$ iff each reverse path from $\Box$ needs to pass through A firstly.*

The node l_{k-1} is the FJD of $\square$ in Fig. 6. We add one unit clause to the inconsistent set in order to break the cycle structure to chain structure. This method creates binary clauses instead of ternary clauses by inference rules.

Rule 5 (Cycle Breaking Rule). *If* $F_1 = \{l_1, \bar{l}_1 \vee l_2, \ldots, \bar{l}_{k-1} \vee l_k, \bar{l}_k \vee l_{k+1}, \ldots, \bar{l}_{k+i-1} \vee l_{k+i}, l_{k-1}, \bar{l}_{k-1} \vee l_{k+i+1}, \bar{l}_{k+i+1} \vee l_{k+i+2}, \ldots, \bar{l}_{k+j-1} \vee l_{k+j}, \bar{l}_{k+i} \vee \bar{l}_{k+j}\} \cup F'$, *then* $F_2 = \{\square, l_1 \vee \bar{l}_2, \ldots, l_{k-1} \vee \bar{l}_k, \ldots, l_{k+i-1} \vee \bar{l}_{k+i}, l_{k-1} \vee \bar{l}_{k+i+1}, \ldots, l_{k+j-1} \vee \bar{l}_{k+j}, l_{k+i} \vee l_{k+j}\} \cup F'$ *is equivalent to* F_1. $(0 \leqslant i < j, k \geqslant 2)$.

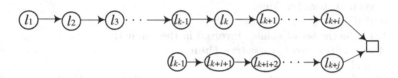

Fig. 7. Implication graph of $F_1 - F'$

Proof. (a) Repeatedly apply Lemma 3 on the clauses from node l_1 to l_{k+i} in Fig. 7. Get the equivalent set $\{l_1 \vee \bar{l}_2, l_2 \vee \bar{l}_3, \ldots, l_{k-2} \vee \bar{l}_{k-1}, l_{k-1} \vee \bar{l}_k, \ldots, l_{k+i-1} \vee \bar{l}_{k+i}, l_{k+i}\}$

(b) Similarly, apply the Lemma 3 on the clauses from node l_{k-1} to l_{k+j} in Fig. 7. Get the equivalent set $\{l_{k-1} \vee \bar{l}_{k+i+1}, l_{k+i+1} \vee \bar{l}_{k+i+2}, \ldots, l_{k+j-1} \vee \bar{l}_{k+j}, l_{k+j}\}$

(c) Prove $\varphi_1 = \{l_{k+i}, l_{k+j}, \bar{l}_{k+i} \vee \bar{l}_{k+j}\}$ is equivalent to $\varphi_2 = \{\square, l_{k+i} \vee l_{k+j}\}$ $\xi(\varphi_1) = (1 - l_{k+i}) + (1 - l_{k+j}) + l_{k+i} l_{k+j} = 1 + (1 - l_{k+i}) + (1 - l_{k+j}) = \xi(\varphi_2)$. So, F_2 is equivalent to F_1.

Cycle breaking rule breaks cycle structure to chain structure by adding a clause c_r to the inconsistent set. Clause c_r should be a unit clause having the same literal with the first joint node in implication graph. Cycle breaking rule is suitable for any length of cycle structure even if the lengths of two paths in the cycle are not equal. It is easy to prove that the length n of cycle creates n binary clauses in Rule 5 by breaking the cycle structure.

3.2   A New Lower Bound with Cycle Breaking Rule

Algorithm upUnderestimation describes a new lower bound method which adopts chain rule and cycle breaking rule to improve LB. In Algorithm 2, UB is an upper bound for the number of unsatisfied clauses by the best complete assignment found so far; $F - \varphi$ is to remove the inconsistent set φ from the formula F in order to get disjoint inconsistent sets. UQ is a queue containing these unit clauses obtained from input instance or the simplified clauses by branch literals. Note that $existUnCl(F - \varphi, FJD)$ returns *true* if it find the unit clause needed by cycle breaking rule. Otherwise, return *false*.

When computing the underestimated LB ($nConflicts$) with unit propagation, Algorithm 2 propagates each unit clause in queue UQ until one empty

---

Algorithm 2. upUnderestimation(F, UB)

Input: a CNF formula F and an upper bound UB
Output: the number of disjoint inconsistent set $nConflicts$
1. $nConflicts = 0$;
2. $F_c \leftarrow F$;
3. **while** unit clause queue $UQ \neq \emptyset$ **do**
4.   get unit clause c_i from UQ;
5.   **if** $unitPropagation(c_i, F_c) == \square$ **then**
6. $nConflicts + +$;
7.   **if** $\#emptyClauses(F) + nConflicts \geqslant UB$ **then**
8.   **return** backTracking;
9.   **end if**
10.   Let φ be the set of clauses involved in the conflict;
11.   **if** $chainStructure(\varphi) == true$ **then**
12.   apply chain rule;
13.   **else if** $cycleStructure(\varphi) == true$ and $existUnCl(F - \varphi, FJD) == true$ **then**
14.   apply cycle breaking rule;
15.   **end if**
16. $F_c \leftarrow F - \varphi$;
17.   **else**
18. $F_c \leftarrow$ the simplified formula F_c returned by $unitPropagation(c_i, F_c)$;
19.   **end if**
20. **end while**
21. **return** $nConflicts$;

---

clause is derived or UQ becomes empty. Once empty clause appears, estimated value of LB increases by 1. If LB $\geqslant$ UB, Algorithm 2 prunes the sub-tree. After getting the inconsistent set φ, algorithm will analyze its structure and choose the suitable inference rule to apply.

4   Evaluation of New Lower Bound

We compared three solvers: ahms-ls-1.55, maxsatz2013f, Brmaxsat. For all these algorithms, either the source code or the executable was provided by the respective author. All the solvers work on the same machine with Centos operation system, Intel E5 2.00 GHz CPU and memory 62.7 GB.

- ahmaxsat-ls-1.55 is a BnB Max-SAT solver which applies Max-SAT resolution on inconsistent sets. It did outstanding performance on random and crafted Max-SAT instances in tenth Max-SAT Evaluation in 2015 [23].
- maxsatz2013f is a BnB Max-SAT solver. It is one of the best solvers on random and crafted instances in eighth Max-SAT Evaluation in 2013.
- Brmaxsat is implemented on top of maxsatz2013f by replacing its lower bounding function by the upUnderestimation function. Brmaxsat is implemented in C programming language and compiled with GCC.

Experiments evaluated the breaking cycle rule on various types of Max-2SAT (see Table 1) and Max-3SAT (see Table 3) instances. All instances come from the 2015 Max-SAT Evaluation website [23]. All the solvers work on the limited CPU time of 1800 s for each instance. Once the solver can't solve this instance in limited time, the process will be killed. The better solver is the one which solves maximum instances or has the shorter time if the solvers solve the same number of instances.

The #*inst* in Table 1 is the number of instances in the instance class. The data format of other columns in Table 1 is A(B). B is the number of instances which find the best solution successfully in limited time and A is the sum of CPU time (in second). Brmaxsat can solve 95 % random instances in Table 1 in 1800 s and reduces 30 %–70 % in running time than ahmaxsat-ls-1.55 for weighted partial instances classes.

Table 1. Experimental results of Max-2SAT

Instance class(#inst)	ahmaxsat-ls-1.55	maxsatz2013f	Brmaxsat
Unweighted random			
Max2SAT/120v(50)	1245.3(50)	3134(50)	**723(50)**
Max2SAT/140v(50)	3727.5(50)	13640(50)	**2865(50)**
Min2SAT/160v(48)	23.9(48)	74(48)	**4(48)**
Min2SAT/200v(48)	153.3(48)	633(48)	**70(48)**
Abrame-habet/120v(45)	8865.3(42)	12224(37)	**9165(45)**
Abrame-habet/140v(45)	12148.8(35)	12615(27)	**14970(44)**
Abrame-habet/160v(45)	7979.5(22)	6356 (14)	**16924(30)**
Total solved(331)	295	274	**315**
Solved radio	89 %	82 %	**95 %**
Unweighted crafted			
maxcut-140-630-0.7(50)	**4307.9(50)**	19572(50)	4560(50)
maxcut-140-630-0.8(50)	**3203.7(50)**	17675(50)	3462(50)
Maxcut/dimacs-mod(62)	1838.1(52)	4768(52)	**1364(52)**
Maxcut/spinglass(5)	205.3(3)	53(3)	**39(3)**
Average time	61.64	271.40	**60.80**
Weighted partial			
Wmax2sat/100v(40)	**669.7(40)**	976(40)	740(40)
Wmax2sat/120v(40)	686.19(40)	1037(40)	**445(40)**
Wmax2sat/140v(40)	1532.9(40)	2864(40)	**1248(40)**
WPmax2sat/hi(30)	182.1(30)	156(30)	**70(30)**
WPmax2sat/lo(30)	20.69(30)	11(30)	**1(30)**
WPmax2sat/me(30)	90.6(30)	46(30)	**28(30)**
Average time	12.48	19.96	**9.89**

Table 2. Experimental results of some Max-2SAT instance

Instance name	ahmaxsat-ls-1.55	maxsatz2013f	Brmaxsat
s2v140c1200_2.cnf	120(76521)	269(660145)	**43(94649)**
s2v160c2000_3.cnf	716.1(251201)	*unsolved*	**439(518123)**
s2v160c2500_1.cnf	*unsolved*	*unsolved*	**1493(1457241)**
V150_C3500_H150_9.wcnf	3.4(189)	3(427)	**1(155)**
maxcut_140_630_0.8_6.wcnf	326.9(126718)	1576(2718631)	**223(374789)**
s2v140c1600_8.wcnf	45.6(3698)	37(6245)	**21(4779)**

Table 3. Experimental results of Max-3SAT instances

Instance class(#inst)	ahmaxsat-ls-1.55	maxsatz2013f	Brmaxsat
Unweighted			
Max3SAT/70v(50)	**9851.9(50)**	16675(50)	13815(50)
Max3SAT/80v(50)	**7744.8(50)**	10781(50)	9092(50)
Abrame-habet/70v (45)	**15117.2(42)**	16274(39)	18105(41)
Total solved(145)	**142**	139	141
Weighted			
Abrame-habet/w70v(45)	**18430.3(40)**	14936 (34)	17488(38)
Abrame-habet/w90v(49)	**11938.3(28)**	7953 (23)	9907(25)
Abrame-habet/w110v(50)	**16266.1(33)**	5928 (26)	6307(29)
Wmax2sat/hi(40)	**2607.5(40)**	4765(40)	3069(40)
Total solved (184)	**141**	123	132

In Table 2, the first column is the instance name. The data format of other columns is A(B) and A is the running time, B is the number of branch nodes for each instance. *unsolved* means that algorithm can't solve this instance in limited time.

The number of branch nodes in Brmaxsat is more than in ahmaxsat-ls-1.55 but the running time is less. This case is sound because Max-SAT resolution applied on each inconsistent set in ahmaxsat-ls-1.55 costs much time because of the lots of temporary clause.

Table 3 is the experimental results of Max-3SAT instances. The data A(B) from second column to fourth column means that A is the sum of running time and B is the number of instances which find the best solution successfully in limited time. Table 3 showed that cycle breaking rule also has a positive effect on Max-3SAT instances, although the improvement is not as well as on Max-2SAT instances. In fact, Brmaxsat is maxsatz2013f reinforced with cycle breaking rule, and solves significantly more Max-3SAT instances than maxsatz2013f.

Sometimes, there are not enough unit clauses for cycle breaking rule. If the unit clause required by cycle breaking rule didn't exist, algorithm didn't apply inference rule on these inconsistent set. So, cycle breaking rule has a greater effect on the LB in Max-3SAT instances when algorithm reaches closely to the leaf nodes.

5   Conclusions

Applying Max-SAT resolution on inconsistent sets costs much time because lots of temporary k-ary clauses ($k \geqslant 2$) are created and difficult to deal with. Inference rules can't also be directly applied on those inconsistent sets containing a length n of cycle ($n > 3$) because new $2(n-2)$ ternary clauses have a bad effect on unit propagation. Cycle breaking rule creates n binary clauses by adding a unit clause to the inconsistent set. These binary clauses not only help to constitute more disjoint inconsistent sets but also help to create new unit clauses by Rule 1. Experiment result shows that cycle breaking rule improves LB efficiently, especially for Max-2SAT.

References

1. Cook, S.A.: The complexity of theorem proving procedures. In: Proceedings of the 3rd Annual ACM Symposium on Theory of Computing, Shaker Heights, pp. 151–158 (1971)
2. Zhang, L., Madigan, C.F., Moskewicz, M.H., Malik, S.: Efficient conflict driven learning in a boolean satisfiability solver. In: Proceedings of IEEE/ACM International Conference on Computer-Aided Design, ICCAD 2001, pp. 279–285 (2001)
3. Ansotegui, C., Bonet, M.L., Levy, J.: SAT-based MaxSAT algorithms. Artif. Intell. **196**, 77–105 (2013)
4. Davies, J., Bacchus, F.: Postponing optimization to speed up MAXSAT solving. In: Schulte, C. (ed.) CP 2013. LNCS, vol. 8124, pp. 247–262. Springer, Heidelberg (2013)
5. Gaspers, S., Sorkin, G.B.: A universally fastest algorithms for Max 2-Sat, Max 2-CSP, and everything in between. J. Comput. Syst. Sci. **78**, 305–335 (2012)
6. Davis, M., Logemann, G., Loveland, D.: A machine program for theorem proving. Commun. ACM **5**(7), 394–397 (1962)
7. Kuegel, A.: Improved exact solver for the weighted MAXSAT problem. In: Workshop Pragmatics of SAT, Edinburgh, Scotland (2010)
8. Argelich, J., Li, C.-M., Manya, F.: An improved exact solver for partial Max-SAT. In: Proceedings of International Conference on Non-convex Programming: Local and Global Approaches, pp. 230–231 (2007)
9. Liu, Y.-L., Li, C.-M., He, K.: Improved lower bounds in MAXSAT complete algorithm based optimizing inconsistent set. Chin. J. Comput. **10**(36), 2087–2096 (2013)
10. Abram, A., Habet, D.: On the resiliency of unit propagation to maxresolution. In: Proceedings of the 24th International Joint Conference on Artificial Intelligence (IJCAI 2015), pp. 268–274 (2015)
11. Anstegui, C., Malitsky, Y., Sellmann, M.: Max-SAT by Improved instance-specific algorithm configuration. In: Proceedings of the 28th National Conference on Artificial Intelligence (AAAI 2014), pp. 2594–2600 (2014)

12. Tompkins, D.A.D., Hoos, H.H.: UBCSAT: an implementation and experimentation environment for SLS algorithms for SAT and MAX-SAT. In: Hoos, H.H., Mitchell, D.G. (eds.) SAT 2004. LNCS, vol. 3542, pp. 306–320. Springer, Heidelberg (2005)
13. Wallace, R., Freuder, E.: Comparative studies of constraint satisfaction and Davis-Putnam algorithms for maximum satisfiability problems. In: Johnson, D., Trick, M. (eds.) Cliques, Coloring and Satisfiability, vol. 26, pp. 587–615. American Mathematical Society, Providence (1996)
14. Shen, H., Zhang, H.: Study of lower bound functions for MAX-2SAT. In: Proceedings of the 19th National Conference on Artificial Intelligence (AAAI 2004), pp. 185–190 (2004)
15. Li, C.-M., Manyà, F., Planes, J.: Exploiting unit propagation to compute lower bounds in branch and bound Max-SAT solvers. In: van Beek, P. (ed.) CP 2005. LNCS, vol. 3709, pp. 403–414. Springer, Heidelberg (2005)
16. Li, C.-M., Manya, F., Planes, J.: New inference rules for Max-SAT. J. Artif. Intell. Res. **30**, 321–329 (2007)
17. Li, C.M., Manyà, F., Mohamedou, N., Planes, J.: Exploiting cycle structures in Max-SAT. In: Kullmann, O. (ed.) SAT 2009. LNCS, vol. 5584, pp. 467–480. Springer, Heidelberg (2009)
18. Li, C.-M., Manya, F., Mohamedou, N., Planes, J.: Resolution-based lower bounds in MaxSAT. Constraints **15**(4), 456–484 (2010)
19. Bonet, M.L., Levy, J., Manya, F.: Resolution for Max-SAT. Artif. Intell. **171**, 606–618 (2007)
20. Luo, C., Cai, S., Wu, W., Jie, Z., Su, K.: CCLS: an efficient local search algorithm for weighted maximum satisfiability. IEEE Trans. Comput. **64**(7), 1830–1843 (2015)
21. Bansal, N., Raman, V.: Upper bounds for MaxSat: further improved. In: Aggarwal, A.K., Pandu Rangan, C. (eds.) ISAAC 1999. LNCS, vol. 1741, pp. 247–258. Springer, Heidelberg (1999)
22. Niedermeier, R., Rossmanith, P.: New upper bounds for maximum satisfiability. J. Algorithms **36**, 63–88 (2000)
23. http://www.maxsat.udl.cat/index.html

On Counting Parameterized Matching and Packing

Yunlong Liu[1] and Jianxin Wang[2(✉)]

[1] Key Laboratory of High Performance Computing and Stochastic Information Processing (Ministry of Education of China), College of Mathematics and Computer Science, Hunan Normal University, Changsha 410081, People's Republic of China
hnsdlyl@163.com
[2] School of Information Science and Engineering, Central South University, Changsha 410083, People's Republic of China
jxwang@mail.csu.edu.cn

Abstract. In this paper, we first show that the complexity of parameterized m-SET PACKING (resp. m-D MATCHING) COUNTING is ♯W[1]-hard by a reduction from parameterized GRAPH (resp. BIPARTITE GRAPH) MATCHING COUNTING ($m \geq 3$). Subsequently, based on the algorithm for 3-D MATCHING COUNTING, we develop fixed-parameter tractable randomized approximation schemes (FPTRAS) for m-SET PACKING COUNTING, m-D MATCHING COUNTING, and BIPARTITE GRAPH MATCHING COUNTING, respectively. Our results indicate that parameterized m-SET PACKING COUNTING and m-D MATCHING COUNTING are typical examples that are ♯W[1]-hard but admit FPTRAS. Furthermore, we show that EDGE DISJOINT SUBGRAPH PACKING COUNTING, i.e., a special subgraph counting problem parameterized by the size of the packing, admits FPTRAS even if some of the counted subgraphs don't have bounded treewidth.

1 Introduction

Counting the number of solutions is one fundamental computation in theoretical computer science. With the development of the parameterized complexity theory, many intractable problems on counting have been studied by parameterized computation approach in recent years [6,7,12,15]. Especially, for the problems that may be not fixed-parameter tractable, it is an important direction to determine their parameterized complexity and/or present fixed-parameter tractable approximation algorithms.

Matching and packing problems form a basic class of NP-hard problems in computational complexity theory. They have also broad application background in resource allocation, code optimization, and computational biology. In this paper, we focus on the counting versions of three parameterized problems: m-SET PACKING, m-D MATCHING, and m-EDGE DISJOINT SUBGRAPH PACKING (m is a constant and $m \geq 3$). We first give some related definitions about them.

This research was supported in part by the National Natural Science Foundation of China under Grants 61572190, 61232001, and 61420106009.

© Springer International Publishing Switzerland 2016
D. Zhu and S. Bereg (Eds.): FAW 2016, LNCS 9711, pp. 125–134, 2016.
DOI: 10.1007/978-3-319-39817-4_13

Let S be a collection of n sets and $P \subseteq S$. P is a *packing* in S if any two sets in P don't intersect each other. The size of P is the number of sets in P. A packing is a *k-packing* if it consists of exactly k sets.

Definition 1. *Parameterized m-SET PACKING COUNTING (p-$\sharp m$-SET PACKING): Given a pair (S,k), where S is a collection of n sets and in each set there are m elements from a universe U, k is the parameter, count the number of distinct k-packings in S.*

Let $T_1, T_2, \ldots, T_m$ be m pairwise disjoint symbol sets. A tuple $(t_1, t_2, \ldots, t_m)$ is called an ordered tuple in $T_1 \times T_2 \times \ldots \times T_m$ if $t_i \in T_i$ $(1 \le i \le m)$. A collection M of ordered tuples is a *matching* if any two tuples in M don't intersect each other. A matching is a *k-matching* if it consists of exactly k tuples.

Definition 2. *Parameterized m-D MATCHING COUNTING (p-$\sharp m$-D MATCHING): Given a pair (S,k), where S is a collection of n ordered tuples and k is the parameter, count the number of distinct k-matchings in S.*

The edge disjoint subgraph packing problem is an extension of the edge disjoint triangle packing problem in [11].

Let $G = (V, E)$ be a simple undirected graph and Y be a connected subgraph of G having m edges. A subgraph packing based on Y is a *k-Y subgraph packing* in G if it is composed of k edge disjoint copies $Y_1, Y_2, \ldots, Y_k$ of Y in G. Obviously, a k-Y subgraph packing in G is also a special subgraph having mk edges.

Definition 3. *Parameterized m-EDGE DISJOINT SUBGRAPH PACKING COUNTING (p-$\sharp m$-EDGE DISJOINT SUBGRAPH PACKING): Given a triple (G,Y,k), where G is a simple undirected graph, Y is a connected subgraph with m edges, and k is the parameter, count the number of distinct k-Y subgraph packings in G.*

The p-$\sharp m$-SET PACKING problem has attracted a lot of attention in recent years. A series of algorithms with running time $O^*(\binom{|U|}{mk/2})$, $O^*(|U|^{\lceil mk/2 \rceil})$, and $O^*(n\binom{|U|}{\lceil mk/2 \rceil})$, respectively, have been developed [2,9,14]. However, its parameterized complexity was not mentioned in these papers. The p-$\sharp m$-D MATCHING problem is the generalization of p-$\sharp 3$-D MATCHING. Although an FPTRAS for p-$\sharp 3$-D MATCHING was presented by Liu et al. [10], its computational complexity was also unknown in [10]. Recently, the complexity of one variant on p-$\sharp m$-D MATCHING, where the parameter is $m+k$, was also discussed by some researchers on the site of Theoretical Computer Science Stack Exchange.

Studies on counting graph matchings have been made with breakthrough progress in recent years. Specifically, the complexity of counting k-matchings on bipartite graphs was proved to be $\sharp W[1]$-hard by Curticapean and Marx [5] in 2014. Meanwhile, counting k-matchings on general graphs was showed to be $\sharp W[1]$-hard by Curticapean [4] in 2013. Besides their own interesting, these results may become the sources for studying other problems.

In this paper, we first generalize the complexity results on counting graph matchings to some problems above[1]. More specifically, we show that the complexity of p-$\sharp m$-SET PACKING (resp. p-$\sharp m$-D MATCHING) is $\sharp W[1]$-hard by a reduction from GRAPH (resp.BIPARTITE GRAPH) MATCHING COUNTING.

Fixed-parameter tractable approximation algorithm is commonly considered to be a practically efficient approach to deal with $\sharp W[1]$-hard problems. Actually, there have been only a few classes of $\sharp W[1]$-hard problems which admit FPTRAS up to now. In this paper, based on the algorithm for p-$\sharp 3$-D MATCHING in [10], we develop FPTRAS for p-$\sharp m$-SET PACKING, p-$\sharp m$-D MATCHING, and BIPARTITE GRAPH MATCHING COUNTING, respectively.

Subgraph counting problems form an important topic studied in the theory of parameterized counting. In particular, exploring FPTRAS has been one main line of research in this area, where the number of vertices in the counted subgraph is usually considered as the parameter. Arvind and Raman [1] firstly introduced FPTRAS to count the number of copies of a k-vertex subgraph with bounded treewidth. Recently, Jerrum and Meeks [7] extended the FPTRAS to count induced subgraphs with property Φ as long as Φ is monotone and every minimal graph with property Φ has bounded treewith. However, for counting subgraphs with unbounded treewidth, many challenges have been faced and some negative results have been obtained recently [13].

In this paper, we show that p-$\sharp m$-EDGE DISJOINT SUBGRAPH PACKING, in which the size of the packing is considered as the parameter, admits FPTRAS even if some of the counted subgraphs (i.e., the subgraph packings) don't have bounded treewidth. This example will help to develop new methods on studying FPTRAS for counting subgraphs with some complicated property.

2   Preliminaries

We introduce some definitions and lemmas employed in this paper.

Definition 4 ([6]). *A parameterized counting problem $F : \sum^* \times N \to N$ is fixed-parameter tractable if there is an algorithm computing $F(x,k)$ in time $f(k)|x|^c$ for some computable function $f : N \to N$ and some constant $c \in N$.*

Definition 5 ([1]). *A parameterized counting problem Q admits a fixed parameter tractable randomized approximation scheme (FPTRAS) if there is an algorithm $\mathcal{A}$ such that for any instance (x,k) of Q, and any positive real number $\epsilon > 0$, $0 < \delta < 1$, the algorithm $\mathcal{A}$ runs in time $f(k)g(|x|, \epsilon, \delta)$, where f is a recursive function, g is a polynomial of $|x|, 1/\epsilon$, and $ln(1/\delta)$, and produces a number h such that $Prob[(1-\epsilon)h_0 \leq h \leq (1+\epsilon)h_0] \geq 1-\delta$, where h_0 is the solution to the instance (x,k).*

Definition 6 ([4]). *Parameterized GRAPH MATCHING COUNTING (p-$\sharp$ MATCHING): Given a simple undirected graph G, a parameter k, count the number of distinct k-matchings in G, where a k-matching is a set of k pairwise disjoint edges.*

---

[1] Note that in these problems, m is a constant rather than a parameter.

Lemma 1 ([4]). *The complexity of p-♯MATCHING is ♯W[1]-hard.*

Definition 7 ([5]). *Parameterized* BIPARTITE GRAPH MATCHING COUNTING *(p-♯2-B* MATCHING*): Given a simple undirected bipartite graph G, a parameter k, count the number of distinct k-matchings in G.*

Lemma 2 ([5]). *The complexity of p-♯2-B* MATCHING *is ♯W[1]-hard.*

Let S be a finite set. A k-coloring f of S is a function mapping S to the set $\{1, 2, \ldots, k\}$. A subset S' of S is colored properly by the k-coloring f if any two elements in S' arenot colored with the same color under f. A family $\mathcal{F}$ of k-colorings of S is a k-color coding scheme if for every subset S' of k elements in S, there is a k-coloring in $\mathcal{F}$ that colors S' properly. The size of the k-color coding scheme $\mathcal{F}$ is the number of k-colorings in $\mathcal{F}$ [3].

Lemma 3 ([3]). *For any finite set U of n elements and any positive integer k ($k \leq n$), there exists a k-color coding scheme $\mathcal{F}$ of size $O(6.4^k n)$ for the set U. Moreover $\mathcal{F}$ can be constructed in time $O(6.4^k n)$.*

3   The Complexity of p-♯m-SET PACKING

To show the computational complexity of p-♯m-SET PACKING, we employ the parameterized parsimonious reduction introduced by Flum and Grohe in [6] and the recent result on p-♯MATCHING in [4].

Definition 8 ([6]). *Let $F : \sum^* \times N \to N$ and $G : \sum^* \times N \to N$ be parameterized counting problems. A parameterized parsimonious reduction from F to G is an algorithm that computes for every instance (x, k) of F an instance (y, l) of G in time $f(k)|x|^c$ such that $l \leq g(k)$ and $F(x, k) = G(y, l)$ for computable function $f, g : N \to N$ and a constant $c \in N$.*

Theorem 1. *The complexity of p-♯m-set packing is ♯W[1]-hard.*

Proof. Firstly, we give a description of the reduction from p-♯MATCHING to p-♯m-SET PACKING as follows.

Let $(G = (V, E), k)$ be an instance of p-♯MATCHING. And assume that $|E| = h$. Now we construct a corresponding instance $(\mathcal{S}, l)$ of p-♯m-SET PACKING. For the initial case, we assume that $\mathcal{S} = \emptyset$. In subsequent steps, we add some sets to $\mathcal{S}$ step by step according to the structure of G. Specifically, for every edge $e_i = (u_i, v_i) \in E$ ($1 \leq i \leq h$), we construct a set $s_i = \{u_i, v_i, w_{i,1}, w_{i,2}, \ldots, w_{i,m-2}\}$ and add it to $\mathcal{S}$, where $w_{i,1}, w_{i,2}, \ldots, w_{i,m-2}$ are some new elements added. Finally we set $l = k$. In the constructed instance $(\mathcal{S}, l)$, there are in total h sets. Obviously, this process can be done in polynomial time. Let $W_i = \bigcup_{j=1}^{j=m-2} \{w_{i,j}\}$, and let $V' = \bigcup_{i=1}^{i=h} W_i$. Obviously, $V' \cap V = \emptyset$ and $W_i \cap W_j = \emptyset$ for any $i \neq j$ ($1 \leq i, j \leq h$).

Next, we argue that the number of k-matchings in (G, k) is exactly equal to the number of k-packings in $(\mathcal{S}, l)$. Let Q_1 be the collection of k-matchings in (G, k), and let Q_2 be the collection of k-packings in $(\mathcal{S}, l)$. We show that

$|Q_1| = |Q_2|$. Obviously, if $Q_1 = \emptyset$ and $Q_2 = \emptyset$, then $|Q_1| = |Q_2|$. Without loss of generality, we may assume that $Q_1 \neq \emptyset$ or $Q_2 \neq \emptyset$ in the following.

On one direction, we assume that $Q_1 \neq \emptyset$. Without loss of generality, we assume that the edge set $E' = \{c_1, c_2, \ldots, e_k\}$ is a k-matching in G, and $s_1, s_2, \ldots, s_k$ are the corresponding k sets in S. By the definition of matching, $Val(e_i) \cap Val(e_j) = \emptyset$ for any $i \neq j$ $(1 \leq i, j \leq k)$, where $Val(e)$ denotes the set of endpoints on the edge e. As shown above, $s_i = Val(e_i) \cup W_i$ and $s_j = Val(e_j) \cup W_j$. Moreover, $W_i \cap W_j = \emptyset$, $W_i \cap Val(e_j) = \emptyset$, and $W_j \cap Val(e_i) = \emptyset$. Thus, $s_i \cap s_j = \emptyset$ for any $i \neq j$ $(1 \leq i, j \leq k)$, which means that $s_1, s_2, \ldots, s_k$ form a k-packing in S. This shows that $Q_2 \neq \emptyset$.

Furthermore, we assume that the edge sets $M = \{e_1, e_2, \ldots, e_k\}$ and $M' = \{e'_1, e'_2, \ldots, e'_k\}$ are two distinct k-matchings in G. According to the proof above, there must exist two corresponding k-packings $P = \{s_1, s_2, \ldots, s_k\}$ and $P' = \{s'_1, s'_2, \ldots, s'_k\}$ in S. Next, we show that $P \neq P'$ if $M \neq M'$. Suppose that $M \neq M'$. There must exist one index $i \in [1, k]$ such that $Val(e_i) \neq Val(e'_i)$. Since $s_i = Val(e_i) \cup W_i$ and $s'_i = Val(e'_i) \cup W'_i$, we conclude that $s_i \neq s'_i$, which means that $P \neq P'$. This process shows that $|Q_1| \leq |Q_2|$.

On the other direction, we assume that $Q_2 \neq \emptyset$. Without loss of generality, we assume that $S' = \{s_1, s_2, \ldots, s_k\}$ is a k-packing in S. In the following, we show that there must exist k corresponding edges in G such that these edges constitute a matching. Firstly, we show that for any set s_i in S' $(i \in [1, k])$, there must be one corresponding edge e_i in the graph G. (1) Suppose that there exist no edge that corresponds to s_i. Then, the elements u_i and v_i in s_i are not adjacent, which is contradicted with the fact that u_i and v_i are the endpoints of one edge in G. (2) Suppose that there are at least two distinct edges that correspond to one set s_i. Since there are at least three endpoints on two edges in any simple graph, there are at least 3 vertices in s_i that belong to V, which is contradicted with the fact that s_i contains only 2 vertices in V. Thus, there must exist a collection E'' containing exactly k edges $e_1, e_2, \ldots, e_k$ in G such that E'' corresponds to the k-packing $\{s_1, s_2, \ldots, s_k\}$ in S. Moreover, by the definition of packing, $s_i \cap s_j = \emptyset$ for any $i \neq j$ $(1 \leq i, j \leq k)$. As shown above, $Val(e_i) = s_i - W_i$ and $Val(e_j) = s_j - W_j$. Therefore, $Val(e_i) \cap Val(e_j) = \emptyset$ for any $i \neq j$ $(1 \leq i, j \leq k)$, which means that the edge set $\{e_1, e_2, \ldots, e_k\}$ constitutes a k-matching in G. This conclusion shows that $Q_1 \neq \emptyset$ and $|Q_2| \leq |Q_1|$.

In short, there exists one parameterized parsimonious reduction from p-$\sharp$ MATCHING to p-$\sharp m$-SET PACKING. The p-$\sharp$MATCHING is $\sharp$W[1]-hard by Lemma 1, therefore, the p-$\sharp m$-SET PACKING is $\sharp$W[1]-hard.   □

4   The Complexity of p-$\sharp m$-D MATCHING

To show the complexity of p-$\sharp m$-D MATCHING, we also employ the complexity result on p-$\sharp 2$-B MATCHING in [5].

Theorem 2. *The complexity of p-$\sharp m$-D MATCHING is $\sharp W[1]$-hard.*

This theorem can be proved exactly on the same lines as Theorem 1. We omit its detailed proof and only describe the construction on the reduction from p-$\sharp$2-B MATCHING to p-$\sharp m$-D MATCHING.

Let (G, k) be an instance of p-$\sharp$2-B MATCHING, in which $G = (V_1 \cup V_2, E)$ is a simple bipartite graph. Assume that $|E| = h$. Now we construct a corresponding instance $(\mathcal{S}, l)$ of p-$\sharp m$-D MATCHING. For the initial case, assume that $\mathcal{S} = \emptyset$. In subsequent steps, for every edge $e_i = (t_{1,i}, t_{2,i}) \in E$, in which $t_{1,i} \in V_1$ and $t_{2,i} \in V_2$, we construct an ordered tuple $\rho_i = (t_{1,i}, t_{2,i}, t_{3,i}, \ldots, t_{m,i})$ and add it to $\mathcal{S}$ $(1 \le i \le h)$. Note that the symbols $t_{3,i}, \ldots, t_{m,i}$ in ρ_i are some new symbols added. Finally, we set $l = k$. In the constructed instance $(\mathcal{S}, l)$, there are in total h ordered tuples $(t_{1,1}, t_{2,1}, \ldots, t_{m,1})$, $(t_{1,2}, t_{2,2}, \ldots, t_{m,2})$, $\ldots$, $(t_{1,h}, t_{2,h}, \ldots, t_{m,h})$. Moreover, $T_1 = V_1$, $T_2 = V_2$, and $T_j = \{t_{j,1}, t_{j,2}, \ldots, t_{j,h}\}$ $(3 \le j \le m)$. Obviously, this process can be done in polynomial time.

The p-$\sharp$2-B MATCHING is $\sharp W[1]$-hard by Lemma 2, therefore, the p-$\sharp m$-D MATCHING is $\sharp W[1]$-hard.

5   The FPTRAS for Considered Problems

For parameterized counting problems, Arvind and Raman [1] first introduced the parameterized version of the Karp-Luby result [8] on this subject. By applying the technique of color-coding, they developed the first fixed-parameter tractable randomized approximation scheme (FPTRAS) for counting k-vertex subgraph with bounded treewidth in a given graph. Along this line, some FPTRAS for solving other problems were also presented in [7,10].

The algorithms we describe in this section are the extensions of FPTRAS for p-$\sharp$3-D MATCHING in [10]. We first sketch its basic idea as follows.

Let $(\mathcal{S}, k)$ be an instance of p-$\sharp$3-D MATCHING and H be the set of all k-matchings in $\mathcal{S}$, where $\mathcal{S}$ is a set of n tripes. Firstly, we apply the technique of color coding improved in [3] to construct a $(2k)$-color coding scheme $\mathcal{F} = \{f_1, \cdots, f_d\}$, where $d = O(6.4^{2k} n)$. For $1 \le i \le d$, let H_i be the subset of H such that H_i consists of all k-matchings in $\mathcal{S}$ that are colored properly by the $(2k)$-coloring f_i. By the definition of color coding scheme, $H = \bigcup_{i=1}^{i=d} H_i$. The main works in the subsequent steps include three parts.

(1) counting the number $|H_i|$ of k-matchings in $\mathcal{S}$, for each i;
(2) random picking, with a uniform probability, a k-matching from the set H_i, for each i;
(3) determining if a given k-matching M is in the set H_i, for each i.
   Part (1) and part (2) can be done by the corresponding procedures, respectively.

Lemma 4 ([10], **Theorem 3**). *There exists a procedure (named local counting procedure) that runs in time $O(2^{2k} kn)$ and returns exactly the number of k-matchings in H_i.*

Lemma 5 ([10], **Theorem 5**). *There exists a procedure (named random sampling procedure) that runs in time $O(2^{2k} kn)$, and returns $\emptyset$ if $H_i = \emptyset$, and returns a k-matching properly colored by f_i with a probability $1/|H_i|$, if $H_i \ne \emptyset$.*

Moreover, part (3) can be done by a trivial procedure in time $O(n)$.

These procedures combined with the Karp-Luby result give an FPTRAS for p-♯3-D MATCHING [10].

By extending the algorithm for p-♯3-D MATCHING, we develop FPTRAS for p-♯m-SET PACKING, p-♯m-EDGE DISJOINT SUBGRAPH PACKING, p-♯m-D MATCHING, and p-♯2-B MATCHING, respectively.

5.1   The FPTRAS for p-♯m-SET PACKING

By Theorem 1, it is unlikely that p-♯m-SET PACKING admits fixed-parameter tractable algorithm, which makes the FPTRAS meaningful for it. Besides this, the FPTRAS for p-♯m-SET PACKING is the base of that for other problems.

Considering the FPTRAS for p-♯m-SET PACKING is an extension of that for p-♯3-D MATCHING, we only describe its extension points and draw the conclusions directly.

Let $(\mathcal{S},k)$ be an instance of p-♯m-SET PACKING and U be the union of all sets in $\mathcal{S}$, where $\mathcal{S}$ is a collection of n sets. First of all, we use mk colors to color the elements in U. By Lemma 3, we obtain a (mk)-color coding scheme $\mathcal{F}= \{f_1, f_2, \ldots, f_q\}$, where $q = O(6.4^{mk}n)$.

The extensions on the local counting procedure and the random sampling procedure are described as follows. (1) We don't need to preprocess the elements in the first column since each element is colored by one color. (2) The dynamic programming subroutine can be implemented by double loops. In the outer loop, we deal with the sets in $\mathcal{S}$; in the inner loop, we deal with the triples of the form (C, h, b) in the storage space $\mathcal{Q}$. (3) The running time of the dynamic programming subroutine is bounded by $O(2^{mk}kn)$, which can be roughly analyzed as follows. For each color set C with mj distinct colors, there exists only one specific triple (C, h, j) in $\mathcal{Q}$. Since the number of combination on choosing mj colors from mk colors is $\binom{mk}{mj}$, the total number of triples in $\mathcal{Q}$ is not greater than $\sum_{j=0}^{j=k} \binom{mk}{mj}$, which can be bounded by 2^{mk}.

These extensions combined with Lemmas 4 and 5 give the following lemmas.

Lemma 6. *Let H_i be the set of all k-packings in S that are properly colored by the (mk)-coloring f_i. The local counting procedure runs in time $O(2^{mk}kn)$ and returns $|H_i|$ correctly.*

Lemma 7. *Let H_i be the set of all k-packings in S that are properly colored by f_i. The sampling procedure runs in time $O(2^{mk}kn)$, and returns $\emptyset$ if $H_i = \emptyset$, and returns a k-packing in H_i with a probability $1/|H_i|$, if $H_i \neq \emptyset$.*

By using Lemmas 6 and 7, we obtain the following conclusion.

Theorem 3. *p-♯m-SET PACKING admits a fixed-parameter tractable randomized approximation scheme. More precisely, for a given instance $(\mathcal{S},k)$, and two positive real number $\epsilon > 0$, $0 < \delta < 1$, the scheme returns a non-negative number R in time $O(12.8^{mk}n^2k^3 ln(2/\delta)/\epsilon^2)$ such that $Prob[(1-\epsilon)R_0 \leq R \leq (1+\epsilon)R_0] \geq 1-\delta$, where R_0 denotes the exact number of k-packings in $\mathcal{S}$.*

5.2 The FPTRAS for p-$\sharp m$-Edge Disjoint Subgraph Packing

p-$\sharp m$-EDGE DISJOINT SUBGRAPH PACKING is also a special subgraph counting problem, where the subgraph pakings counted have different shapes. Some of the subgraph packings have bounded treewidth, however, others don't have bounded treewidth. We take the edge disjoint triangle packing problem [11] as an example for the latter. We consider the case that the k triangles are pairwise jointed by one vertex and the jointed vertices induce a k-cycle, i.e., a cycle with k vertices. This case is denoted by $P_\triangle$. Obviously, the treewidth of $P_\triangle$ is $(k-1)$, which cannot be bounded by a constant. Nevertheless, an FPTRAS for p-$\sharp m$-EDGE DISJOINT SUBGRAPH PACKING can be obtained when the size of the packing is considered as the parameter.

Let (G, Y, k) be an instance of p-$\sharp m$-EDGE DISJOINT SUBGRAPH PACKING, where Y is a connected subgraph with m edges. We solve it by two steps. In the first step, we translate (G, Y, k) into an instance $(\mathcal{S},k)$ of p-$\sharp m$-SET PACKING. To be specific, for each subgraph Z isomorphic to Y in G, we construct a corresponding set in which each element corresponds one edge in Z. By this way, we enumerate all of the subgraphs isomorphic to Y and obtain the collection $\mathcal{S}$ containing all of the constructed sets. In the second step, we directly apply the algorithm for p-$\sharp m$-SET PACKING to the instance $(\mathcal{S},k)$.

Its running time can be analyzed as follows. The number of combination on choosing m edges from the given graph $G = (V, E)$ is $\binom{|E|}{m}$, which can be bounded by $|E|^m$. For each choice, determining if it is isomorphic to the subgraph Y can be done in time $O((2m)!m)$. Thus, the time in the first step can be bounded by $O(|E|^m(2m)!m)$. Moreover, the size of the (mk)-color coding scheme is bounded by $O(6.4^{mk}|E|)$ and the size of $\mathcal{S}$ is bounded by $|E|^m$. By Theorem 3, the second step can be done in time $O(12.8^{mk}|E|^{m+1}k^3 ln(2/\delta)/\epsilon^2)$. Thus, the total time can be bounded by $O(12.8^{mk}|E|^{m+1}k^3 ln(2/\delta)/\epsilon^2)$.

Therefore, we draw the following conclusion.

Theorem 4. p-$\sharp m$-EDGE DISJOINT SUBGRAPH PACKING *admits a fixed-parameter tractable randomized approximation scheme. More precisely, for a given instance* $(G = (V, E), Y, k)$, *and two positive real number* $\epsilon > 0, 0 < \delta < 1$, *the scheme returns a non-negative number R in time* $O(12.8^{mk}|E|^{m+1}k^3 ln(2/\delta)/\epsilon^2)$ *such that* $Prob[(1\text{-}\epsilon)R_0 \leq R \leq (1+\epsilon)R_0] \geq 1\text{-}\delta$, *where R_0 denotes the exact number of k-Y subgraph packings in G.*

5.3 The FPTRAS for p-$\sharp m$-D Matching and p-$\sharp 2$-B Matching

p-$\sharp m$-D MATCHING is the generalization of p-$\sharp 3$-D MATCHING. So, we can obtain an FPTRAS for p-$\sharp m$-D MATCHING by generalizing that for p-$\sharp 3$-D MATCHING.

Based on the algorithm for p-$\sharp 3$-D MATCHING, the algorithm for p-$\sharp m$-D MATCHING includes the following generalized aspects. (1) We use $(m-1)k$ colors to color the symbols in the 2nd to the m-th dimensions of $\mathcal{S}$ and the size of the $((m-1)k)$-color coding scheme is bounded by $O(6.4^{(m-1)k}n)$. (2) In the dynamic programming subroutine, the condition of adding a tuple ρ to an existed matching M is that $C(M) \cap \{cl(Val^2(\rho)), \ldots, cl(Val^m(\rho))\} = \emptyset$, where $C(M)$ denotes the color set colored on M and $cl(Val^i(\rho))$ denotes the color colored on the i-th symbol in ρ $(2 \le i \le m)$. (3) The running time of the dynamic programming subroutine is bounded by $O(2^{(m-1)k}kn)$, which can be roughly analyzed as follows. For each color set C with $(m-1)j$ distinct colors, there is only one specific triple (C, h, j) in Q. Therefore, the total number of triples in Q is not greater than $\sum_{j=0}^{j=k} \binom{(m-1)k}{(m-1)j}$, which can be bounded by $2^{(m-1)k}$.

These generalizations combined with the FPTRAS for p-$\sharp$3-D MATCHING give an FPTRAS for p-$\sharp m$-D MATCHING.

Theorem 5. *p-$\sharp m$-D MATCHING admits a fixed-parameter tractable randomized approximation scheme. More precisely, for a given instance $(\mathcal{S}, k)$, and two positive real number $\epsilon > 0$, $0 < \delta < 1$, the scheme returns a non-negative number R in time $O(12.8^{(m-1)k}n^2k^3ln(2/\delta)/\epsilon^2)$ such that $Prob[(1-\epsilon)R_0 \le R \le (1+\epsilon)R_0] \ge 1-\delta$, where R_0 denotes the exact number of k-matchings in $\mathcal{S}$.*

The complexity of p-$\sharp$2-B MATCHING is $\sharp$W[1]-hard. Although the approximation algorithm for p-$\sharp$MATCHING in [1] can be applied to p-$\sharp$2-B MATCHING, we also present a new algorithm for p-$\sharp$2-B MATCHING, applying the techniques in the algorithm for p-$\sharp m$-D MATCHING.

Let $(G = ((V_1 \cup V_2), E), k)$ be an instance of p-$\sharp$2-B MATCHING. Based on the properties on bipartite graphs, we can take each edge in E as a tuple, in which the two elements correspond to the endpoints of one edge. The main strategy used in p-$\sharp m$-D MATCHING can be also applied to p-$\sharp$2-B MATCHING. Thus, we directly color the vertices in V_2 by using only k colors, and keep the vertices in V_1 uncolored. By Lemma 3, the size of the k-color coding scheme is bounded by $O(6.4^k|V_2|)$.

Moreover, we can simplify some steps in the algorithm for p-$\sharp m$-D MATCHING. Thus, the local counting procedure and the random sampling procedure can be done in time $O(2^kk|E|)$, respectively. Therefore, the total running time is bounded by $O(6.4^k|V_2| \times 2^kk|E| \times k^2ln(2/\delta)/\epsilon^2)=O(3.58^{2k}|V_2||E|k^3ln(2/\delta)/\epsilon^2)$.

Based on the analysis above, we draw the following conclusion.

Theorem 6. *p-$\sharp$2-B MATCHING admits a fixed-parameter tractable randomized approximation scheme. More precisely, for a given instance $(G = (V, E), k)$, and two positive real number $\epsilon > 0$, $0 < \delta < 1$, the scheme returns a non-negative number R in time $O(3.58^{2k}|V||E|k^3ln(2/\delta)/\epsilon^2)$ such that $Prob[(1-\epsilon) R_0 \le R \le (1+ \epsilon) R_0] \ge 1-\delta$, where R_0 denotes the exact number of k-matchings in G.*

6   Conclusions

In this paper, we show that the complexity of p-$\sharp m$-SET PACKING (resp. p-$\sharp m$-D MATCHING) is $\sharp$W[1]-hard by employing the complexity result on p-$\sharp$MATCHING (resp. p-$\sharp$2-B MATCHING). Based on the FPTRAS for $\sharp$3-D MATCHING, we also develop FPTRAS for p-$\sharp m$-SET PACKING, p-$\sharp m$-D MATCHING, and p-$\sharp$2-B MATCHING, respectively. Moreover, we show that the p-$\sharp m$-EDGE DISJOINT SUBGRAPH PACKING problem, which is parameterized by the size of the packing, admits FPTRAS even if some of the counted subgraphs don't have bounded treewidth.

Acknowledgments. We are grateful to the anonymous referees for helpful suggestions that improve the presentation of this paper.

References

1. Arvind, V., Raman, V.: Approximation algorithms for some parameterized counting problems. In: Bose, P., Morin, P. (eds.) ISAAC 2002. LNCS, vol. 2518, pp. 453–464. Springer, Heidelberg (2002)
2. Björklund, A., Husfeldt, T., Kaski, P., Koivisto, M.: Counting paths and packings in halves. In: Fiat, A., Sanders, P. (eds.) ESA 2009. LNCS, vol. 5757, pp. 578–586. Springer, Heidelberg (2009)
3. Chen, J., Lu, S., Sze, S.-H., Zhang, F.: Improved algorithms for path, matching, and packing problems. In: Proceedings of the 18th Annual ACM-SIAM Symposium on Discrete Algorithms (SODA2007), pp. 298–307 (2007)
4. Curticapean, R.: Counting matchings of size k is $\sharp$W[1]-hard. In: Fomin, F.V., Freivalds, R., Kwiatkowska, M., Peleg, D. (eds.) ICALP 2013, Part I. LNCS, vol. 7965, pp. 352–363. Springer, Heidelberg (2013)
5. Curticapean, R., Marx, D.: Complexity of counting subgraphs: only the boundedness of the vertex-cover number counts. In: Proceedings of the 55th Annual IEEE Symposium on Foundations of Computer Science (FOCS 2014), pp. 130–139 (2014)
6. Flum, J., Grohe, M.: The parameterized complexity of counting problems. SIAM J. Comput. **33**(4), 892–922 (2004)
7. Jerrum, M., Meeks, K.: The parameterised complexity of counting connected subgraphs and graph motifs. J. Comput. Syst. Sci. **81**(4), 702–716 (2015)
8. Karp, R.M., Luby, M., Madras, N.: Monte-Carlo approximation algorithms for enumeration problems. J. Algorithms **10**, 429–448 (1989)
9. Koutis, I., Williams, R.: Limits and applications of group algebras for parameterized problems. In: Albers, S., Marchetti-Spaccamela, A., Matias, Y., Nikoletseas, S., Thomas, W. (eds.) ICALP 2009, Part I. LNCS, vol. 5555, pp. 653–664. Springer, Heidelberg (2009)
10. Liu, Y., Chen, J., Wang, J.: On counting 3-D matchings of size k. Algorithmica **54**, 530–543 (2009)
11. Mathieson, L., Prieto, E., Shaw, P.: Packing edge disjoint triangles: a parameterized view. In: Downey, R.G., Fellows, M.R., Dehne, F. (eds.) IWPEC 2004. LNCS, vol. 3162, pp. 127–137. Springer, Heidelberg (2004)
12. McCartin, C.: Parameterized counting problems. Ann. Pure Appl. Logic **138**(13), 147–182 (2006)
13. Meeks, K.: The challenges of unbounded treewidth in parameterised subgraph counting problems. Discrete Appl. Math. **198**, 170–194 (2016)
14. Yu, D., Wang, Y., Hua, Q.-S., Lau, F.C.M.: Exact parameterized multilinear monomial counting via k-layer subset convolution and k-disjoint sum. In: Fu, B., Du, D.-Z. (eds.) COCOON 2011. LNCS, vol. 6842, pp. 74–85. Springer, Heidelberg (2011)
15. Zhang, C., Chen, Y.: Counting problems in parameterized complexity. TsingHua Sci. Technol. **19**(4), 410–420 (2014)

Online Scheduling with Increasing Subsequence Serving Constraint

Kelin Luo[1]([✉]), Yinfeng Xu[1,2], and Xin Feng[1]

[1] School of Management, Xi'an Jiaotong University, Xi'an 710049, China
{luokelin,fengxin.xjtu}@stu.xjtu.edu.cn, yfxu@xjtu.edu.cn
[2] The State Key Lab for Manufacturing Systems Engineering, Xi'an 710049, China

Abstract. This paper studies an online scheduling problem with increasing subsequence serving constraint. Customers requests are released over-list, and the operator has to decide whether or not to accept current request and arrange it to a server immediately. Each server has to process an increasing subsequence requests. There are two online scheduling problems in this paper. The first problem is to find a schedule which occupies the minimal servers if the operator accepts all requests. The second problem is to find a schedule which accepts the maximal requests if the operator has just one server. In this paper, we propose two optimal algorithms, Double-Greedy Algorithm and Partition Algorithm, for the above two problems, respectively.

Keywords: Online scheduling · Increasing subsequence · Online strategy · Competitive ratio

1 Introduction

Instant delivery is a rising industry which has been developed rapidly in this century [1] because we have to satisfy people's substantial demand as soon as possible. For example, online ordering take-out has been attracting increasingly wide attention in the Internet business [2]. However, take-out service is a tough problem. The restaurant owner not only needs to consider take-out as soon as possible, but also considers hiring the less courier for reducing cost. The restaurant owner has to balance the above concerns. Another instant delivery example is taxi booking, such as Uber, Didi, and so on. It is easy to know the possible request locations, however, we do not know whether these requests occur, or when they occur. Thus it is hard to allocate servers (couriers, drivers, etc.) by following a fixed strategy. In many practical problems, requests do not reveal themselves until they come, and the requests come via an online fasion. On observing a request, the decision maker needs to make an irrevocable decision on whether to accept or reject the current request and arrange this request to a server, with the overall objective which satisfies some constraints.

As we known, whether a request could be accepted by a server depends on the server's current location. For example, a taxi driver has taken a passenger

D. Zhu and S. Bereg (Eds.): FAW 2016, LNCS 9711, pp. 135–144, 2016.
DOI: 10.1007/978-3-319-39817-4_14

to a supermarket 15 miles away from a downtown, meanwhile, another passenger requests to go to the cinema which is only 5 miles away from the downtown. The driver may reject the trip to cinema because he has to drive 10 miles without any revenue. Thus, he is likely to accept the ride near the supermarket. Motivated by the instance, we may label all the request locations from 1 to n based on their increasing subsequence serving constraints. The increasing subsequence serving constraint can express various factors, such as the serving time and the distance between two request locations.

In the *online scheduling with increasing subsequence serving constraint* problem, there are N requests and M servers. Based on the location constraint, every request has been given a label, so there are n labels (i.e., 1 through n). The request on some locations may be released more than once, so we do not know the actual requests sequence length in advance. A schedule for the requests is feasible if each server processes an increasing subsequence requests. In other words, if request r_i occurs, we could arrange this request to an active(allocated) server whose label of last request is not more than r_i; otherwise, we will reject this request or allocate a new server for this request. All research comes from two sources.

The first one is online scheduling problem. In practice, requests are released from the customer to the operator one by one. Due to the uncertainty of future requests, the operator has no information about a request until he receives it. There are various algorithms for online scheduling problem with serving dealine constraints [3,4]. These results mainly focused on serving cost but ignored the constraint of customers in make-to-order environment. Most researchers studied scheduling problem with minimizing makespan or minimizing total completion time. Kaminsky and Lee [5] showed an online model to minimize the sum (or average) of serving time, and demonstrated that heuristics are effective. To the best of our knowledge, few researchers consider scheduling with an increasing subsequence constraint.

The second one is the longest increasing subsequence problem. The static longest increasing subsequence problem has been studied for many years. Fredman et al. [6] and Albert et al. [7] investigated this problem on a line. Deorowicz [8] studied this problem in a circle. Then, some researchers studied the longest increasing subsequence problem from a dynamic view. After observing a sequence of independent non-negative random variables, the decision maker has to decide whether to accept the current variable immediately for selecting a longest increasing subsequence in the sequence. Many researchers [9–11] paid attention to the mean and standard deviation analysis. Recently, Arlotto et al. [12] solved this problem with a continuous distribution and obtained a central limit theorem. However, few researchers connected this problem with online scheduling problem. Nagarajan and Sviridenko [13] firstly combined the permution flow shop scheduling problem with the increasing subsequence constraint, and assumed that there are m machines and n jobs. They came up with a randomized algorithm with $O(min\{\sqrt{m}, \sqrt{n}\})$ approximation. In this paper, we apply the competitive analysis to the online scheduling problem with increasing subsequence serving constraint.

There are two different objectives, both of which are useful for take-out and taxi allocation. The first objective is to minimize the active(allocated) servers for arranging all requests; if we only have one server, the second objective is to maximize the requests that we accept to serve. The requests are released over-list. We should arrange all these request immediately after it released. Our contributions are as follows:

- we study an online problem with increasing subsequence serving constraint and analyze it from a more proper perspective, i.e., the competitive analysis;
- we propose algorithms that achieve good performance for solving two different online scheduling problems, respectively.

The rest of this paper is organized as follows. Section 2 introduces two online increasing subsequence scheduling problems and some preliminaries. In Sect. 3, we present an optimal algorithm for the first problem in Sect. 3.1; and also present an optimal algorithm for the second problem with competitive analysis in Sect. 3.2. Final conclusions and remarks are given in Sect. 4.

2   Preliminaries

Consider an online scheduling problem with increasing subsequence serving constraint. Suppose time is divided into discrete time slots $T = \{1, 2, 3, \cdots\}$. A set of requests $R = \{r_1, r_2, r_3, \cdots\}$ are released over-list. We also abuse r_i to denote its label when there is no ambiguity. A set of servers $S = \{s^1, s^2, s^3, \cdots\}$ are allocated in the process. Let $S_i = \{s_i^1, s_i^2, \cdots\}$ and $S_{*,i} = \{s_{*,i}^1, s_{*,i}^2, \cdots\}$ be the active servers allocated by an online algorithm and by optimal offline adversary(OPT) after r_i occurred, respectively. Let s_i^j denote the last request arranged to server s^j by an online algorithm after request r_i is released. and let $s_{*,i}^j$ denote the last request arranged to server $s_{*,i}^j$ by OPT after request r_i is released.

Definition 1 *((Strict) Increasing Subsequence).* For a sequence $R = \{r_1, r_2, \cdots, r_m\}$, R is called an increasing subsequence if $\forall k \in [1, m], r_k \geq r_{k-1}$; R is called a strict increasing subsequence if $\forall k \in [1, m], r_k > r_{k-1}$.

Definition 2 *(Mapping).* Saying, $S_{*,i}$ is a mapping of S_i means that $\forall s_i^{j_1} \in S_i$, we can find a one-to-one mapping item $s_{*,i}^{j_2} \in S_{*,i}$, which satisfies $s_i^{j_1} \leq s_{*,i}^{j_2}$. For example, $S_i = \{5, 4, 3\}$ and $S_{*,i} = \{4, 5, 5, 1\}$. Then we say $S_{*,i}$ is a mapping of S_i because $5 \leq 5$, $4 \leq 4$ and $3 \leq 5$.

Min-OIS Problem. Notice that some companies state that they do not reject any requests [14], although sometimes they need many servers. These companies need to minimize the number of servers when they arrange requests. We denote this problem as a Minimum Online Scheduling with Increasing Subsequence (Min-OIS) problem.

Max-OIS Problem. Considering the limited resources of a company in this paper, such as servers, drivers, and etc., we assume there is only one server available, so companies could reject some requests. Those companies need to maximize the accepted requests in the process. We denote this problem as a Maximum Online Scheduling with Increasing Subsequence (Max-OIS) problem.

2.1   Competitive Ratio

We use the competitive analysis (see [15]) to measure the performance of online scheduling for the problem under consideration. Translated into our problem terminology, for an arbitrary order sequence R, $A(R)$ be the objective value of schedule produced by an online algorithm A, and $OPT(R)$ be that obtained by an optimal offline scheduler OPT who has the information of all requests in advance.

We say that the online algorithm for a Min objective problem is ρ_A-competitive if $A(R) \leq \rho_A OPT(R) + \varepsilon$ holds for any R where $\rho_A \geq 1$ is some constant and ε is an arbitrary positive number. ρ_A is also called the competitive ratio of the online algorithm. The online algorithm for a Max objective problem is ρ_A-competitive if $\rho_A A(R) + \varepsilon \geq OPT(R)$ holds for any R where $\rho_A \geq 1$ is some constant and ε is an arbitrary positive number. ρ_A is also called the competitive ratio of the online algorithm. Clearly, $\rho_A \geq 1$.

Let ON be the complete set of online strategies for the problem. The lower bound w of competitive ratio is defined as $w = inf_{A \in ON} \rho_A$. We say A is an optimal online strategy if $\rho_A = w$.

3   Min-OIS Problem and Max-OIS Problem

In this section, we propose online algorithms for two online scheduling problems, Min-OIS problem and Max-OIS problem, and conduct the competitive analysis.

3.1   Min-OIS Problem

Considering the special increasing subsequence constraint of Min-OIS problem, we use Double-Greedy Algorithm(DGA) to arrange requests to optimize schedule with minimized servers. Once a request is released, we consider giving priority to the nearest server, if there exists. Otherwise, we will allocate a new server.

Lemma 1. *For a schedule obtained by Double-Greedy Algorithm, the last request of each server is different.*

It is a straightforward lemma. If a new request's label is equal to the last request of an active server, we will schedule this new request to that active server, because we arrange all requests following the rule: scheduling the next request to the active server whose last request's label is positive nearest to this request's label.

Algorithm 1. Double-Greedy Algorithm

Step 0: Initialize. Set $i = 1$, $j = 1$, $s_i^j = r_i$.

Step 1: Repeat for $i = i + 1, i + 2, \cdots$

Step 2: If $r_i \geq min\{s_{i-1}^1, s_{i-1}^2, s_{i-1}^3, \cdots, s_{i-1}^j\}$, go to Step 3;
otherwise, go to Step 4.

Step 3: Arrange r_i to server s^k $(k \leq j)$ which is positive nearest to r_i, let $s_i^k = r_i$,
and let the other servers $s_i^1 = s_{i-1}^1, s_i^2 = s_{i-1}^2, \cdots, s_i^j = s_{i-1}^j$, go to Step 5.

Step 4: Arrange r_i to a new server s^{j+1}, $s_i^{j+1} = r_i$, and let $s_i^1 = s_{i-1}^1, s_i^2 = s_{i-1}^2, \cdots$,
$s_i^j = s_{i-1}^j$. Let $j = j + 1$, go to Step 5.

Step 5: If all the requests have already been arranged, the game terminates;
otherwise, go to Step 1.

Note: s^k is positive nearest to r_i means that: $min \mid s_{i-1}^k - r_i \mid (s_{i-1}^k - r_i \geq 0)$.

Theorem 1. *The Double-Greedy Algorithm for Min-OIS problem is 1-competitive, which is optimal.*

Proof. Consider an arbitrary input sequence $R = \{r_1, r_2, \cdots\}$. Let $|S_i|$ and $|S_{*,i}|$ denote the number of servers, which are allocated to serve requests. We use mathematical induction to prove this theorem.

Basic Step: For $i = 1$, the schedule produced by OPT is the same as that produced by DGA. We know $|S_i| = |S_{*,i}| = 1$, $s_i^1 = s_{*,i}^1 = r_1$. So $|S_i| \leqslant |S_{*,i}|$, and $S_{*,i}$ is a mapping of S_i.

Inductive Step: Now we assume when $i = k$(for some $k \in Z^+$), $|S^i| \leqslant |S^{*,i}|$, and $S_{*,i}$ is a mapping of S_i are true. We assume that for $i = k+1$, $|S_i| \leqslant |S_{*,i}|$, and $S_{*,i}$ is a mapping of S_i.

For $i = k + 1$, there are three different cases as following.

Case 1. if $r_i < min\{s_{i-1}^1, s_{i-1}^2, s_{i-1}^3, \cdots, s_{i-1}^j\}$, and the schedule produced by OPT is the same as that produced by DGA. We have $|S_i| = |S_{i-1}| + 1 = |S_{*,i}| = |S_{*,i-1}| + 1$, and $S_{*,i}$ is still a mapping of S_i.

Case 2. if $r_i \geq min\{s_{i-1}^1, s_{i-1}^2, s_{i-1}^3, \cdots, s_{i-1}^j\}$, and the schedule produced by OPT is the same as that produced by DGA, they both arrange this request to the server whose last request's label is positive nearest to r_i. We have $|S_i| = |S_{i-1}| = |S_{*,i}| = |S_{*,i-1}|$, and $S_{*,i}$ is still a mapping of S_i.

Case 3. if $r_i \geq min\{s_{i-1}^1, s_{i-1}^2, s_{i-1}^3, \cdots, s_{i-1}^j\}$, and the schedule produced by OPT is different with that produced by DGA. There are two different cases.

Case 3.1 The OPT allocates a new server to this request, so we have $|S_i| = |S_{i-1}| < |S_{*,i}| = |S_{*,i-1}| + 1$, and $S_{*,i}$ is still a mapping of S_i.

Case 3.2 The OPT arranges this request to a server $(eg.s^k)$ whose last request's label is not positive nearest to r_i and $s_{*,i-1}^k \leq r_i$, so we have $|S_i| = |S_{i-1}| \leq |S_{*,i}| = |S_{*,i-1}|$, and $S_{*,i}$ is still a mapping of S_i.

According to Case $1, 2, 3$, we still have $|S_i| \leqslant |S_{*,i}|$, and $S_{*,i}$ is a mapping of S_i. Hence, for each $k \in Z^+$), it follows that $|S_k| \leqslant |S_{*,k}| \Longrightarrow |S_{k+1}| \leqslant |S_{*,k+1}|$.

In conclusion, for all input sequences, we have $|S_i| \leqslant |S_{*,i}|$, and $S_{*,i}$ is a mapping of S_i. Thus, the servers used by DGA is not more than the servers by offline adversary's schedule for serving any request sequence,

$$C/C_{OPT} = 1.$$

From above all, the Double-Greedy Algorithm for Min-OIS problem is 1-competitive, i.e., it is an optimal algorithm. $\square$

3.2 Max-OIS Problem

In this subsection, we prove that there is no constant competitive ratio can be obtained for Max-OIS problem by constructing a special instance. Then we studied this problem with strict increasing subsequence constraint and proposed an optimal algorithm for this problem.

Lemma 2. *For Max-OIS problem, there is no deterministic on-line algorithm that can achieve a constant competitive ratio.*

Proof. Given an arbitrary order sequence R, let σ and σ^* be the schedules produced by Partition Algorithm and by OPT, respectively. Let $|\sigma|$ and $|\sigma^*|$ denote the number of the accepted requests. There are two different cases as following.

Case 1. If an online algorithm rejects all the requests except for $r_i = 1$, the offline adversary will release infinite requests greater than $k, \forall k \in Z^+$ requests, i.e., the offline adversary released requests $\{r_{i+1} = 2, r_{i+2} = 2, \cdots, r_{i+k} = 2, \cdots\}$.

$$C_{OPT}/C \geq k, \forall k \in Z^+.$$

Case 2. If an online algorithm accepts a request $r_i \neq 1$, the offline adversary will release infinite requests greater than $k, \forall k \in Z^+$ requests, i.e., the offline adversary released requests $\{r_{i+1} = r_i - 1, r_{i+2} = r_i - 1, \cdots, r_{i+k} = r_i - 1, \cdots\}$.

$$C_{OPT}/C \geq k, \forall k \in Z^+.$$

From the two cases above, it can be concluded that there is no deterministic on-line algorithm that can achieve a constant competitive ratio for Max-OIS problem. $\square$

Since we proved that the lower bound of Max-OIS problem is infinite, we will restrict this problem to be a new online scheduling problem with strict increasing subsequence constraint. We denote this problem as a Maximum Online Scheduling with Strict Increasing Subsequence (Max-OSIS) problem. We will propose an online Partition Algorithm(PA) to obtain a preferable serving schedule for this problem. Before a request is released, we divide the whole requests into two parts, acceptable part and rejectable part, based on the previous schedule.

Algorithm 2. Partition Algorithm

Step 0: Initialize. Set $i = 0$, $s_i^1 = 0$.

Step 1: Repeat for $i = i + 1, i + 2, \cdots$

Step 2: If $r_i > s_i^1$, go to Step 3;

   otherwise, reject i_{th} request, $s_i^1 \leq s_{i-1}^1$, go to Step 4.

Step 3: If $r_i \leq max\{\lceil \frac{n+s_{i-1}^1}{2} \rceil, \lceil \frac{n+r_i}{2} \rceil\}$, accept i_{th} request, $s_i^1 = r_i$, go to Step 4;

   otherwise, reject i_{th} request, $s_i^1 = s_{i-1}^1$, go to Step 4.

Step 4: If all the requests have already been arranged, go to Step 5;

   otherwise, go to Step 1.

Step 5: If $s_i^1 > 0$, the game terminates;

   otherwise, let the server accept the final request, the game terminates.

Note: n represents the total request labels.

Theorem 2. *The Partition Algorithm(PA) for Max-OSIS problem is $\lceil \frac{n}{2} \rceil$-competitive.*

Proof. Considering an arbitrary order input sequence R, let σ and σ^* be the schedules produced by Partition Algorithm and by OPT, respectively. Let $|\sigma|$ and $|\sigma^*|$ denote the number of the accepted requests. There are two different cases as following.

Case 1. If PA accepts no request before the final request arrived, it means that the released request is at least labeled $\lceil \frac{n}{2} \rceil$. PA only accepts the final request. The offline adversary at most accepts $|\sigma^*| = \lceil \frac{n}{2} \rceil$ requests, and $\sigma^* = \{r_1 = \lceil \frac{n}{2} \rceil, r_2 = \lceil \frac{n}{2} \rceil + 1, \cdots, r_{\lceil \frac{n}{2} \rceil} = n\}$.

$$C_{OPT}/C \leq \lceil \frac{n}{2} \rceil.$$

Case 2. If PA could accept at least one request before the final request arrived, we denote the first request arranged by PA as r_k. We know that $0 < r_k \leq \frac{n}{2}$. Then,

Case 2.1 if PA could not accept one more request, it means that the offline adversary could not release a request which is labeled greater than r_k.

   Thus, PA will reject all the rest of the requests. The schedule produced by PA is $\sigma = \{r_k\}$. In the optimal schedule, there are not more than $\lceil \frac{n}{2} \rceil$ requests that are arranged, $\sigma^* = \{1, 2, \cdots, r_k\}$. Due to $r_k \leq \lceil \frac{n}{2} \rceil$,

$$C_{OPT}/C \leq \lceil \frac{n}{2} \rceil.$$

Case 2.2 if PA could accept one more request, it means that the offline adversary will release a request r_j $(j > k)$, $r_j > r_k$.

   Thus, the schedule produced by PA is $\sigma = \{r_k, r_j\}$. PA will serve 2 requests. The offline adversary at most accept $|\sigma^*| = r_k + n - \frac{n+r_k}{2} = \frac{n+r_k}{2} \leq \frac{3}{4}n$ requests, and $\sigma^* = \{r_{k+1} = 1, r_{k+2} = 2, \cdots, r_{r_k} = r_k; r_{r_k+1} = \lceil \frac{n+r_k}{2} \rceil, r_{r_k+2} = \lceil \frac{n+r_k}{2} \rceil + 1, \cdots, r_j = n\}$.

$$C_{OPT}/C \leq \frac{3}{8}n.$$

According to Case 1, 2, we obtain $\rho_{PA} = max\{\lceil \frac{n}{2} \rceil, \frac{3}{8}n\} = \lceil \frac{n}{2} \rceil$. Thus, the Partition Algorithm for Max-OSIS problem is $\lceil \frac{n}{2} \rceil$-competitive.   □

Theorem 3. *The Partition Algorithm is the optimal algorithm for Max-OSIS problem.*

Proof. Considering an arbitrary order input sequence R, there are three different cases as following.

Case 1. If an online algorithm accepts the first request $r_1 = k$, the offline adversary will release a set of requests which are labeled $\{r_2 = 1, r_3 = 2, r_4 = 3, \cdots, r_{k+1} = k\}$. The schedule produced by the online algorithm is $\sigma = \{r_1\}$. In the optimal schedule, $\sigma^* = \{r_2, r_3, r_4, \cdots, r_{k+1}\}$ are arranged.

$$C_{OPT}/C = k.$$

Case 2. If the online algorithm rejects the first request, the offline adversary will release a request which is labeled $r_2 = k + 1$.

Case 2.1. If the online algorithm accepts this requests, the offline adversary will release a set of requests which are labeled $\{r_3 = 1, r_4 = 2, r_5 = 3, \cdots, r_{k+3} = k+1\}$. The schedule produced by the online algorithm is $\sigma = \{r_2\}$. In the optimal schedule, $\sigma^* = \{r_3, r_4, r_5, \cdots, r_{k+3}\}$ are arranged.

$$C_{OPT}/C = k + 1.$$

Case 2.2. If the online algorithm rejects all the previous requests before r_j, $j \geq 2$, the offline adversary will release a request which is labeled $\{r_{j+1} | r_j < r_{j+1} < n, r_{j+1} \in Z^+\}$.

If the online algorithm accepts request r_{j+1}, the offline adversary will release a set of requests which are labeled $\{r_{j+2} = 1, r_{j+3} = 2, r_{j+4} = 3, \cdots, r_{k+2j} = k + j - 1\}$. The schedule produced by the online algorithm is $\sigma = \{r_{j+1}\}$. In the optimal schedule, $\sigma^* = \{r_{j+2}, r_{j+3}, r_{j+4}, \cdots, r_{k+2j}\}$ are arranged.

$$C_{OPT}/C = k + j - 1.$$

Case 3. If the online algorithm rejects all requests until the last request is labeled as $\{r_j = n | r_j \in Z^+\}$, the online algorithm will be strictly required to arrange the last request. The schedule produced by the online algorithm is $\sigma = \{r_j\}$. In the optimal schedule, $\sigma^* = \{r_1 = k, r_2 = k + 1, \cdots, r_j = n\}$ are arranged.

$$C_{OPT}/C = n - k + 1.$$

From the three cases above, it can be conclued that there is no online algorithm which can achieve a competitive ratio less than $Min\{k, k + 1, \cdots, n - 1, n - k + 1\}$. For $0 < k < n, k \in Z^+$, $Max\, Min\{k, n - k + 1\} = \lceil \frac{n}{2} \rceil$. Thus, the Partition Algorithm is the optimal algorithm for Max-OSIS problem.   □

4   Conclusions

We investigate the following two online scheduling problems with increasing subsequence serving constraint. The conditions we use are independent of the optimal objective value, the length of request sequence, and the distribution of input sequence. As for the first problem Min-OIS, we assume that the servers are infinite and we should not reject any requests. We come up with the Double-Greedy Algorithm and prove it is optimal. In the other problem Max-OIS, we prove that its lower bound is infinite. For Max-OSIS problem, we prove a lower bound of Max-OSIS problem is $\lceil \frac{n}{2} \rceil$ and propose the Partition Algorithm, whose competitive ratio is equal to the lower bound.

A straightforward application of the algorithms is to be injected into the software of request assignment system, such as taxi booking system and food delivery system, because the requests appear uncertainly.

There are many questions for further research. One important question is whether there is a scheduling algorithm which is the most efficient for specific server company. In addition, as we show in this paper, we do not consider the balance between the size of requests and the size of servers, which is left to be explored in the future. Also, if some companies allow the number of servers to fluctuate over time, online scheduling problem with increasing subsequence constraint will also be an interesting direction for further research.

Acknowledgments. This research is partially supported by the NSFC (grant number 61221063), and by the PCSIRT (grant number IRT1173) and by China Postdoctoral Science Foundation(grant numbers 2014M550503 and 2015T81040).

References

1. Howell, G.A.: What is lean construction. In: Proceedings IGLC (1999)
2. Moodie, D.R.: Due date demand management: negotiating the trade-off between price and delivery. Int. J. Prod. Res. **37**(5), 997–1021 (1999)
3. Hall, N.G., Posner, M.E.: Earliness-tardiness scheduling problems, I: weighted deviation of completion times about a common due date. Oper. Res. **39**(5), 836–846 (1991)
4. Portougal, V., Trietsch, D.: Setting due dates in a stochastic single machine environment. Comput. Oper. Res. **33**(6), 1681–1694 (2006)
5. Kaminsky, P., Lee, Z.H.: Effective on-line algorithms for reliable due date quotation and large-scale scheduling. J. Sched. **11**(3), 187–204 (2008)
6. Fredman, M.L.: On computing the length of longest increasing subsequences. Discrete Math. **11**(1), 29–35 (1975)
7. Albert, M.H., Golynski, A., Hamel, A.M., et al.: Longest increasing subsequences in sliding windows. Theor. Comput. Sci. **321**(2), 405–414 (2004)
8. Deorowicz, S.: An algorithm for solving the longest increasing circular subsequence problem. Inf. Process. Lett. **109**(12), 630–634 (2009)
9. Odlyzko, A.M., Rains, E.M.: On longest increasing subsequences in random permutations. Contemp. Math. **251**, 439–452 (2000)

10. Arlotto, A., Gans, N., Steele, J.M.: Markov decision problems where means bound variances. Oper. Res. **62**(4), 864–875 (2014)
11. Romik, D.: The Surprising Mathematics of Longest Increasing Subsequences. Cambridge University Press, Cambridge (2015)
12. Arlotto, A., Nguyen, V.V., Steele, J.M.: Optimal online selection of a monotone subsequence: a central limit theorem. Stochast. Process. Appl. **125**, 3596–3622 (2015)
13. Nagarajan, V., Sviridenko, M.: Tight bounds for permutation flow shop scheduling. Math. Oper. Res. **34**(2), 417–427 (2009)
14. Ausiello, G., Feuerstein, E., Leonardi, S., et al.: Algorithms for the on-line travelling salesman 1. Algorithmica **29**(4), 560–581 (2001)
15. Borodin, A., El-Yaniv, R.: Online Computation and Competitive Analysis. Cambridge University Press, Cambridge (1998)

Notes on the $\frac{6}{5}$-Approximation Algorithm for One-Sided Scaffold Filling

Jingjing Ma and Haitao Jiang[(✉)]

School of Computer Science and Technology, Shandong University, Jinan 250101,
People's Republic of China
majingjing_sdu@163.com, htjiang@sdu.edu.cn

Abstract. We focus on designing algorithm for One-sided Scaffold
Filling. Jiang et al. proposed a non-oblivious local search algorithm for
this problem recently. We can give an example to show that this algorithm cannot approximate One-sided Scaffold Filling to $\frac{6}{5}$. In this paper,
we propose a new objective function based local search algorithm for
One-sided Scaffold Filling, and give the accurate proof to show that its
approximation ratio is $\frac{6}{5}$.

1 Introduction

With the development of biological sequencing technology, there has been an
increasing trend that genomes are being published in scaffold form [9]. A scaffold
has usually been viewed as a *draft genome* with some gene fragments missed.
The draft genomes are often used to make analysis and interpretations, which is
tentative and prone to error and leads to particular problems in genomic analysis.
Thus the scaffold filling problem is motivated by extracting whole genomes from
scaffolds via computation [12].

Muñoz *et al.* pioneered to propose the one-sided scaffold filling problem, and
devised a polynomial time algorithm to fill a signed permutation by minimizing
some rearrangement distance [12]. Jiang *et al.* showed that it could be solved in
polynomial time to fill a permutation by minimizing the breakpoint distance as
its objective [8], no matter whether the permutation is signed or not. On the
other hand, there exist trivial reductions which can show it NP-Hard to fill an
unsigned permutation by minimizing a distance of rearrangement like reversal
or translocation [17,18].

Whichever a genome similarity measure can be used as an optimization objective in filling a scaffold with duplicated genes. Since those genome similarity measures such as exemplar distance [13], minimum common string partition (MCSP)
distance [3] and maximum common string number [1,10,11] are themselves computationally difficult, NP-Hard namely [2–7], it must be computationally difficult to use one of them as an optimization objective in filling a scaffold with
duplicated genes. Although the breakpoint distance or the adjacency number
is computationally easy for genomes with duplicated genes, it has been shown
that One-Sided Scaffold Filling is NP-Hard under breakpoint distance [8], which

© Springer International Publishing Switzerland 2016
D. Zhu and S. Bereg (Eds.): FAW 2016, LNCS 9711, pp. 145–157, 2016.
DOI: 10.1007/978-3-319-39817-4_15

means also NP-Hard under adjacency number. Two-Sided Scaffold Filling under breakpoint distance or adjacency number is NP-Hard consequently, because One-Sided Scaffold Filling is a special case of it.

Scaffold filling for genomes with duplicated genes has been demonstrated to admit approximation algorithms with constant performance ratios, if adjacency number is used as the maximization objective. For One-Sided Scaffold Filling, Jiang et $al.$ designed a greedy $\frac{4}{3}$-approximation algorithm [10,11]. Liu et $al.$ then improved the performance ratio to $\frac{5}{4}$ by the so called local improvement following behind greedy method [14]. One can look up in [15] for how to approximate Two-Sided Scaffold Filling to a constant performance ratio. Recently, Jiang et $al.$ proposed a $\frac{6}{5}$-approximation algorithm by a non-obvious local search, but there are some errors in it [16]. We can give an example of instance ($A = xaymbnam1cd23e4bc35e6f5$, $B = xymn123456m3aeb5c$, $B' = xyman123cb45e6fm3aeb5cd$, $B^* = xaymbn1cd23e45f6m3aeb5c$,$B'$ is obtained by algorithm, and B^* is the optimal), which shows the algorithm cannot approximate the problem to $\frac{6}{5}$, but $\frac{11}{9}$.

In this paper, we present a new algorithm for One-Sided Scaffold Filling, which can achieve a performance ratio $\frac{6}{5}$ accurately. This will be done by non-obvious local search technique, which benefits from a new objective function expressed by a weighted number summation of those missed gene strings with one, two, three and four genes, other than the intuitional adjacency numbers. By a bipartite conflict graph, we prove that the obtained scaffold comes true with as many increased adjacencies as needed to approximate One-Sided Scaffold Filling to $\frac{6}{5}$.

2   Preliminaries

Let Σ be an alphabet in which each element represents a gene family. A $scaffold$ on Σ is a gene $sequence$ whose genes each is an occurrence of a gene family in Σ. Let $S = s_1\ s_2\ \cdots\ s_n$ be a scaffold on Σ, then all genes in S form a multi-set which will be denoted as $c(S) = \{s_1, ..., s_n\}$. A substring in S with k genes is referred to as an k-$string$. A 2-string in S is particularized as a $pair$. Let $P[S]$ be the pair set of S. Then $P[S] = \{s_1s_2, s_2s_3, ..., s_{n-1}s_n\}$. A pair is in S, if and only if it is in $P[S]$.

Let $A = a_1a_2\ \cdots\ a_n$ and $B = b_1\ b_2\ \cdots\ b_m$ be two scaffolds on Σ, $P[A]$ and $P[B]$ the pair sets of A and B respectively. To identify a pair in A (resp. B) as an $adjacency$ or $breakpoint$, we have to set a maximum matching between $P[A]$ and $P[B]$.

A pair $a_ia_{i+1} \in P[A]$ and a pair $b_jb_{j+1} \in P[B]$ form a $match$ (a_ia_{i+1}, b_jb_{j+1}) between $P[A]$ and $P[B]$, if $a_ia_{i+1} = b_jb_{j+1}$ or $a_ia_{i+1} = b_{j+1}b_j$. Let $\mathcal{R}$ be the set of all matches between $P[A]$ and $P[B]$. Then a subset $R \subseteq \mathcal{R}$ is a $maximum$ $matching$ between $P[A]$ and $P[B]$ if, (1) any pair in A or B does not occur in two or more matches in R; (2) the cardinality of R is maximized over all subsets of $\mathcal{R}$ which subjects to (1). The $adjacency$ and $breakpoint$ can be formalized as,

Definition 1. *Let R be a maximum matching between $P[A]$ and $P[B]$. Then a pair in A (resp. B) is an **adjacency** relative to B with respect to R, if it forms a match in R with a pair in B (resp. A), otherwise, a **breakpoint**.*

As a simple consequence of Definition 1, A must have as many adjacencies as B has with respect to whichever maximum matching between $P[A]$ and $P[B]$. Since so, we usually address a pair in A or B as an adjacency or a breakpoint without mentioning the maximum matching with respect to which the pair serves as an adjacency or a breakpoint. For convenience, we denote by $a(A, B)$ the set of adjacencies in A (or B), $b_A(A, B)$ and $b_B(A, B)$ the set of breakpoints in A and B respectively. For example, let $A = a\ c\ d\ a\ b\ c\ d$, $B = a\ b\ d\ c\ a\ c\ d$. Then $P[A] = \{ac, cd, da, ab, bc, cd\}$, $P[B] = \{ab, bd, dc, ca, ac, cd\}$. If setting a maximum matching between $P[A]$ and $P[B]$ as $R = \{(ac, ca), (cd, dc), (ab, ab), (cd, cd)\}$, then with respect to R,

$$a(A, B) = \{ac, cd, ab, cd\}$$
$$b_A(A, B) = \{da, bc\}$$
$$b_B(A, B) = \{bd, ac\}$$

An *insertion* of a gene into a scaffold refers to the operation to insert the gene between two genes of a pair, onto the left side of the first gene or the right side of the last gene in the scaffold. Let $S = s_1 \ldots s_i\ s_{i+1} \ldots s_n$ be a scaffold on Σ and x a gene of a symbol in Σ. Then inserting x between s_i and s_{i+1} transforms S into $S' = s_1 \ldots s_i\ x\ s_{i+1} \ldots s_n$, $1 \le i \le n-1$, inserting x onto the left (resp. right) side of s_1 (resp. s_n) transforms S into $S' = x\ s_1 \ldots s_n$ (resp. $S' = s_1 \ldots s_n$ x). We denote by $S + x$ the set of scaffolds produced by inserting x into S. For a gene set X, we denote by $S + X$ the set of scaffolds produced by inserting all genes in X into S.

A scaffold has usually been considered with some genes missed. Muñoz et al. pioneered the scaffold filling problem which suggested to get a seemingly no-gene-missed scaffold by filling a scaffold with those genes in an already existed scaffold but it, where the adjacency number has been most commonly used to measure how good a resulting scaffold is. Thus the one-sided scaffold filling problem (abbr. OSSF) uses a scaffold for reference, asks to find a scaffold by filling a given scaffold with those genes in the reference scaffold but the given one, such that the number of adjacencies in the scaffold relative to the reference one is maximized. It can be formalized as,

Instance: Two scaffolds A, B on Σ with $c(B) - c(A) = \varnothing$.

Objective: Find a scaffold $B' \in B + (c(A) - c(B))$, such that $|a(B', A)|$ is maximized over all scaffolds in $B + (c(A) - c(B))$.

Let $X = c(A) - c(B)$. A scaffold in $B + X$ is optimal, if its adjacencies relative to A is number maximized over all scaffolds in $B + X$. Moreover, a gene is *missed* if it is in $X = c(A) - c(B)$.

To avoid inserting a missed gene onto the leftmost or rightmost side of a scaffold, we add a special gene # to both ends of A and B, and accept that any gene other than the first and the last in A or B is not $\#$. Thus, we will treat the instance of OSSF as $A = a_0\ a_1\ ...\ a_n\ a_{n+1}$, $B = b_0\ b_1\ ...\ b_m\ b_{m+1}$, where $a_0 = a_{n+1} = b_0 = b_{m+1} = \#$.

3   How to Approximate the Scaffold Filling

Let $Y \subseteq X$, $B' \in B+Y$. Then those genes in Y must occur in B' in the form of gene strings. A gene string in $B' \in B+Y$ is *missed*, if each gene in it is missed, whereas neither of that one on the left side of the string's first gene and that one on the right side of the string's last gene is missed. A missed string in B' can always be thought of formed by inserting the string into B once as a whole. In [14], it has been shown that one can always fill B with genes to increase at least as many adjacencies as those genes which have been inserted into B. We formally state it as,

Theorem 1. *Let $Y \subseteq X$, $B' \in B+Y$. Then in polynomial time, a group of string insertions can be found to insert all genes in $X - Y$ into B', such that each of them can insert a string to increase at least as many adjacencies as those genes the string has.*

3.1   A Sufficient Condition for $\frac{6}{5}$-Approximation

Let $Y \subseteq X$, $B' \in B+Y$. A missed k-string in B' is *good*, if removing (all genes of) the string from B' transforms B' into B^+, such that $|a(B',A)| - |a(B^+,A)| = k+1$. Else if $|a(B',A)| - |a(B^+,A)| = k$, the missed k-string is not good. *inserting a good i-string*, say I, into B' refers to that the insertion of the missed i-string I into B' transforms B' into B^+, such that the string turns into good in B^+. A string is missed by default, if it is mentioned for inserting into or removing from a scaffold.

Let $B^* \in B + X$ be optimal. Let in B^*, b_i^* be the number of good i-strings, c_i^* the number of missed i-strings which are not good. Then,

Lemma 1.

$$|a(B^*,A)| = |a(B,A)| + \sum_{i \geq 1}(i+1)b_i^* + \sum_{i \geq 1} ic_i^*$$

$$= |a(B,A)| + |X| + \sum_{i \geq 1} b_i^*. \tag{1}$$

Proof. Inserting a good k-string increases just $k+1$ adjacencies in B^*. By Theorem 1 moreover, inserting an i-string into a scaffold increases either $i+1$ (if good) or i adjacencies for the scaffold (otherwise). This leads to the first equation of (1). The second equation follows from that all the missed strings in B^* together have $|X|$ genes.

Let $B'' \in B + X$. By Theorem 1 again, each of those string insertions to transform B into B'' can be assumed to have increased at least as many adjacencies as the genes the string has. Since so, we can present on what a scaffold in $B + X$ can play the role of what we cry. Let in B'', b_i be the number of good i-strings, c_i the number of missed strings which are not good, then a sufficient condition to approximate OSSF to what we cry can be stated as,

Lemma 2. If $6\sum_{i\geq 1} b_i \geq 4b_1^* + 3b_2^* + 2b_3^* + b_4^*$, then $\frac{|a(B^*,A)| - |a(B,A)|}{|a(B'',A)| - |a(B,A)|} \leq \frac{6}{5}$.

Proof. By Lemma 1, the adjacencies in B^* can be bounded by,

$$|a(B^*, A)| = |a(B, A)| + |X| + \sum_{i\geq 1} b_i^*$$

$$= |a(B, A)| + |X| + b_1^* + b_2^* + b_3^* + b_4^* + \sum_{i\geq 5} b_i^*$$

$$\leq |a(B, A)| + |X| + b_1^* + b_2^* + b_3^* + b_4^* + \frac{1}{5}(|X| - b_1^* - 2b_2^* - 3b_3^* - 4b_4^*)$$

$$\leq |a(B, A)| + \frac{6}{5}(|X| + \frac{4}{6}b_1^* + \frac{3}{6}b_2^* + \frac{2}{6}b_3^* + \frac{1}{6}b_4^*)$$

The adjacencies in B'' can also be bounded by $|a(B'', A)| \geq |a(B, A)| + |X| + \sum_{i\geq 1} b_i$. Thus if $6\sum_{i\geq 1} b_i \geq 4b_1^* + 3b_2^* + 2b_3^* + b_4^*$, then

$$\frac{|a(B^*, A)| - |a(B, A)|}{|a(B'', A)| - |a(B, A)|} \leq \frac{\frac{6}{5}(|X| + \frac{4}{6}b_1^* + \frac{3}{6}b_2^* + \frac{2}{6}b_3^* + \frac{1}{6}b_4^*)}{|X| + \sum_{i\geq 1} b_i} \leq \frac{6}{5}. \tag{2}$$

Then the remainder effort is to find a scaffold in $B + X$ to meet what Lemma 2 asks.

3.2 The Objective Function of Local Search

To approximate OSSF to $\frac{6}{5}$, it suffices to find a scaffold in $B + X$ with as many good strings as Lemma 2 asks. Local search always asks for an objective function to quantify how optimal its solutions are. Thus it is not unusual to set the local search objective function with a weighted summation of the good i-string numbers for $i = 1, 2, 3, 4$. Let $B' \in B + Y$ with $Y \subseteq X$ be the scaffold the local search algorithm should maintain during it runs, b_i the number of good i-strings in B'. Then the objective function can be generalized as,

$$D(B') = w_1 b_1 + w_2 b_2 + w_3 b_3 + w_4 b_4. \tag{3}$$

Strategically, our algorithm always repeats to find and substitute some missed strings in B' with new ones, so that the object function value can be increased, until no string substitution we allow holds for B'. A *string substitution* for B' refers to a series of operations which remove some strings from B' to transform B' into B^+, then insert some other strings into B^+, where if those strings removed from B' are $I_1, ..., I_k$, then a gene should be in $c(I_1) \bigcup ... \bigcup c(I_k) \bigcup (X - Y)$, if it is in a string inserted into B^+. Actually, a removed string as well as an inserted string must be good, whenever it happens in a substitution. For the sake of enough good strings of length one, two, three and four, we technically set $w_1 = 1$, $w_2 = \frac{1}{3}$, $w_3 = \frac{1}{10}$, $w_4 = \frac{1}{41}$.

3.3 What Good String Substitutions Are Allowed

A string substitution for B' is *adoptable*, if it transforms B' into B^+, such that $D(B^+) - D(B') > 0$. In regard to improving a scaffold to get a larger objective function value, we are interested in just four kinds of adoptable string substitutions. Note again that a string must be good, and has at most 4 genes, whenever it is removed from or inserted into a scaffold.

(1) Insert a good string

An string insertion is what we must accept, if it inserts a string with all genes in $X - Y$ into B' to increase $D(B')$ by a positive amount. In what follows, *good-insertion*(B, B', X, I) will be used as a subroutine to find a string I with $c(I) \subseteq X - Y$, whose insertion into B' can increase $D(B')$. Later, no string can be inserted into B' to increase $D(B')$.

(2) Add a good string

If removing a good string from B' transforms B' into B^+, then the insertions into B^+ of two or more strings can transform B^+ into B^{++} with $D(B^{++}) > D(B')$. This suggests a kind of substitutions of one string with two or more ones, where we just accept those adoptable substitutions of two substituting one. Thus in what follows, the subroutine named as *good-string*(B, B', X, T, I) will be used to find a good string I in B', and a set T of two strings with genes in $c(I) \cup (X - Y)$ whose insertions into B' with I removed can increase $D(B')$. In the situation good-string(B, B', X, T, I) returns true, I and T will be found such that each member in T turns into good in the scaffold with the strings in T inserted into that B' with I removed. Later, no two strings can substitute a good string in B' to increase $D(B')$.

(3) Add a shorter good string

If removing some good strings from B' transforms B' into B^+, then the insertion into B^+ of another string, if shorter than each of those removed from B', can transform B^+ into B^{++} with $D(B^{++}) > D(B')$. This suggests a kind of substitutions of at least one string with one, where we accept those adoptable substitutions of one substituting 1, 2, 3 or 4 strings. In what follows, the subroutine named as *shorter-string*(B, B', X, I, T) will be used to find a good string set T in B', and a string I with $c(I) \subseteq \bigcup_{J \in T} c(J) \cup (X - Y)$, whose insertion into that B' with all strings in T removed can increase $D(B')$. In the situation shorter-string(B, B', X, I, T) returns true, T and I can be found such that I will turn into good in the scaffold with it inserted into that B' with all strings in T removed. Later, no one string can substitute one, two, three or four good strings in B' to increase $D(B')$.

(4) More efforts to add a shorter good string

A string can substitute a good string of the same length as it to transform B' into a scaffold with equal objective function value to $D(B')$. If fortunate, another string substitution so as mentioned in (3) may come true for that B' with the former substitution done. Based on this observation, we just accept those adoptable string substitutions each of which includes two sub substitutions, where the first must happen to two good strings of the same length, and the second must be so as mentioned in (3). Specifically, the subroutine named as $more\text{-}shorter(B, B', X, T, T')$ will be used to find in B' a good string set $T' = \{I'\} \cup T^+$ with $T^+ \neq \emptyset$, and a string set $T = \{I, I^+\}$ with $c(I) \subseteq (X - Y) \cup c(I')$, $|c(I)| = |c(I')|$, $c(I^+) \subseteq (X - ((Y - c(I'))$ $\cup c(I))) \cup (\bigcup_{J \in T^+} c(J))$, such that I can substitute I' without decreasing $D(B')$, if this substitution transforms B' into B^+, then I^+ can substitute all strings in T^+ to increase $D(B^+)$. As previously stated, in the situation $more\text{-}shorter(B, B', X, T, T')$ returns true, T and T' can be found such that the strings in T will turn into good in that scaffold with the strings in T inserted into B' with all strings in T' removed.

The algorithm always repeats to find and implement a string substitution so as mentioned in (1), (2), (3), (4) to improve B'. The string substitutions will come into effects in the order as they have been presented before. Only when no substitution mentioned in (i) can be found, can those substitutions mentioned in $(i+1)$ be used to improve B', $i \leq 3$. If for B', no string substitution as mentioned in (1), (2), (3), (4) can be found, those genes in $X - Y$ will be inserted into B' by the algorithm proposed in [14] without decreasing the number of good strings. In summation, the algorithm is given as Scaffold-Filling(A, B) formally.

Algorithm 1. Scaffold-Filling(A, B)

Input: two scaffolds A, B, $c(B) \subseteq c(A)$
Output: $B' \in B + c(A) - c(B)$
1: $X \leftarrow c(A) - c(B)$; $B' \leftarrow B$; $D \leftarrow -1$; $D' \leftarrow 0$.
2: **while** ($D' \neq D$) **do**
3:    $D \leftarrow D'$;
4:    If (good-insertion(B, B', X, I) = true), insert I into B'.
5:    Else if (good-string(B, B', X, I, T) = true), substitute I with the strings in T for B'.
6:    Else if (shorter-string(B, B', X, I, T) = true), substitute the strings in T with I for B'.
7:    Else if (more-shorter(B, B', X, T, T') = true), substitute the strings in T' with the strings in T for B'.
8:    $D' \leftarrow D(B')$.
9: **end while**
10: **return** ($B'' \leftarrow$ inserting all genes in $c(A) - c(B')$ into B' by the the algorithm in [15]).

---

4   Why Is the Performance Ratio $\frac{6}{5}$

Let B^* be an optimal scaffold in $B + X$, B'' a scaffold returned by Scaffold-Filling(A, B). We show $\frac{|a(B^*,A)|-|a(B,A)|}{|a(B'',A)|-|a(B,A)|} \leq \frac{6}{5}$ for B'' in this section.

Let B' be the scaffold Scaffold-Filling(A, B) maintains when it runs into the while loop end at Line 9. Then B' subjects to that, there are no four kinds of string substitution what we allowed holds adoptable for B', and there are at least as many good strings in B'' as in B'. So we can arrive at $\frac{|a(B^*,A)|-|a(B,A)|}{|a(B'',A)|-|a(B,A)|}$ $\leq \frac{6}{5}$, if B' has at least $\frac{1}{6}$ $(4b_1^* + 3b_2^* + 2b_3^* + b_4^*)$ good strings. Actually, B' can be shown to have at least $\frac{1}{6}$ $(4b_1^* + 3b_2^* + 2b_3^* + b_4^*)$ good strings each with at most 4 genes.

A good string, say I in B', *destroys* a good string, say I^* in B^*, if I shares a gene with I^* or occurs between the same two genes in B as those I^* occurs between. A given good string in B' can make no other good strings than those in B^* it destroys fail to occur in B'.

Lemma 3. *Let $Y \subseteq X$, $B' \in B+Y$. Then a good i-string in B' can destroy at most $i + 1$ good strings in B^*; a good i-string in B^* can be destroyed by at most $i + 1$ good strings in B'.*

Proof. Let $b_j b_{j+1}$ be a breakpoint in B, I a good i-string in B' which occurs between b_j and b_{j+1}. Since I has i genes, it can share genes with at most i good strings in B^*. Moreover, at most one good string in B^* can occur between b_j and b_{j+1}. Totally, at most $i + 1$ good strings in B^* could be destroyed by I. For the same reason, a good i-string in B^* can be destroyed by at most $i + 1$ good strings in B'. □

To help compare the good string number of B' with that of B^*, we set a bipartite graph $G = (L, R, E)$, where a vertex in L (resp. R) corresponds to a good i-string in B' (resp. B^*) with $i \leq 4$, an edge is set between a vertex u in L and a vertex v in R, if the good string u corresponds to destroys the good string v corresponds to. Thus let S_i (resp. S_i^*) be the set of vertices corresponding to those good i-strings in B' (resp. B^*), then $L = \bigcup_i^4 S_i$, $R = \bigcup_i^4 S_i^*$, $E = \{(u,v) \mid u$ (resp. v) corresponds to a good string s (resp. s^*) in B' (resp. B^*), s destroys $s^*\}$. For example, let A and B as two scaffolds. Given an optimal scaffold B^* and an arbitrary scaffold B' in $B + X$ as follows,

$$A = \natural x_1 a y_1 12 x_2 34 y_2 567 x_3 cdef y_3 aabb x_4 x_1 1 y_1 x_2 5 a y_2 y_3 246 dx_4 37 c x_5 m\natural.$$
$$B = \natural x_1 y_1 x_2 y_2 x_3 y_3 x_4 x_5 x_1 5 y_1 6 x_2 dy_2 1a23 y_3 45 x_4 mc\natural.$$
$$X = \{1, 2, 3, 4, 5, 6, 7, a, a, a, b, b, c, d, e, f\}.$$
$$B' = \natural x_1 \underline{1} y_1 x_2 \underline{5a} y_2 x_3 \underline{bbaa} y_3 \underline{246} dx_4 \underline{37c} x_5 x_1 5 y_1 6 x_2 dy_2 1a23 \underline{ef} y_3 45 x_4 mc\natural.$$
$$B^* = \natural x_1 a y_1 \underline{12} x_2 \underline{34} y_2 \underline{567} x_3 \underline{cdef} y_3 \underline{aabb} x_4 x_5 x_1 5 y_1 6 x_2 dy_2 1a23 y_3 45 x_4 mc\natural.$$

Then the good string set in B' and B^* respectively is $L = \{1, 5a, 246d, 37c\}$, $R = \{a, 12, 34, 567, cdef, aabb\}$, which will be used as the vertices of the graph.

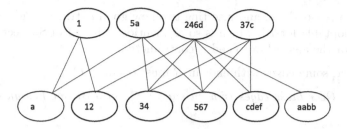

Fig. 1. An example for bipartite graph

We present the bipartite graph as follows in Fig. 1 to show how the good strings in B' destroy those in B^*.

By Lemma 2 and the fact that $|S_i| = b_i$, $|S_i^*| = b_i^*$, it suffices to show

$$6|S_1| + 6|S_2| + 6|S_3| + 6|S_4| \geq 4|S_1^*| + 3|S_2^*| + 2|S_3^*| + |S_4^*|, \qquad (4)$$

for arriving at $\frac{|a(B^*,A)| - |a(B,A)|}{|a(B'',A)| - |a(B,A)|} \leq \frac{6}{5}$.

Since a good string is always of length at most four, a vertex of G is incident with at most five edges by Lemma 3. To simplify the proof of Inequality (4), we try to delete some edges of G to reduce it into one with each vertex in R incident with at most 2 edges. Concretely, those edges of G will be considered for deletion one by one in such a principle as: an edge between $u(\in L)$ and $v(\in R)$ should be deleted, if $d(v) > 2$, and the length of good string u corresponds to is the maximal among all the good strings corresponded by all the vertices in L connected by v.

A graph reduced from G is *simplified*, if none of its edges could be deleted by this principle. Later, let $G' = (L, R, E')$ be a simplified graph reduced from G with $L = \bigcup_i^4 S_i$, $R = \bigcup_i^4 S_i^*$, $E' \subseteq E$. A vertex of G' turns to have at most $2°$, if it is in R.

Since G' is a subgraph of G, both a vertex and an edge of G' are treated to be of G. In what follows, we concentrate on G' to show Inequality (4). Although a vertex in R has at most $2°$ for G', no isolated vertex can occur in R. That is,

Lemma 4. *For G', a vertex in R is incident with at least one edge.*

Proof. If a vertex r in R is incident with no edge, it must be incident with no edge of G. Thus the good string in B^* r corresponds to has not been destroyed by any good strings in B'. An string insertion so as mentioned in (1) in Subsect. 3.3 holds true for B'. The condition for the algorithm to end is contradicted. □

Some edges with ends in L can be excluded from G'. That is,

Lemma 5. *If two edges have two respective 1 degree ends in S_i^* and S_j^*, where $i \leq j$, they cannot share one end in S_i.*

Proof. Let $(v, u), (v, u')$ be two edges with $v \in S_i$, $u \in S_i^*$, $u' \in S_j^*$, where $i \leq j$. If u and u' both have one degree, then the substitution of removing the good

i-string v corresponds to, and inserting two new good i-strings u, u' correspond to holds adoptable for B', which is so as mentioned in (2) in Subsect. 3.3. The condition for the algorithm to end is contradicted. □

Moreover, some edges with ends in R must occur in G'.

Lemma 6. *Over those edges with an end in S_j^*, at least one has an end in S_i with $i \leq j$.*

Proof. A vertex in R is incident with at least one edge by Lemma 4. Let $v \in S_j^*$. If it happens contrary, each edge with v as an end has an end in S_i with $i > j$, then by the expression of $D(B')$, a substitution of removing at least one good string then inserting the j-string v corresponds to must hold adoptable for B', which suggests a substitution so as mentioned in (3) in Subsect. 3.3. The condition for the algorithm to end is contradicted. □

Let (l,r) be an edge with $l \in S_i$, $r \in S_j^*$, $i \leq j$. Then an edge sharing the end l with (l,r) can be excluded conditionally. That is,

Lemma 7. *If an edge with end r has an end in S_k with $k > j$, then no edge with end l can have a 1 degree end in S_i^*.*

Proof. If G' has an edge (l, r'), where $r' \in S_i^*$ has one degree, then r' must have 1 degree in G, which means the string r' corresponds to can substitute the string l corresponds to for B' with $D(B')$ unchanged. Since $k > j$, each edge of G with end r has an end in S_l with $l \geq k > j$, even if it does not occur in G'. Thus if the substitution of removing the string l corresponds to and inserting the string r' corresponds to transforms B' into B^+, those good strings the vertices incident with the edges which share the end r correspond to, can be substituted with the string r corresponds to for B^+ to increase $D(B^+)$. The condition for the algorithm to end is contradicted because this has suggested a string substitution so as mentioned in (4) in Subsect. 3.3. □

We turn to bound the number of vertices in S_j^* for j with $1 \leq j \leq 4$. Since a vertex in S_j^* has either one or two degrees, let $X_j^k = \{v \mid v \in S_j^*, d(v) = k \leq 2\}$. It follows Lemma 4 that $|S_j^*| = |X_j^1| + |X_j^2|$, $1 \leq j \leq 4$. Since the end of an edge, if in S_j^*, is in either X_j^1, or X_j^2, we divide those edges with ends in S_j^* into two sets, which are respectively formalized as, $K_{i,j} = \{(u,v) \mid u \in S_i, v \in S_j^*, d(v) = 1\}$ and $M_{i,j} = \{(u,v) \mid u \in S_i, v \in S_j^*, d(v) = 2\}$. An edge in $M_{i,j}$ should be specialized, if given an integer d, it shares an end in S_j^* with an edge with another end in S_d. Thus let $M_{i,j}^d = \{(u,v) \mid u \in S_i, w \in S_d, v \in S_j^*, (w,v) \in E', u \neq w\}$. By Lemma 6, we have $|M_{i,j}| = \sum_{d=1}^j |M_{i,j}^d|$, if $i > j$. Those 1 degree vertices in S_1^* can be number bounded by the in-S_1 vertex number minus a numerical value made of those edges with ends in S_1. That is,

Lemma 8.

$$|X_1^1| \leq |S_1| - \frac{1}{2}(|K_{1,2}| + |K_{1,3}| + |K_{1,4}| + |M_{2,1}| + |M_{3,1}|$$
$$+ |M_{4,1}| + |M_{3,2}^1| + |M_{4,2}^1| + |M_{4,3}^1|). \tag{5}$$

Proof. It follows Lemma 6 that $|X_1^1| = |K_{1,1}|$.

An edge in $K_{1,1}$ must have an end in S_1. On the other hand, a vertex in S_1 is incident with at most one edge in $K_{1,1}$ by Lemma 5. By Lemma 5 again, a vertex in S_1 cannot be incident with any edge in $K_{1,1}$, if it happens incident with an edge in $K_{1,2}$, $K_{1,3}$ or $K_{1,4}$.

By Lemma 7 moreover, a vertex in S_1 cannot be incident with any edge in $K_{1,1}$, if it happens incident with an edge in $M_{1,1} \cup M_{1,2} \cup M_{1,3}$, which shares an end with someone in $M_{2,1} \cup M_{3,1} \cup M_{4,1} \cup M_{3,2}^1 \cup M_{4,2}^1 \cup M_{4,3}^1$. Thus we have $|K_{1,2}| + |K_{1,3}| + |K_{1,4}| + |M_{2,1}| + |M_{3,1}| + |M_{4,1}| + |M_{3,2}^1| + |M_{4,2}^1| + |M_{4,3}^1|$ edges whose ends in S_1 all fail to be incident with any edges in $K_{1,1}$. Finally, the lemma follows from that a vertex in S_1 has at most $2°$. $\qquad\square$

Those ends-in-S_1^* edges can also help provide a numerical value bound on those in-S_1^* vertices. That is,

Lemma 9.

$$|X_1^1| + 2|X_1^2| \leq 2|S_1| + |M_{2,1}| + |M_{3,1}| + |M_{4,1}| - (|K_{1,2}| + |K_{1,3}| + |K_{1,4}|$$
$$+ |M_{1,2}| + |M_{1,3}| + |M_{1,4}|). \tag{6}$$

Proof. Let E^+ be the edge set in which each member has an end in S_1^*. Then by Lemma 4, $|E^+| = |X_1^1| + 2|X_1^2|$. By Lemma 6, $|X_1^1| = |K_{1,1}|$, $2|X_1^2| = |M_{1,1}| + |M_{2,1}| + |M_{3,1}| + |M_{4,1}|$. An edge in $K_{1,1}$ and $M_{1,1}$ must have an end in S_1. By Lemma 3, a vertex in S_1 has at most two degrees. Those edges in $K_{1,2}$, $K_{1,3}$, $K_{1,4}$, $M_{1,2}$, $M_{1,3}$, $M_{1,4}$ all fall outside E^+, although each of them has an end in S_1. Thus $|K_{1,1}| + |M_{1,1}| \leq 2|S_1| - (|K_{1,2}| + |K_{1,3}| + |K_{1,4}| + |M_{1,2}| + |M_{1,3}| + |M_{1,4}|)$. This leads to the lemma inequality. $\qquad\square$

In the same way as for Lemmas 8 and 9, we further bound the vertex numbers in S_2^*, S_3^*, S_4^* respectively by the following six lemmas.

Lemma 10.

$$|X_2^1| \leq |S_2| + |K_{1,2}| - \frac{1}{3}(|K_{2,3}| + |K_{2,4}| + |M_{3,2}^2| + |M_{4,2}^2| + |M_{4,3}^2|). \tag{7}$$

Lemma 11.

$$|X_2^1| + 2|X_2^2| \leq 3|S_2| + |K_{1,2}| + |M_{1,2}| + |M_{3,2}| + |M_{4,2}|$$
$$- (|K_{2,3}| + |K_{2,4}| + |M_{2,1}| + |M_{2,3}| + |M_{2,4}|) \tag{8}$$

Lemma 12.

$$|X_3^1| \leq |S_3| + |K_{1,3}| + |K_{2,3}| - \frac{1}{4}(|K_{3,4}| + |M_{4,3}^3|) \tag{9}$$

Lemma 13.

$$|X_3^1| + 2|X_3^2| \leq 4|S_3| + |K_{1,3}| + |K_{2,3}| + |M_{1,3}| + |M_{2,3}| + |M_{4,3}| - (|K_{3,4}|$$
$$+ |M_{3,4}| + |M_{3,1}| + |M_{3,2}|) \tag{10}$$

Lemma 14.

$$|X_4^1| \leq |S_4| + |K_{1,4}| + |K_{2,4}| + |K_{3,4}| \tag{11}$$

Lemma 15.

$$|X_4^1| + 2|X_4^2| \leq 5|S_4| + |K_{1,4}| + |K_{2,4}| + |K_{3,4}| + |M_{1,4}| + |M_{2,4}| + |M_{3,4}|$$
$$-(|M_{4,1}| + |M_{4,2}| + |M_{4,3}|) \tag{12}$$

Theorem 2. $6|S_1| + 6|S_2| + 6|S_3| + 6|S_4| \geq 4|S_1^*| + 3|S_2^*| + 2|S_3^*| + |S_4^*|.$

Proof. The lemma inequality will come out of adding those inequalities of Lemma from 8 to 15 with technical weights, then by $((5) + (6)) \times 2 + ((7) + (8)) \times \frac{3}{2} + ((9) + (10)) + ((11) + (12)) \times \frac{1}{2}$, we have

$$4|S_1^*| + 3|S_2^*| + 2|S_3^*| + |S_4^*| \leq 6|S_1| + 6|S_2| + 5|S_3| + 3|S_4|$$
$$+ \frac{1}{2}|M_{41}| + \frac{1}{2}|M_{42}^2| + \frac{1}{4}|M_{43}^3|$$

Finally, the lemma inequality follows from $|M_{4,1}| + |M_{4,2}^2| + |M_{4,3}^3| \leq 5|S_4|$. $\square$

Looking from Theorem 2 back to Lemma 1, we come true to show that Scaffold-Filling(A, B) can always output a scaffold $B'' \in B + (c(A) - c(B))$ with approximation ratio no more than $\frac{6}{5}$.

5   Conclusion

We have presented a local search algorithm which can approximate One-sided Scaffold Filling to $\frac{6}{5}$. It is interesting and open that if this problem can be approximated to a smaller performance ratio. On the other hand, although the problem is proved Max-SNP-Hard, it awaits a real value found to reject any polynomial time algorithm to approximate the problem within it, which seems also interesting.

Acknowledgments. This research is partially supported NSF of China under grant 61472222,61202014.

References

1. Angibaud, S., Fertin, G., Rusu, I., Thevenin, A., Vialette, S.: On the approximability of comparing genomes with duplicates. J. Graph Algorithms Appl. **13**(1), 19–53 (2009)
2. Blin, G., Fertin, G., Sikora, F., Vialette, S.: The EXEMPLAR BREAKPOINT DISTANCE for non-trivial genomes cannot be approximated. In: Das, S., Uehara, R. (eds.) WALCOM 2009. LNCS, vol. 5431, pp. 357–368. Springer, Heidelberg (2009)
3. Cormode, G., Muthukrishnan, S.: The string edit distance matching problem with moves. In: Proceedings of 13th ACM-SIAM Symposium on Discrete Algorithms (SODA 2002), pp. 667–676 (2002)

4. Chen, Z., Fowler, R., Fu, B., Zhu, B.: On the inapproximability of the exemplar conserved interval distance problem of genomes. J. Comb. Optim. **15**(2), 201–221 (2008)
5. Chen, Z., Fu, B., Zhu, B.: The approximability of the exemplar breakpoint distance problem. In: Cheng, S.-W., Poon, C.K. (eds.) AAIM 2006. LNCS, vol. 4041, pp. 291–302. Springer, Heidelberg (2006)
6. Goldstein, A., Kolman, P., Zheng, J.: Minimum common string partition problem: hardness and approximations. In: Fleischer, R., Trippen, G. (eds.) ISAAC 2004. LNCS, vol. 3341, pp. 484–495. Springer, Heidelberg (2004)
7. Jiang, M.: The zero exemplar distance problem. In: Tannier, E. (ed.) RECOMB-CG 2010. LNCS, vol. 6398, pp. 74–82. Springer, Heidelberg (2010)
8. Jiang, H., Zheng, C., Sankoff, D., Zhu, B.: Scaffold filling under the breakpoint distance. In: Tannier, E. (ed.) RECOMB-CG 2010. LNCS, vol. 6398, pp. 83–92. Springer, Heidelberg (2010)
9. Huson, D.H., Reinert, K., Myers, E.W.: The greedy path-merging algorithm for Contig scaffolding. J. ACM **49**(5), 603–615 (2002)
10. Jiang, H., Zhong, F., Zhu, B.: Filling scaffolds with gene repetitions: maximizing the number of adjacencies. In: Giancarlo, R., Manzini, G. (eds.) CPM 2011. LNCS, vol. 6661, pp. 55–64. Springer, Heidelberg (2011)
11. Jiang, H., Zheng, C., Sankoff, D., Zhu, B.: Scaffold filling under the breakpoint and related distances. IEEE/ACM Trans. Comput. Biol. Bioinform. **9**(4), 1220–1229 (2012)
12. Muñoz, A., Zheng, C., Zhu, Q., Albert, V., Rounsley, S., Sankoff, D.: Scaffold filling, contig fusion and gene order comparison. BMC Bioinform. **11**, 304 (2010)
13. Sankoff, D.: Genome rearrangement with gene families. Bioinformatics **15**(11), 909–917 (1999)
14. Liu, N., Jiang, H., Zhu, D., Zhu, B.: An improved approximation algorithm for scaffold filling to maximize the common adjacencies. IEEE/ACM TCBB **10**(4), 905–913 (2013)
15. Liu, N., Zhu, D.: The algorithm for the two-sided scaffold filling problem. In: Chan, T.-H.H., Lau, L.C., Trevisan, L. (eds.) TAMC 2013. LNCS, vol. 7876, pp. 236–247. Springer, Heidelberg (2013)
16. Jiang, H., Ma, J., Luan, J., Zhu, D.: Approximation and nonapproximability for the one-sided scaffold filling problem. In: Xu, D., Du, D., Du, D. (eds.) COCOON 2015. LNCS, vol. 9198, pp. 251–263. Springer, Heidelberg (2015)
17. Caprara, A.: Sorting by reversals is difficult. In: Proceedings of RECOMB 1997, pp. 75–83 (1997)
18. Zhu, D., Wang, L.: On the complexity of unsigned translocation distance. Theor. Comput. Sci **352**(1–3), 322–328 (2006)

Nonlinear Dimension Reduction by Local Multidimensional Scaling

Yuzhe Ma[1], Kun He[1(✉)], John Hopcroft[2], and Pan Shi[1]

[1] Huazhong University of Science and Technology, Wuhan, China
brooklet60@hust.edu.cn
[2] Cornell University, New York, USA

Abstract. We propose a neighbourhood-preserving method called LMB for generating a low-dimensional representation of the data points scattered on a nonlinear manifold embedded in high-dimensional Euclidean space. Starting from an exemplary data point, LMB locally applies the classical Multidimensional Scaling (MDS) algorithm on small patches of the manifold and iteratively spreads the dimension reduction process. Differs to most dimension reduction methods, LMB does not require an input for the reduced dimension, as LMB could determine a well-fit dimension for reduction in terms of the pairwise distances of the data points. We thoroughly compare the performance of LMB with state-of-the-art linear and nonlinear dimension reduction algorithms on both synthetic data and real-world data. Numerical experiments show that LMB efficiently and effectively preserves the neighbourhood and uncovers the latent embedded structure of the manifold. LMB also has a low complexity of $O(n^2)$ for n data points.

Keywords: Dimension reduction · Nonlinear manifold · Neighbourhood-preserving · Local multidimensional scaling

1 Introduction

Many high-dimensional data in real-world applications can be modeled as points distributed on a low-dimensional manifold. Dimension reduction is widely used to obtain a compact representation of the high-dimensional data to reduce the redundancy and preserve the principal properties of the neighbourhood structure or the global structure. In the last decades, various linear and nonlinear methods have been proposed for the dimension reduction [1–3]. Due to the curvature and distortion of the manifold, however, traditional linear dimension reduction methods often fail to yield a neighbourhood-preserving representation of the data points in a low-dimensional space, and they are unable to uncover the global structure of the manifold. Researchers then shift their attention to nonlinear methods to construct the low-dimensional manifolds from high-dimensional space [4–6]. We propose a new dimension reduction algorithm and demonstrate its efficiency on both synthetic and real-world data. We will first highlight several linear and nonlinear popular methods and show the difference of our algorithm.

© Springer International Publishing Switzerland 2016
D. Zhu and S. Bereg (Eds.): FAW 2016, LNCS 9711, pp. 158–171, 2016.
DOI: 10.1007/978-3-319-39817-4_16

Linear methods require the representation to be a linear combination of the original variables. There are two lines of works. One line projects the data points onto the most significant subspaces, such as Principal Component Analysis (PCA) [7] and Multidimensional Scaling (MDS) [8]. The linear projection may cause the data points far from each other in high-dimensional space be close in the subspace and fail to preserve the neighbourhood. Another line seeks to extract the principal variables underlying the original data by using the statistical analysis, including Factor Analysis (FA) [9], Linear Discriminant Analysis [10], Canonical Correlation Analysis [11], and Independent Component Analysis (ICA) [12].

Kernel PCA [13] and Isomap [6,14] are two classical nonlinear methods. Kernel PCA maps the original data onto an inner-product feature space so that the data in the feature space can be separated linearly. The nonlinear mapping may distort the data structure and not well preserve the neighbourhood as measured by the distances in the original space. Isomap is an extension of MDS, which approximates the geodesic distance by the length of the shortest path confined to the embedded manifold and then utilizes MDS to obtain a compact representation. Isomap could preserve the geodesic distance globally but it does not efficiently preserve the neighbourhood.

In order to well preserve the neighbourhood, several nonlinear local methods have been proposed. As a typical local nonlinear method, Locally Linear Embedding (LLE) [5] constructs a local geometric structure of the data points and reduces the data points into a low-dimensional space that could best preserve the local geometries [15]. LLE maps the data into a single global coordinate system and enforces the representation variables to have identity variance-covariance matrix. LLE is extended to Hessian-based Locally Linear Embedding (HLLE) [16], which works with the Hessian matrix of the graph and is able to handle a wider range of the data set. Other nonlinear local methods include Local Tangent Space Alignment (LTSA) [17] and principal curves [18].

As small patches on a manifold could be approximately regarded as local hyperplanes, applying linear methods on small patches could well preserve the local structure of the neighbourhood. Following this observation, and based on the efficient linear method MDS [8], we propose a local nonlinear method called Local Multidimensional scaling with Breath-first search (LMB). LMB iteratively applies MDS on small patches of the manifold and spreads the local reduction process via Breath-First Search (BFS). To better understand LMB, we briefly discuss MDS in the appendix.

Compared with the global nonlinear methods, LMB works on the local patches of the manifold and better preserves the neighbourhood. By spreading the local reduction process via BFS, LMB could also reveal the manifold structure like Isomap [6] does. Compared with the local method LLE [5], LMB better preserves the neighbourhood and tries not to change the Euclidean distances of the data points. Another strength of LMB is that it automatically determines the dimension of the manifold in terms of the pairwise distances of the data points, so it does not need a priori parameter for the reduced dimension. Also, LMB is very fast in dealing with ten thousands of data points within several minutes, as LMB does not require the decomposition of a large matrix which indicates the pairwise distances of the data points. Numerical experiments and comparisons with other state-of-the-art methods show that LMB is more efficient in preserving the neighbourhood on both synthetic and real-world data.

2   The Proposed LMB Algorithm

2.1   Determine the Dimension of the Embedded Manifold

A d-dimensional manifold possesses the property that a local region on the manifold is approximately a d-dimensional hyperplane. The local observation of small regions on the manifold provides insight for the dimension of the embedded manifold. This intuitive perceptive can be demonstrated if the embedded structure is smooth and the data is well-sampled. For simplicity, we show cases for dimension $d \leq 2$. It is similar for higher-dimensional manifolds. A one-dimensional manifold embedded in an m-dimensional space is a curve, which can be represented by a vector-valued function:

$$f(s) = (f_1(s), f_2(s), ..., f_m(s)).$$

Suppose we choose a point $v_0 = f(s_0)$, and a neighbouring point of v_0 is $f(s_0 + \delta_s)$:

$$v_{nei} = f(s_0 + \delta_s) = (f_1(s_0 + \delta_s), f_2(s_0 + \delta_s), ..., f_m(s_0 + \delta_s)).$$

If the curve is differentiable at the point v_0, v_{nei} can be approximated by:

$$v_{nei} \approx (f_1(s_0) + f_1'(s_0)\delta_s, f_2(s_0) + f_2'(s_0)\delta_s, ..., f_m(s_0) + f_m'(s_0)\delta_s). \quad (1)$$

All neighbouring points of v_0 can be approximated by Eq. (1), only δ_s varies for these points. Locate v_0 at the origin, the local observation near the point v_0 is derived by measuring the relative location of its neighbouring points:

$$v_r = v_{nei} - v_0 \approx (f_1'(s_0), f_2'(s_0), ..., f_m'(s_0))\delta_s.$$

The vectors are almost along the same direction $(f_1'(s_0), f_2'(s_0), ..., f_m'(s_0))$, which indicates that the neighbouring points are approximately in a one-dimensional space. If $d = 2$, the vector-valued function is $f(s) = (f_1(x,y), f_2(x,y), ..., f_m(x,y))$ and the neighbouring points are almost in the plane spanned by $(\frac{\partial f_1}{\partial x}, \frac{\partial f_2}{\partial x}, ..., \frac{\partial f_m}{\partial x})$ and $(\frac{\partial f_1}{\partial y}, \frac{\partial f_2}{\partial y}, ..., \frac{\partial f_m}{\partial y})$.

For each point i, find its k-nearest points, and define i to be the father point of this neighbourhood. We execute MDS on the pairwise distances of these $k+1$ points to obtain their coordinates X_i. The column vectors of X_i are approximately in a d-dimensional hyperplane, and the first d eigenvalues of $X_i^T X_i$ are significant. Therefore we define the **local dimension** of point i as d if the first

d eigenvalues satisfy Eq. (2), where r is a constant close to 1 (such as 0.95) and S is the summation of all the eigenvalues.

$$\lambda_1 + \lambda_2 + ... + \lambda_d > rS \tag{2}$$

The dimension d of the embedded manifold is determined by the integer closest to the average local dimension of all the data points. Table 1 illustrates the response of d against different values of r.

2.2 Dimension Reduction by LMB

Based on the determined dimension of the embedded manifold d, and the Euclidean pairwise distances of the data points, LMB generates a compact representation of the original data in a d-dimensional space.

By performing BFS from the starting point to get its k-nearest neighbors and executing MDS on this neighbourhood, we place the first patch in a d-dimensional space with its center at the origin, and do translation to move the starting point to the origin. For the second neighbourhood, its father point belongs to the neighbourhood of the starting point. Local MDS on the second neighbourhood generates the second patch with its center at the origin. We translate the second patch such that its father point coincides with the same point in the first patch, and then do rotation or reflection (**orthogonal transformation**) on the second patch by fixing its father point as the pivot, such that other intersected points of these two patches coincide as much as possible. Continue the process for other patches and finally merge all the patches together in the d-dimensional space. A point is called **reduced** if we already obtain its coordinates in d-dimensional space, and a point is called **explored** if all its k-nearest neighbors are reduced.

2.3 Orthogonal Transformation

In Fig. 1, P_1 and P_2 are the father points of the two patches, and C_1 and C_2 are the patch centers. Points in patch ① are reduced. P_2 belongs to patch ① and therefore is also reduced. By executing local MDS on patch ②, we obtain the coordinates for patch ②, as shown in ③. Do translation to fix P_2 at the origin, as shown in ④. Then rotate or reflect patch ② by fixing P_2 as the pivot. After the orthogonal transformation, patch ② is rotated to the right direction and could be merged with patch ① correctly.

The orthogonal transformation is determined according to the distance constraint between $s_q(q = 1, 2, 3, ..., l)$ and the neighbouring points of i. The d-dimensional representation of points 1, 2, 3, ..., k are as shown in Eq. (3), where U is the orthogonal matrix needs to be determined and x_i is the d-dimensional representation of i.

$$x_j = Ux_{ij} + x_i, j = 1, 2, ..., k, k \geq d. \tag{3}$$

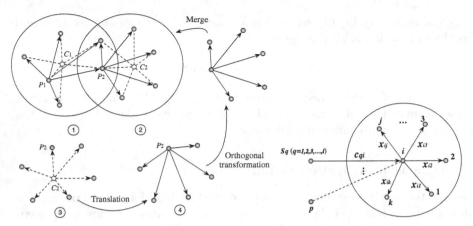

Fig. 1. Merge the patches

Fig. 2. Points 1, 2, 3, ..., k are the nearest neighbors of i. Point p is the father point of i and $s_q(q = 1, 2, 3, ..., l)$ are $l(l \geq d)$ reduced points close to i. x_{ij} is the coordinates of point j with i at the origin and c_{qi} is the vector from s_q to i.

The vector from s_q to j is $Ux_{ij} + c_{qi}$. The length of this vector should equal the distance between s_q and j.

$$\| Ux_{ij} + c_{qi} \| = d_{s_q j}, j = 1, 2, ..., k. \tag{4}$$

Expanding the left hand side of Eq. (4) leads to

$$(Ux_{ij} + c)^T (Ux_{ij} + c) = d_{s_q j}^2.$$

$$c_{qi}^T Ux_{ij} + c_{qi}^T c_{qi} + x_{ij}^T U^T Ux_{ij} + x_{ij}^T U^T c_{qi} = d_{s_q j}^2.$$

The matrix U to be determined is orthogonal and $c_{qi}^T Ux_{ij} = x_{ij}^T U^T c_{qi}$, and we have

$$c_{qi}^T Ux_{ij} = \frac{d_{s_q j}^2 - c_{qi}^T c_{qi} - x_{ij}^T x_{ij}}{2} = L_{qj}(1 \leq q \leq l, 1 \leq j \leq k). \tag{5}$$

Let C be $(c_{1i}, c_{2i}, ..., c_{li})$, X_i be $(x_{i1}, x_{i2}, ..., x_{ik})$, and L be an $l \times k$ matrix obtained from the right hand side of Eq. (5). We have

$$C^T UX_i = L. \tag{6}$$

Patches on the manifold are not strictly d-dimensional hyperplanes and there is actually a small angle between adjacent patches, thus Eq. (6) has no strict solution. However, with the pseudo-inverse, we can obtain a very accurate solution, as shown in Eq. (7).

$$U = (C^T)^\dagger L(X_i)^\dagger. \tag{7}$$

The U calculated by Eq. (7) satisfies the distance constraint between patches very well, but it is not strictly orthogonal and thus it may distort the inner structure of patches. To protect the inner structure of the patches, we adjust U to the nearest orthogonal matrix U' by Eq. (8) [19]. As U better preserves the distances between patches while U' better preserves the inner structure of the patches, we need to maintain a trade-off between U and U'. When U is far from an orthogonal matrix, that is, when $\|U^T U - I\|_F > 0.05$, adjust U to U' and use U' for Eq. (3), otherwise, use U for Eq. (3).

$$U' = U(U^T U)^{-\frac{1}{2}}. \tag{8}$$

2.4 Algorithm Description

LMB first determines the dimension d of the embedded manifold, and yields the local dimension for each data point $Ld_{1 \times n}$. $Ld_{1 \times n}$ is used to select the starting point for the BFS, which will be discussed in Sect. 4.1. The main body of LMB implements the dimension reduction based on d and the pairwise distances of the data points to generate a compact d-dimensional representation for the original data.

Algorithm 1. Dimension reduction by LMB

Input:
  Dimension d of the embedded manifold, and pairwise distances $D_{n \times n}$ of the data points;
Output:
  A compact d-dimensional representation $X_{d \times n}$ with the starting point v_1 at the origin;
1: Let $D_c = D$. Find the k-nearest neighbors of the starting point v_1 according to D_c. Extract the pairwise distances of these $k+1$ points from D, and locally apply MDS to obtain the coordinates. Subtract all these coordinates by the coordinates of v_1 to fix v_1 at the origin.
2: For each reduced but unexplored point v_i, find its k-nearest neighbors and apply MDS to obtain the coordinates of its neighbors. Subtract the coordinates of the neighbors by the coordinates of v_i to fix v_i at the origin.
3: Select l reduced point close to v_i and determine the orthogonal transformation U.
4: If $\|U^T U - I\|_F > 0.05$, adjust U by Eq. (8).
5: Calculate the d-dimensional representation of the neighbouring points of v_i by Eq. (3).
6: Update the distances between reduced points and the neighbouring points of v_i in D_c by the distances in d-dimensional space.
7: Recursively do steps 2 to 6 until the BFS process ends.
8: **return** X

The pairwise distances between the reduced points are updated in D_c and it takes $O(n^2)$ time. We choose the neighbouring points according to D_c, not

D, as some manifolds (such as the cylinder) are closed. In the experiment, the cylinder is split by LMB along the side and then unrolled into a rectangle, as illustrated in Fig. 4(h). To achieve the result, we enforce the neighbourhoods on one side of the split contain no points on the other side. However, points on opposite sides of the split are close in high-dimensional space, and if we select the neighbouring points according to D, the neighbourhood on one side of the split will contain points on the other side. To avoid this problem, we update the distances between reduced points in D_c by the distances in d-dimensional space, and choose neighbouring points according to D_c. Though neighbourhoods across the split can not be preserved well, there are only a small portion of points across the split and such cases only happen when the manifold is closed. The local MDS searches for the k-nearest neighbouring points, which costs $O(logn)$ time. Local MDS is performed at most n times and thus LMB costs $O(n^2 + nlogn)$ time.

3   Experiments

We do experiments on three synthetic data set, a cone, a cylinder, a Swiss roll, and two real data set, a half earth data and a sculpture face data. Table 1 gives the dimension of the embedded manifold determined by LMB. Based on the Local Continuity Meta Criteria (LCMC) [20], Table 2 lists the top four algorithms for different data set.

Table 1. Dimension of the embedded manifold

R	Cone	Cylinder	Swiss roll	Half earth	Sculpture face
0.95	2	2	2	2	7
0.96	2	2	2	2	7
0.97	2	2	2	2	8
0.98	2	2	2	2	8
0.99	2	2	2	2	10

Table 2. The top 4 methods

Rank	Cone	Cylinder	Swiss roll	Half earth	Sculpture face
1	LMB	LMB	LMB	LTSA	LMB
2	Isomap	LTSA	Isomap	Isomap	LTSA
3	MDS	HLLE	HLLE	LMB	Isomap
4	LTSA	MDS	MDS	MDS	HLLE

In Fig. 3, MDS reduces the cone into a triangle as it just projects the cone onto the plane. LLE, HLLE, and LTSA also yield a triangle but with some distortion. LMB spreads the cone into a circular sector, which preserves the neighbourhood much better than other algorithms. For the cylinder, LMB unrolls it into a rectangle with length of 6π and width of 10. The size corresponds with the initial cylinder. MDS, HLLE, and LTSA also yield a rectangle but not exactly the rectangle unrolled from the initial cylinder. In Fig. 5, LMB and Isomap yield the same rectangle for the Swiss roll data, but when measured by the LCMC criteria, LMB performs better (Table 3(c)).

Figure 6(a) shows a half earth data containing about 7500 points, and some of the data points are cities distributed in America. In the result of MDS, Isomap, LTSA, and LMB, countries could be easily recognized and North America and South America are described well in a 2D plane.

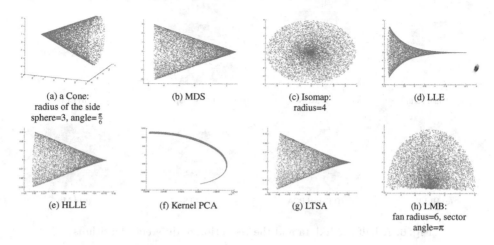

| (a) a Cone: radius of the side sphere=3, angle=$\frac{\pi}{6}$ | (b) MDS | (c) Isomap: radius=4 | (d) LLE |

| (e) HLLE | (f) Kernel PCA | (g) LTSA | (h) LMB: fan radius=6, sector angle=π |

Fig. 3. An 8000-point cone and the reduction of different algorithms

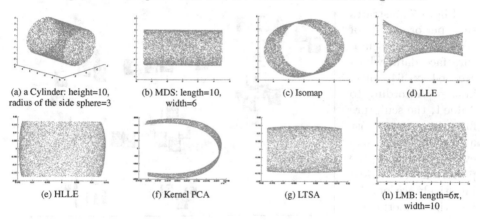

| (a) a Cylinder: height=10, radius of the side sphere=3 | (b) MDS: length=10, width=6 | (c) Isomap | (d) LLE |

| (e) HLLE | (f) Kernel PCA | (g) LTSA | (h) LMB: length=6π, width=10 |

Fig. 4. A 10000-point cylinder and the reduction of different algorithms

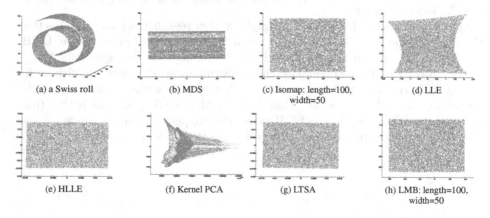

| (a) a Swiss roll | (b) MDS | (c) Isomap: length=100, width=50 | (d) LLE |

| (e) HLLE | (f) Kernel PCA | (g) LTSA | (h) LMB: length=100, width=50 |

Fig. 5. A 15000-point Swiss roll and the reduction of different algorithms

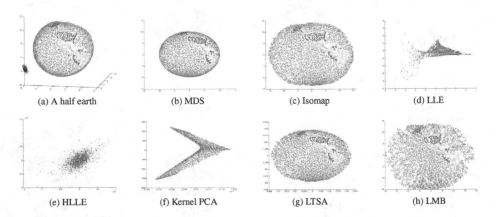

(a) A half earth   (b) MDS   (c) Isomap   (d) LLE

(e) HLLE   (f) Kernel PCA   (g) LTSA   (h) LMB

Fig. 6. A half earth data and the reduction of different algorithms

Figure 7 illustrates the performance of LMB on a sculpture face data, which are 64×64 pixel images. According to Table 1, the sculpture face data is at least a 7D manifold. However, to show the performance of LMB intuitively, we let $d = 2$ for the experiments and comparisons with other algorithms. In Fig. 7, faces sharing

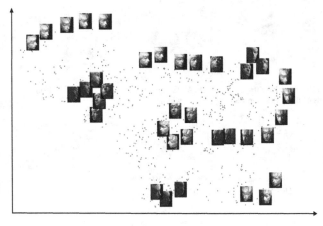

Fig. 7. The 2D representation of the sculpture face data

similar poses or with similar lighting angles gather together, which indicates that the neighbourhood of the high-dimensional face data is well preserved by LMB. Most of the images on the left, face right while images on the right, face left.

In Tables 3 and 4, we compare the performance of different algorithms by the LCMC criteria, which counts the number of points that remain inside the neighbourhood after reduction. Result shows that LMB performs better than most of the other algorithms. LMB ranks the toppest on cone, cylinder, Swiss roll and the sculpture face data, and top three for the half earth data.

Table 3. Comparison on the synthetic data and the sculpture face data

(a) Cone

k'	LMB	MDS	Isomap	LLE	HLLE	LTSA	kernel PCA
10	**0.977**	0.466	0.782	0.386	0.441	0.447	0.031
12	**0.978**	0.468	0.784	0.388	0.443	0.449	0.034
15	**0.977**	0.472	0.786	0.393	0.448	0.452	0.039
20	**0.977**	0.479	0.789	0.396	0.455	0.458	0.044
avg	**0.977**	0.471	0.785	0.391	0.447	0.452	0.037

(b) Cylinder

k'	LMB	MDS	Isomap	LLE	HLLE	LTSA	kernel PCA
10	**0.985**	0.470	0.378	0.451	0.481	0.481	0.018
12	**0.986**	0.470	0.379	0.453	0.482	0.483	0.020
15	**0.985**	0.471	0.382	0.455	0.483	0.484	0.022
20	**0.985**	0.474	0.385	0.457	0.485	0.484	0.024
avg	**0.985**	0.471	0.381	0.454	0.483	0.483	0.021

(c) Swiss roll

k'	LMB	MDS	Isomap	LLE	HLLE	LTSA	kernel PCA
10	**0.983**	0.401	0.956	0.776	0.776	0.779	0.019
12	**0.985**	0.401	0.959	0.778	0.779	0.783	0.020
15	**0.985**	0.403	0.963	0.779	0.782	0.784	0.021
20	**0.988**	0.406	0.966	0.781	0.785	0.786	0.023
avg	**0.985**	0.403	0.961	0.779	0.781	0.783	0.021

(d) Sculpture face data

k'	LMB	MDS	Isomap	LLE	HLLE	LTSA	kernel PCA
10	**0.453**	0.290	0.393	0.311	0.348	0.416	0.047
12	**0.461**	0.305	0.413	0.321	0.368	0.433	0.054
15	**0.462**	0.375	0.433	0.335	0.394	0.455	0.063
20	0.464	0.359	0.456	0.350	0.430	**0.483**	0.074
avg	**0.460**	0.332	0.424	0.329	0.385	0.447	0.060

Table 4. Comparison on the half earth data. The left table takes all the data points into account, while the right table only considers the city points.

(a) On all data points

k'	LMB	MDS	Isomap	LLE	HLLE	LTSA	kernel PCA
10	0.677	0.648	0.831	0.605	0.003	**0.836**	0.036
12	0.687	0.648	0.830	0.605	0.003	**0.836**	0.038
15	0.702	0.651	0.833	0.605	0.004	**0.837**	0.039
20	0.718	0.655	0.835	0.606	0.005	**0.838**	0.043
avg	0.694	0.651	0.832	0.605	0.004	**0.837**	0.039

(b) On city points

k'	LMB	MDS	Isomap	LLE	HLLE	LTSA	kernel PCA
10	0.868	0.863	**0.928**	0.677	0.010	0.912	0.065
12	0.877	0.867	**0.928**	0.680	0.012	0.913	0.070
15	0.885	0.879	**0.937**	0.682	0.014	0.919	0.075
20	0.893	0.889	**0.941**	0.692	0.018	0.924	0.085
avg	0.881	0.875	**0.934**	0.683	0.014	0.917	0.074

4 Parameter Study of LMB

4.1 The Starting Point

According to Table 1, the cone, cylinder, and Swiss roll are 2-dimensional manifolds embedded in a 3-dimensional space, and the neighbourhoods on these manifolds are locally 2-dimensional. However, the neighbourhoods on the cone are not consistently 2-dimensional. For the points near the apex, the local dimension is three. We call those points **singular points** as the local dimension does not conform with the dimension of the embedded manifold. Figure 8 shows two different results of LMB on the cone data with different starting points (in red bold).

The cone should be spread as a circular sector with angle of π, but the second result is different from what we would expect. The reason lies in the selection of the starting point. Although most of the points on the cone are locally two-dimensional, there are some singular points near the apex of the cone which are locally three-dimensional. The neighbourhood near the apex could not be preserved well in the plane by local MDS. If LMB starts from a point near the apex, the neighbourhood of the starting point would contain those singular points. We expect all the singular points to be distributed in a circular sector of angle π. However, local MDS would spread the singular points in a sphere and generates the unexpected result.

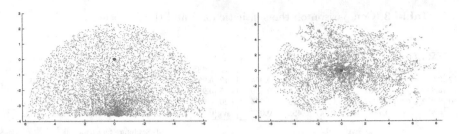

Fig. 8. The LMB reduction result for different starting points (Color figure online)

In order to avoid this problem, the starting point should be far from the singular points, which could be achieved by analysing the local dimension of the data points $Ld_{1 \times n}$. If a very large neighbourhood (such as $\frac{n}{3}$) of a point i has no singular points, i could be chosen as the starting point. When BFS starts from any point far from the apex, LMB yields the expected circular sector.

4.2 The Size of the Neighbourhood

The neighbourhood size k should not be too large so that the neighbourhood is not far from a local hyperplane, and MDS could reduce the neighbourhood accurately. We do experiments on the synthetic and real-world data with different k, and we use the 10-nearest neighbourhood to obtain the LCMC value. We also use the 12, 15, and 20-nearest neighbourhoods to obtain the LCMC value, and the results are almost the same.

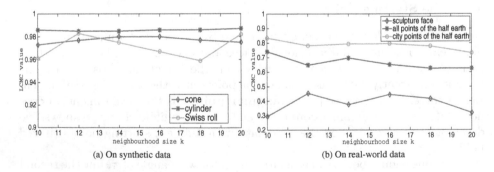

(a) On synthetic data   (b) On real-world data

Fig. 9. The LCMC value against different neighbourhood size

According to Fig. 9(a), the size of the neighbourhood has little influence on the performance of LMB for the synthetic data. The real data, however, is a little more influenced by the size of the neighbourhood. It is probably because the local structure of the real data deviates from a hyperplane to some degree or the data points are not dense enough. We let $k = 12$ for the final experiments in this paper.

5   Conclusions and Future Work

We propose a new nonlinear dimension reduction algorithm called Local Multi-dimensional Scaling with BFS (LMB) for generating a compact representation of a d-dimensional manifold embedded in a high-dimensional space. We locally employ the classical linear algorithm MDS on the k-nearest neighbourhoods and spread the local reduction process via BFS. Differs to most dimension reduction methods, LMB determines the dimension of the embedded manifold such that the reduction process bases only on the Euclidean pairwise distances.

LMB reveals three advantages when comparing with other state-of-the-art dimension reduction algorithms. First, LMB better preserves the neighbour-hood as measured by the LCMC criteria. Second, LMB also uncovers the global structure of the manifold as Isomap does. Besides, LMB does not involve the decomposition of large matrix and it has a lower computational complexity of $O(n^2)$ for n data points scattered on the manifold. LMB could be used for the pre-processing of the image data or other real-world data and it well serves the information retrieval since it well preserves the neighbourhood. Besides, LMB could also be used for uncovering the embedded manifold or for reconstructing the data points in its original space.

We make a thorough parameter study on LMB. We discuss the selection of the starting point for the spreading process, and a point with a comparatively large smooth neighbourhood is selected to avoid some singular points on the manifold. We then test the impact of different neighbourhood size on the performance of LMB. Experiments show that the neighbourhood size has little influence on synthetic data and a little more influence on real data. We address the following issues for future research.

- The neighbourhood size affects a little on the performance of LMB on real data. It is probably because the neighbourhoods of real data deviates from a hyperplane or the points are not dense enough. We will investigate further on the causes and develop possible solutions, such as adjusting the local structure of the real data.
- The closed manifold is split by LMB and the neighbourhoods across the split can not be preserved well. A possible solution is to duplicate points along the split. Take the cylinder as an example, we may copy the points on the left edge to the right of the right edge such that the neighbourhoods of the points on the right edge can also be preserved.
- We do not need to split the unclosed manifolds and the intermediate parameter D_c can be omitted. We will see if we could determine whether a manifold is closed in $O(nlogn)$ time, in this way the time complexity of LMB will be reduced to $O(nlogn)$.

Acknowledgments. This work is supported by National Natural Science Foundation (61472147) and National Science Foundation of Hubei Province (2015CFB566).

Appendix: A Brief Introduction of the MDS Algorithm

Let n be the size of the data and D be the matrix of Euclidean pairwise distances. MDS generates the coordinates X for all the data points with their center at the origin. Each column of X represents a data point. MDS first calculates the matrix $X^T X$ by Eq. (9). The diagonal entries of H are $1 - \frac{1}{n}$ and the rests are $-\frac{1}{n}$.

$$X^T X = -\frac{1}{2} H D^2 H. \tag{9}$$

$X^T X$ is positive semi-definite, and can decomposed as Eq. (10), which leads to Eq. (11).

$$X^T X = V \sigma^2 V^T = V \sigma^T \sigma V^T = (\sigma V^T)^T \sigma V^T. \tag{10}$$

$$X = \sigma V^T. \tag{11}$$

If the data points are distributed in an m-dimensional space, the first m diagonal entries of σ are non-zero. Extracting the first d entries of σ and the corresponding eigenvectors generates an approximation of the original data in a lower d-dimensional space. The approximation is very accurate if the first d entries are the most significant ones and the rests are close to zero. If MDS is applied on the k-nearest neighbourhoods on a d-dimensional manifold, σ is a $k+1$ by $k+1$ matrix and the first d entries are significant while the rests are almost zero. However, If MDS is applied on all the data points of a manifold, σ is a n by n matrix, and due to the global geometry of the manifold, the first m ($m > d$) entries of $\sigma_{n \times n}$ are significant. Therefore, only the first d dimensions are not enough to well represent the data points, and MDS on the whole nonlinear manifold may result in an inaccurate approximation in the d-dimensional space.

References

1. Sorzano, C.O.S., Vargas, J., Montano, A.P.: A survey of dimensionality reduction techniques (2014). arXiv preprint arXiv: 1403.2877
2. Sarveniazi, A.: An actual survey of dimensionality reduction. Am. J. Comput. Math. 4(5), 55–72 (2014)
3. van der Maaten, L.J., Postma, E.O., van den Herik, H.J.: Dimensionality reduction: a comparative review. J. Mach. Learn. Res. 10(1–41), 66–71 (2009)
4. Jiang, X., Gao, J., Hong, X., Cai, Z.: Gaussian processes autoencoder for dimensionality reduction. In: Tseng, V.S., Ho, T.B., Zhou, Z.-H., Chen, A.L.P., Kao, H.-Y. (eds.) PAKDD 2014, Part II. LNCS, vol. 8444, pp. 62–73. Springer, Heidelberg (2014)
5. Roweis, S.T., Saul, L.K.: Nonlinear dimensionality reduction by locally linear embedding. Science 290(5500), 2323–2326 (2000)
6. Tenenbaum, J.B., De Silva, V., Langford, J.C.: A global geometric framework for nonlinear dimensionality reduction. Science 290(5500), 2319–2323 (2000)
7. Abdi, H., Williams, L.J.: Principal component analysis. Wiley Interdisc. Rev. Comput. Stat. 2(4), 433–459 (2010)
8. Borg, I., Groenen, P.J.: Modern Multidimensional Scaling: Theory and Applications. Springer Science & Business Media, New York (2005)

9. Brown, T.A.: Confirmatory Factor Analysis for Applied Research. Guilford Publications, New York (2015)
10. Li, M., Yuan, B.: 2d-lda: a statistical linear discriminant analysis for image matrix. Pattern Recogn. Lett. **26**(5), 527–532 (2005)
11. Hardoon, D.R., Shawe-Taylor, J.: Convergence analysis of kernel canonical correlation analysis: theory and practice. Mach. Learn. **74**(1), 23–38 (2009)
12. Hyvärinen, A., Oja, E.: Independent component analysis: algorithms and applications. Neural Netw. **13**(4), 411–430 (2000)
13. Schölkopf, B., Smola, A., Müller, K.-R.: Kernel principal component analysis. In: Gerstner, W., Hasler, M., Germond, A., Nicoud, J.-D. (eds.) ICANN 1997. LNCS, vol. 1327. Springer, Heidelberg (1997)
14. Bernstein, M., De Silva, V., Langford, J.C., Tenenbaum, J.B.: Graph approximations to geodesics on embedded manifolds, Technical report, Department of Psychology, Stanford University (2000)
15. Saul, L., Roweis, S.: "Think globally, fit locally: unsupervised learning of nonlinear manifolds," Technical report MS CIS-02-18 (2002)
16. Donoho, D.L., Grimes, C.: Hessian eigenmaps: locally linear embedding techniques for high-dimensional data. Proc. Natl. Acad. Sci. **100**(10), 5591–5596 (2003)
17. Zhang, Z., Zha, H.: Nonlinear dimension reduction via local tangent space alignment. In: Liu, J., Cheung, Y., Yin, H. (eds.) IDEAL 2003. LNCS, vol. 2690. Springer, Heidelberg (2003)
18. Hastie, T.: Principal curves and surfaces, Technical report, DTIC Document (1984)
19. Horn, B.K., Hilden, H.M., Negahdaripour, S.: Closed-form solution of absolute orientation using orthonormal matrices. JOSAA **5**(7), 1127–1135 (1988)
20. Lee, J.A., Verleysen, M. et al.: Quality assessment of nonlinear dimensionality reduction based on k-ary neighborhoods. In: FSDM, pp. 21–35 (2008)

Mechanism Design for One-Facility Location Game with Obnoxious Effects

Lili Mei, Deshi Ye$^{(\boxtimes)}$, and Guochuan Zhang

College of Computer Science, Zhejiang University, Hangzhou, China
{meilili,yedeshi,zgc}@zju.edu.cn

Abstract. In classic obnoxious facility games [5,6,12], each agent i has a *private* location x_i on a closed interval $[0, 1]$ and one facility y is planned to build on the interval according to the bids of all the agents. In this paper we consider obnoxious effects among the game by introducing two thresholds d_1 and d_2 into utility functions, where $0 \leq d_1 \leq d_2 \leq 1$. Let $d(y, x_i) = |y - x_i|$ be the distance between agent i and facility y. The utility function of agent i is 0 if $d(y, x_i)$ is at most d_1; 1 if $d(y, x_i)$ is at least d_2; otherwise a linear increasing function between 0 and 1. Each agent aims to get a largest possible utility while the social welfare is to maximize the sum of all the agents' utilities.

The classic obnoxious facility game is a special case of our problem when $d_1 = 0$ and $d_2 = 1$. We show that if $d_1 = d_2$, a mechanism that outputs the leftmost optimal facility location is strategy-proof. If $d_1 \geq \frac{1}{2}$, we show the problem cannot have any bounded deterministic strategy-proof mechanism. By further detailed analysis, if the thresholds d_1, d_2 are restricted to some ranges, we design strategy-proof mechanisms and provide the approximation ratios with respect to d_1 and d_2.

1 Introduction

Approximate mechanism design without money for facility location problem was first advocated by Procaccia and Tennenholtz [15]. A set of strategic agents have different locations, and a mechanism is a function that outputs the facility location based on the locations reported by the agents. Each agent prefers to be as close to the facility as possible. Recently mechanism design on obnoxious facility game was proposed by Cheng et al. [5], where each agent prefers to stay far away from the facility. In the obnoxious facility game, agent i's utility is simply the distance to the facility. The social welfare is to maximize sum of all the agents' utilities.

Anyway, in previous studies, the facilities may be classified to be either purely desirable where they should be close to the users, or purely undesirable which means they should be as far away as possible. This classification is not always

Research was partially supported by the Nature Science Foundation of Zhejiang Province (NO. LQ15A010001).

D. Zhu and S. Bereg (Eds.): FAW 2016, LNCS 9711, pp. 172–182, 2016.
DOI: 10.1007/978-3-319-39817-4_17

true in some practical scenario, as being pointed out by Brimberg and Juel [3] that some type of facilities have undesirable effects in the real life. For example, a government plans to build a garbage dump, a chemical plant, or a nuclear reactor on a street. When the facility is close to a resident within a fixed range, it is totally unacceptable for her. Similarly, if it is already far away enough, small increases in the distance cannot change the obnoxious effects.

In this work, taking the obnoxious effects into account, we consider mechanism design on obnoxious facility game with two thresholds d_1 and d_2 which are two constants in $[0, 1]$ and $d_1 \le d_2$. And agent i's utility function $u_i(y, x_i)$ is defined as follows. To state conveniently, we let $d(y, x_i)$ denote the distance between agent i and facility y. The function $u_i(y, x_i)$ is 0 if $d(y, x_i)$ is at most d_1; 1 if $d(y, x_i)$ is at least d_2; otherwise a linear increasing function between 0 and 1. The social welfare is to maximize the sum of all the agents' utilities. We wish to design mechanisms that are strategy-proof, i.e., preventing any agent from lying to benefit. At the same time, the proposed mechanisms are expected to have small approximation ratios with respect to the optimal social welfare.

1.1   Our Results

We show that the optimization version of this problem can be solved in polynomial time. It is easy to verify that the leftmost optimal facility location is strategy-proof when $d_1 = d_2$. Our main work is to deal with the approximate mechanisms for different kinds of two thresholds d_1 and d_2, where $0 \le d_1 < d_2 \le 1$. First of all, we show that any deterministic strategy-proof mechanisms are unbounded when $d_1 \ge 1/2$. That motivated us to design 2-approximated randomized strategy-proof mechanisms. Moreover, we show that no randomized mechanism can achieve an approximation ratio less than $4/3$ for this case.

To address the case when $d_1 < 1/2$, we propose a majority mechanism that motivated by Cheng et al. [5] and show its approximation ratio with respect to d_1 and d_2. Furthermore, we show that the mechanism is the best possible for some d_1 and d_2. We also provide a family of improved deterministic mechanisms for $d_1 < d_2 \le 1/2$ with $(1 + \frac{1}{k})$-approximation ratios, where k is an integer such that $\frac{1}{2(k+1)} < d_2 \le \frac{1}{2k}$.

1.2   Related Works

Mechanism design for facility location games has a considerable amount of work in the literature. Procaccia and Tennenholtz [15] first studied mechanism design for facility location games. In the setting, the utility of each agent is the distance from the facility to agent's location and each agent attempts to minimize the utility. They considered two objective functions, minimizing the sum of all the agents utilities (minSum) and the maximum utility (minMax). They gave some lower bounds and upper bounds for 1-facility and 2-facility. They also considered an extended model – multiple locations per agent. Subsequently, Alon et al. [1, 2] extended the randomized and deterministic mechanisms for 1-facility on other networks. Lu *et al.* [13,14] successively improved some results for 2-facility on general metric networks under the minSum objective function. Feldman and Wilf [9] then considered the deterministic and randomized mechanisms for the objective function of minimizing the sum of squares of all the agents utilities (minSOS). The number of facilities which is more than three was considered by Fotakis *et al.* [11] and Escoffier *et al.* [7]. Recently, Filos-Ratsikas *et al.* [10] considered a facility game that each agent prefers two locations which they named double-peaked preferences.

Mechanism design for the obnoxious facility games was first studied by Cheng *et al.* [5]. They considered deterministic and randomized strategy-proof mechanisms for maxSum objective function on the line. Cheng et al. extended the results on the other networks in [6]. Ibara and Nagamochi [12] gave the characterization of strategy-proof mechanisms for the obnoxious facility game for the maxSum objective function. Subsequently, Cheng *et al.* [4] investigated the obnoxious facilities with a bounded service range. Recently, Ye *et al.* [17] gave some results for the objectives of maxSum and maxSOS. They also considered the extended model of multiple locations per agent for maxSum and maxSOS objectives.

Zhang and Li [18] studied the weighted version for both facility games and obnoxious facility games. Moreover, they considered the facility location game with one threshold with respect to the utility function, an optimal deterministic mechanism was provided. Zou and Li [19] considered the problem where two preferences of agents, staying close to and staying away from the facility, exist. They called this problem the facility location game with dual preference. They gave a deterministic mechanism. They also considered the two-opposite-facility location game with limited distance. Here, two-opposite-facility location means one facility is that each agent want to stay close to; the other one is opposite. Feigenbaum and Sethuraman [8] considered randomized mechanisms for the facility location game with dual preference.

On the optimization aspect, our work is related to approximation algorithm for semi-desirable k-facility location problem [3,16].

2   Preliminaries

Let $N = \{1, 2, \ldots, n\}$ be a set of agents. All agents are located on a closed interval $I = [0, 1]$. Each agent i has a location $x_i \in I$. We refer to the set $\mathbf{x} = (x_1, x_2, \ldots, x_n) \in I^n$ as the location profile. Let d_1 and d_2 be the given two thresholds, which are constants in $[0, 1]$ and $0 \leq d_1 \leq d_2 \leq 1$.

A *deterministic mechanism* is a function $f : I^n \to I$ that maps a given location profile $\mathbf{x}$ to a facility location $f(\mathbf{x})$. If the facility is located at y, the distance between x_i and y is denoted by $d(y, x_i) = |y - x_i|$. We define agent i's utility function $u_i(y, x_i)$ as below,

$$u_i(y, x_i) = \begin{cases} 0 & \text{if } d(y, x_i) \leq d_1, \\ \frac{d(y,x_i)-d_1}{d_2-d_1} & \text{if } d_1 < d(y, x_i) \leq d_2, \\ 1 & \text{otherwise.} \end{cases}$$

A *randomized mechanism* is a function $f : I^n \to \Delta I$ that maps a given location profile to probability distributions over I. If $f(\mathbf{x}) = P$, where P is a probability distribution, the utility of agent i is the expected utility, *i.e.*, $u_i(P, x_i) = E_{y \sim P}[u_i(y, x_i)]$.

A mechanism f is *strategy-proof* (SP) in this setting if no agent can benefit from misreporting a location deliberately, regardless of the strategies of the other agents, *i.e.*, for all $\mathbf{x} \in I^n$, for all $i \in N$, and for all $x_i' \in I$, we have $u_i(f(\mathbf{x}), x_i) \geq u_i(f(x_i', \mathbf{x}_{-i}), x_i)$, where $\mathbf{x}_{-i} = (x_1, \ldots, x_{i-1}, x_{i+1}, \ldots, x_n)$ is the location profile without agent i.

Given a location profile $\mathbf{x} \in I^n$, the *social welfare* of a deterministic mechanism f is denoted by $sw(f(\mathbf{x}), \mathbf{x})$. Given a profile $\mathbf{x}$ and the facility location y, the social welfare is given by $sw(y, \mathbf{x}) = \sum_{i \in N} u_i(y, x_i)$. Moreover, the *social welfare* of a distribution P is $sw(P, \mathbf{x}) = E_{y \sim P}[sw(y, \mathbf{x})]$.

To evaluate a mechanism's performance, we use the standard worst-case approximation notion. We say that a mechanism f is ρ-approximate, if for any profile $\mathbf{x}$,

$$sw(OPT(\mathbf{x}), \mathbf{x}) \leq \rho \cdot sw(f(\mathbf{x}), \mathbf{x}),$$

where $OPT(\mathbf{x})$ is an optimal facility placement for profile $\mathbf{x}$, *i.e.*, $OPT(\mathbf{x}) \in argmax_{y \in I} sw(y, \mathbf{x}))$.

3   Hardness of Two Thresholds

Before jumping into the detailed study on strategy-proof mechanisms, we first consider an optimal location for any location profile. The following lemma shows that the optimal location can be found on the boundaries of agents that is related to the threshold d_2.

Lemma 1. *Given a location profile $\mathbf{x}$, there exists an optimal location y^* such that $y^* \in \mathcal{O} = \{0, 1, x_i \pm d_2 \in I | i \in N\}$ for maximizing the social welfare. (Due to the space limitation, the proof can be found in the full version of the paper.)*

By Lemma 1, we can establish an optimal algorithm.

Algorithm 1. *Given a location profile $\mathbf{x}$, the facility location is $argmax_{y \in \mathcal{O}} sw(y, \mathbf{x})$. If there is a tie, choose the leftmost y.*

Remark. Using the similar analysis as Theorem 12 in [18], we can get that Algorithm 1 is a strategy-proof mechanism when $d_1 = d_2$.

We presently turn to consider the utility function with two thresholds. We first show that for $d_1 \geq \frac{1}{2}$, no deterministic strategy-proof mechanism has a bounded approximation ratio.

3.1   Lower Bounds for $d_1 \geq \frac{1}{2}$

Theorem 1. *Let $N = \{1, 2, \cdots, n\}$, where $n \geq 2$. If $\frac{1}{2} \leq d_1 < d_2 \leq 1$, any deterministic strategy-proof mechanisms do not have a bounded approximation ratio.*

Proof. We first deal with the case where $N = \{1, 2\}$, and then extend the proof to an arbitrary n agents.

Consider the location profile $\mathbf{x}$ where $x_1 = 1 - d_2$ and $x_2 = d_2$. Since $\frac{1}{2} < d_2 \le 1$, we can get that $0 \le d(0, x_1) = x_1 = 1 - d_2 < \frac{1}{2} < d_1$. Hence, the first agent at x_1 can only get positive utility in $(1 - d_2 + d_1, 1]$. By the symmetry of agent 1 and 2, the second agent at x_2 can only get positive utility in $[0, d_2 - d_1)$. Since $d_1 \ge \frac{1}{2}$ and $d_2 \le 1$, we can get that $d_2 - d_1 \le \frac{1}{2}$, which implies that $d_2 - d_1 \le 1 - d_2 + d_1$. The utility functions of two agents are illustrated by Fig. 1.

From Fig. 1, we can easily see that the optimal facility is at 0 or 1 and the social welfare is 1. Let f be a strategy-proof mechanism, and let $f(\mathbf{x}) = f_y \in [0, 1]$. If $f_y \in [d_2 - d_1, 1 - d_2 + d_1]$, $sw(f_y, \mathbf{x}) = 0$ and the approximation ratio is already unbounded. Hence, we can assume that $f_y \in I \backslash [d_2 - d_1, 1 - d_2 + d_1]$. Since the symmetry of two agents, without loss of generality, we can assume that $f_y \in (1 - d_2 + d_1, 1]$.

Consider another location profile $\mathbf{x}' = (x_1' = f_y - d_1 - \epsilon, x_2 = d_2)$, where $0 < \epsilon \ll d(f_y, 1 - d_2 + d_1)$. Due to $f_y \le 1$ and $0 < \epsilon \ll d(f_y, 1 - d_2 + d_1)$, we can have that $1 - d_2 < x_1' \le 1 - d_1$. Moreover, By $d_1 \ge \frac{1}{2}$, then we can get $x_1 = 1 - d_2 < x_1' < d_1$, i.e., x_1' is in the right of x_1 and only has positive utility in $(f_y - \epsilon, 1]$. See Fig. 1 for an illustration of the utility functions of x_1' in the whole interval. Let y^* be the optimal facility location of the profile $\mathbf{x}'$. From Fig. 1, we can easily see that $y^* = 0$ and $sw(y^*, \mathbf{x}') = 1$. We consider the facility location of mechanism f. Let $f(\mathbf{x}') = f_{y'}$. We claim that $f_{y'} = f_y$. By strategyproofness, $f_{y'}$ cannot be in $[0, f_y)$, otherwise agent 1 in $\mathbf{x}'$ can misreport to x_1 and benefit; Similarly, $f_{y'}$ cannot be in $(f_y, 1]$, otherwise agent 1 in $\mathbf{x}$ can misreport to x_1' and benefit. Hence, $f_{y'} = f_y$. From Fig. 1, we can see that $sw(f_y, \mathbf{x}') = u_1(f_y, x_i') = \frac{\epsilon}{d_2 - d_1}$. It follows that the approximation ratio is at least $\frac{d_2 - d_1}{\epsilon}$, which is unbounded when ϵ trends to 0.

In order to extend this result to n agents, we locate all the agents $N \backslash \{1, 2\}$ at $\frac{1}{2}$ in each one of the profiles described above. $\qquad\square$

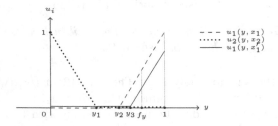

Fig. 1. The utility functions of x_1, x_2 and x_1'. Here y denotes the facility location. $y_1 = d_2 - d_1$, $y_2 = 1 - d_2 + d_1$ and $y_3 = x_1' + d_1$. To display clearly, we separate the lines on the horizontal axis (The following figures are the same.).

4   Approximation Strategy-Proof Mechanisms

In this section, we first show that the mechanism given by Cheng [5] that was designed for $d_1 = 0, d_2 = 1$ is also strategy-proof in our setting. We call that mechanism *Majority Mechanism*. We show the approximation ratio of Majority Mechanism in our setting.

4.1   Majority Mechanism

Mechanism 1. *Given a location profile* $\mathbf{x} \in I^n$, *let* n_l *denote the number of agents in* $[0, 1/2]$ *and* n_r *be the number of the rest agents. The mechanism selects the position 0 as the facility if* $n_l \leq n_r$ *and otherwise the position 1 will be the facility.*

In Majority Mechanism, only two points 1 and 0 are regarded as the candidate facilities. The intuition why we do not use the other optimal locations $x_i \pm d_2$ as candidates is that all other positions depend on the bids of the agents, which might be manipulated by agents.

Theorem 2. *Majority Mechanism is strategy-proof. If* $d_1 < \frac{1}{2}$ *and* $d_2 > \frac{1}{2}$, *the approximation ratio is* $1 + \frac{2(d_2 - d_1)}{1 - 2d_1}$; *if* $d_1 < d_2 \leq 1/2$, *the approximation ratio is* 2.

Since the utility function is nondecreasing with respect to the distance, it is easy to show the strategyproofness. Due to the space limitation, we only prove the approximation ratio for the case $d_1 < \frac{1}{2}$ and $d_2 > \frac{1}{2}$.

Proof. Without loss of generality, we assume Majority Mechanism locates the facility at 0 which indicates that $n_l \leq n_r$. It is obvious that if the optimal facility is also located at 0. Note that the social welfare of Majority Mechanism is at least $\frac{\frac{1}{2} + \epsilon - d_1}{d_2 - d_1} n_r$, which can be gotten with the profile that all the agents in $[0, 1/2]$ are at 0 and the rest agents are at $\frac{1}{2} + \epsilon$, where $0 < \epsilon \ll d_2 - \frac{1}{2}$ is a small number. In the rest of the proof, to avoid many notations, we abuse ϵ to denote a small number between 0 and $d_2 - \frac{1}{2}$. Let y^* denote the optimal facility location. To state conveniently, we let $sw(y, \mathbf{x}_L) = \sum_{x_i \in [0,1/2]} u_i(y, x_i)$ and $sw(y, \mathbf{x}_R) = \sum_{x_i \in (1/2,1]} u_i(y, x_i)$. According to Lemma 1, case studies on the optimal facility locations are given as below.

Case 1: $y^* = 1$. In this case, one can check that the worst case instance is that all the agents in $[0, 1/2]$ are at 0 and all the agents in $(1/2, 1]$ are at $1/2 + \epsilon$. And we denote the worst case instance as I_w. Hence, the approximation ratio is

$$\frac{sw(1, I_w)}{sw(0, I_w)} = \frac{n_l + \frac{1/2 - \epsilon - d_1}{d_2 - d_1} n_r}{\frac{1/2 + \epsilon - d_1}{d_2 - d_1} n_r} < 1 + \frac{2(d_2 - d_1)}{1 - 2d_1} \frac{n_l}{n_r} \leq 1 + \frac{2(d_2 - d_1)}{1 - 2d_1}. \quad (1)$$

Case 2: The optimal facility is at $x_l + d_2$, where x_l is agent l's location. By $d_2 > 1/2$, we can easily get that the optimal facility is in $(1/2, 1]$. Then we consider the worst case instance $\mathbf{x}$. We first claim that all the agents in $[0, 1/2]$ are at 0. Otherwise, we move agents in $(0, 1/2]$ to 0 and the approximation ratio cannot be better after the movements. Similarly, we move all the agents in $(1/2, x_l + d_2]$ to $\frac{1}{2} + \epsilon$. For the agents in $(x_l + d_2, 1]$, they all get the utility 1 at 0 since $d(0, x_l + d_2) \geq d_2$. It implies that move all these agents to 1 do not increase the utilities. Analogously, we construct the worst case instance for all the agents in $(1/2, 1]$, and the approximation ratio does not decrease. We still use $\mathbf{x}$ to denote the profile after modifications. Then,

$$sw(y^*, \mathbf{x}_R) \leq \frac{\frac{1}{2} - \epsilon - d_1}{d_2 - d_1} n_r. \tag{2}$$

The Inequality holds since for any agent i in $(1/2, 1]$, $d(y^*, x_i) \leq \frac{1}{2} - \epsilon$. The approximation ratio for the profile $\mathbf{x}$ is at most

$$\frac{sw(y^*, \mathbf{x})}{sw(0, \mathbf{x})} \leq \frac{n_l + sw(y^*, \mathbf{x}_R)}{\frac{\frac{1}{2} + \epsilon - d_1}{d_2 - d_1} n_r} \leq 1 + \frac{2(d_2 - d_1)}{1 - 2d_1}. \tag{3}$$

Case 3: The optimal facility is at $x_l - d_2$ for some $l \in N$. The optimal facility location is in $[0, 1/2)$. For agents in $(1/2, 1]$, The social welfare $sw(y^*, \mathbf{x}_R) \leq n_r$, since the maximum value of the utility of each agent in this interval is 1. Meanwhile, we get that $sw(y^*, \mathbf{x}_L) \leq \frac{\frac{1}{2} - d_1}{d_2 - d_1} n_l$ since $d(y^*, x_i) \leq \frac{1}{2}$. The approximation ratio is at most

$$\frac{sw(y^*, \mathbf{x})}{sw(0, \mathbf{x})} \leq \frac{n_r + \frac{\frac{1}{2} - d_1}{d_2 - d_1} n_l}{\frac{\frac{1}{2} + \epsilon - d_1}{d_2 - d_1} n_r} < 1 + \frac{2(d_2 - d_1)}{1 - 2d_1} \frac{n_l}{n_r} \leq 1 + \frac{2(d_2 - d_1)}{1 - 2d_1}.$$

The approximation ratio is tight. Given a location profile $\mathbf{x} = (x_1 = 0, x_2 = \frac{1}{2} + \epsilon)$, the optimal facility is at 1 and social welfare is $1 + \frac{\frac{1}{2} - \epsilon - d_1}{d_2 - d_1}$. Majority Mechanism locates the facility at 0 and the social welfare is $\frac{\frac{1}{2} + \epsilon - d_1}{d_2 - d_1}$, and then the approximation ratio is at least $1 + \frac{2(d_2 - d_1)}{1 - 2d_1}$ when $\epsilon \to 0$. □

4.2   Lower Bounds for $d_1 < \frac{1}{2}$ and $d_2 > \frac{1}{2}$

For this case, the approximation ratio of Majority Mechanism is $1 + \frac{2(d_2 - d_1)}{1 - 2d_1}$. In the following theorem, we show that the mechanism is best possible if $d_1 + d_2 \geq 1, d_2 - d_1 \leq \frac{1}{2}$ and $d_1 < \frac{1}{2}$. It is interesting to consider the lower bounds for the rest scenarios.

Theorem 3. *Let $N = \{1, 2, \cdots, n\}$, where $n \geq 2$. If $d_1 + d_2 \geq 1, d_2 - d_1 \leq \frac{1}{2}$ and $d_1 < \frac{1}{2}$, any deterministic strategy-proof mechanisms cannot have an approximation ratio less than $1 + \frac{2(d_2 - d_1)}{1 - 2d_1}$ for maximizing the social welfare. (Due to the space limitation, the proof can be found in the full version of the paper.)*

4.3   A Family of Strategy-Proof Mechanisms for $d_2 \leq 1/2$

For the case $d_2 \leq 1/2$, we present a family of mechanisms for $\frac{1}{2(k+1)} < d_2 \leq \frac{1}{2k}$, where k is a positive integer.

We try to divide the interval into several sub-intervals. In each sub-interval, we can find a facility location that all the agents in the rest sub-intervals have utility of 1. We establish the following mechanism.

To describe the mechanism, we define the following notations. Let k be an integer such that $\frac{1}{2(k+1)} < d_2 \leq \frac{1}{2k}$. We divide $I = [0,1]$ into $k+1$ sub-intervals. The length of the 1st and the last one are both $\frac{1}{2k}$. The length of the rest sub-intervals are all $\frac{1}{k}$. We denote the ith sub-interval by I_i, for all $i = 1, \cdots, k+1$. Formally, $I_1 = [0, \frac{1}{2k}]$; $I_i = (\frac{2i-3}{2k}, \frac{2i-1}{2k}]$, $i = 2, \cdots, k$; $I_{k+1} = (\frac{2k-1}{2k}, 1]$. Let n_i denote the number of agents in sub-interval I_i, for all $i = 1, \cdots, k+1$.

Mechanism 2. *Given any location profile* **x**, *compute* $l = \underset{i=1,\cdots,k+1}{argmin}\{n_i\}$. *If there is a tie, take the one with minimum index. Return* $\frac{l-1}{k}$.

Theorem 4. *When* $0 \leq d_1 < d_2 \leq \frac{1}{2}$, *Mechanism 2 is a strategy-proof* $(1 + \frac{1}{k})$-*approximated mechanism for the social welfare, where k is an integer such that* $\frac{1}{2(k+1)} < d_2 \leq \frac{1}{2k}$. *(Due to the space limitation, the proof can be found in the full version of the paper.)*

4.4   Randomized Mechanisms for $d_1 \geq 1/2$

In Sect. 3.1 we have shown no deterministic strategy-proof mechanism can achieve a bounded approximation ratio. One question arises whether randomized mechanisms can help to reduce the approximation ratio. In the case $d_1 \geq 1/2$, we found a nice property of optimal facility locations other than those results in Lemma 1.

Proposition 1. *Given a location profile* **x**, *if* $\frac{1}{2} \leq d_1 < d_2 \leq 1$, *there exists an optimal solution located at either* 0 *or* 1 *for maximizing the social welfare. (Due to the space limitation, the proof can be found in the full version of the paper.)*

Due to Proposition 1, we propose a randomized mechanism which outputs 0 and 1 with probability $\frac{1}{2}$, respectively. It is easy to show that the mechanism is strategy-proof and 2-approximate. Moreover, we show the following theorem.

Theorem 5. *Let* $N = \{1, 2, \cdots, n\}$, *where* $n \geq 2$. *If* $\frac{1}{2} \leq d_1 < d_2 \leq 1$, *any randomized strategy-proof mechanism which only take 0 and 1 be candidates cannot achieve an approximation ratio less than 2 for maximizing the social welfare. (Due to the space limitation, the proof can be found in the full version of the paper.)*

We now turn to consider the lower bounds of any randomized strategy-proof mechanisms.

Theorem 6. *Let $N = \{1, 2, \cdots, n\}$, where $n \geq 2$. If $\frac{1}{2} \leq d_1 < d_2 \leq 1$, any randomized strategy-proof mechanism cannot attain an approximation ratio less than $\frac{4}{3}$.*

Proof. With the same reason as in the proof of Theorem 1, we only need to deal with the case where $N = \{1, 2\}$.

Consider the location profile $\mathbf{x}$ where $x_1 = 1 - \frac{d_1 + d_2}{2}$ and $x_2 = \frac{d_1 + d_2}{2}$. When $\frac{1}{2} \leq d_1 < d_2 \leq 1$, we can have that $1 < d_1 + d_2 < 2$, which immediately gets that $0 < x_1 < 1/2 \leq d_1$. Hence x_1 only has positive utility in $(1 - \frac{d_2 - d_1}{2}, 1]$. Since agent 1 and 2 are symmetric, agent 2 can only have positive utility in $[0, \frac{d_2 - d_1}{2})$. Meanwhile, we have that $d_2 - d_1 < 1$, which implies $\frac{d_2 - d_1}{2} < 1 - \frac{d_2 - d_1}{2}$. See Fig. 2 for an illustration of the utility functions of x_1 and x_2. From Fig. 2, we can know that the optimal facility is at 0 or 1 and the social welfare is $frac12$.

Let f be a strategy-proof randomized mechanism and $f(\mathbf{x}) = P$. Note that $sw(P, \mathbf{x}) = u_1(P, x_1) + u_(P, x_2) \leq \frac{1}{2}$. Without loss of generality, we can assume that $u_2(P, x_2) \leq \frac{1}{4}$.

Consider another location profile $\mathbf{x}'$ where $x_1 = 1 - \frac{d_1 + d_2}{2}$ and $x_2' = d_2$. It is easy to see that x_2' can only have positive utility in $[0, d_2 - d_1)$. The utility functions of x_1 and x_2' are illustrated in Fig. 2. Let y^* denote the optimal facility location of $\mathbf{x}'$. From Fig. 2, we can see that $y^* = 0$ and $sw(y^*, \mathbf{x}') = 1$.

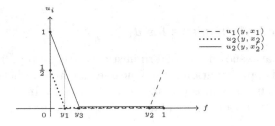

Fig. 2. The utility functions of x_1, x_2 and x_2'. Here y denotes the facility location;. $y_1 = \frac{d_2 - d_1}{2}$, $y_2 = 1 - \frac{d_2 - d_1}{2}$ and $y_3 = d_2 - d_1$.

Let $f(\mathbf{x}') = P'$ be the probability distribution of profile $\mathbf{x}'$. Formally, we let y denote the randomized variable of the facility location for the profile $\mathbf{x}'$ and let $p(y)$ be the probability density function. Now we turn to consider the social welfare of P' for profile $\mathbf{x}'$. We denote $p = \int_{\frac{d_2 - d_1}{2}}^{1} p(y) dy$. The social welfare for mechanism f of the profile $\mathbf{x}'$ is

$$sw(P', \mathbf{x}') = \int_0^{\frac{d_2 - d_1}{2}} u_2(y, x_2') p(y) dy + \int_{\frac{d_2 - d_1}{2}}^{1} sw(y, \mathbf{x}') p(y) dy$$

$$\leq \int_0^{\frac{d_2 - d_1}{2}} u_2(y, x_2') p(y) dy + p \max_{\frac{d_2 - d_1}{2} \leq y \leq 1} \{sw(y, \mathbf{x}')\}$$

$$= \int_0^{\frac{d_2-d_1}{2}} u_2(y,x_2')p(y)dy + \frac{p}{2}$$

The last equation holds since $\max_{\frac{d_2-d_1}{2} \leq y \leq 1}\{sw(y,\mathbf{x}')\}$ is the larger one of $u_2(\frac{d_2-d_1}{2},x_2')$ and $u_1(1,x_1)$ and $u_2(\frac{d_2-d_1}{2},x_2') = u_1(1,x_1) = \frac{1}{2}$. Moreover, if $y \leq \frac{d_2-d_1}{2}$, $u_2(y,x_2') = 1 - \frac{y}{d_2-d_1}$ and $u_2(y,x_2) = 1 - \frac{y}{d_2-d_1}$. Then we can get that $u_2(y,x_2') = u_2(y,x_2) + \frac{1}{2}$. Using this equality, we get that

$$sw(P',\mathbf{x}') \leq \int_0^{\frac{d_2-d_1}{2}} u_2(y,x_2')p(y)dy + \frac{p}{2} = \int_0^{\frac{d_2-d_1}{2}} u_2(y,x_2)p(y)dy + \frac{1}{2}$$

$$= u_2(P',x_2) + \frac{1}{2} \leq u_2(P,x_2) + \frac{1}{2} \leq \frac{3}{4}.$$

The last but one inequality holds by the strategyproofness. The last inequlity holds since x_2 only has positive utility in $[0, \frac{d_2-d_1}{2})$. We subsequently get the lower bound is at least $\frac{sw(y^*,\mathbf{x}')}{sw(P',\mathbf{x}')} \geq \frac{4}{3}$, which completes our proof.   □

5 Conclusion

In this work, we introduce obnoxious effect into obnoxious facility game problem by giving two thresholds. The problem with two thresholds is much harder than one threshold, because it is unbounded if $d_1 \geq 1/2$, while the optimal solution is strategy-proof for one threshold. We show *Majority Mechanism* is the best possible deterministic strategy-proof mechanism for almost all cases. Next we propose an improved mechanism, or give a randomized mechanism to reduce the approximation ratio. There remains many open questions. It is interesting to attain some lower bounds which we do not get in this paper or consider randomized mechanisms other than $d_1 \geq \frac{1}{2}$. It is also interesting to extend our model to another network instead of a line.

References

1. Alon, N., Feldman, M., Procaccia, A.D., Tennenholtz, M.: Strategy proof approximation of the minimax on networks. Math. Oper. Res. (MOR) **35**(3), 513–526 (2010)
2. Alon, N., Feldman, M., Procaccia, A.D., Tennenholtz, M.: Strategyproof approximation for location on networks (2009). arXiv.org/abs/0907.2049
3. Brimberg, J., Juel, H.: A bi-criteria model for locating a semi-desirable facility in the plane. Eur. J. Oper. Res. (EJOR) **106**(1), 144–151 (1998)
4. Cheng, Y., Han, Q., Zhang, G.: Obnoxious facility game with a bounded service range. In: Chan, T.-H.H., Lau, L.C., Trevisan, L. (eds.) TAMC 2013. LNCS, vol. 7876, pp. 272–281. Springer, Heidelberg (2013)
5. Cheng, Y., Yu, W., Zhang, G.: Mechanisms for obnoxious facility game on a path. In: Wang, W., Zhu, X., Du, D.-Z. (eds.) COCOA 2011. LNCS, vol. 6831, pp. 262–271. Springer, Heidelberg (2011)

6. Cheng, Y., Yu, W., Zhang, G.: Strategy-proof approximation mechanisms for an obnoxious facility game on networks. Theor. Comput. Sci. (TCS) **35**(3), 513–526 (2011)
7. Escoffier, B., Gourvès, L., Kim Thang, N., Pascual, F., Spanjaard, O.: Strategy-proof mechanisms for facility location games with many facilities. In: Brafman, R. (ed.) ADT 2011. LNCS, vol. 6992, pp. 67–81. Springer, Heidelberg (2011)
8. Feigenbaum, I., Sethuraman, J.: Strategy proof mechanisms for one-dimensional hybrid and obnoxious facility location models. In: Workshops at the 29th AAAI Conference on Artificial Intelligence (2015)
9. Feldman, M., Wilf, Y.: Strategyproof facility location and the least squares objective. In: The 14th ACM Conference on Electronic Commerce (EC), pp. 873–890 (2013)
10. Filos-Ratsikas, A., Li, M., Zhang, J., Zhang, Q.: Facility location with double-peaked preferences. In: The 29th Conference on Artificial Intelligence (AAAI) (2015)
11. Fotakis, D., Tzamos, C.: Strategy proof facility location for concavecost functions. In: The 14th ACM Conference on Electronic Commerce (EC), pp. 435–452 (2013)
12. Ibara, K., Nagamochi, H.: Characterizing mechanisms in obnoxious facility game. In: Lin, G. (ed.) COCOA 2012. LNCS, vol. 7402, pp. 301–311. Springer, Heidelberg (2012)
13. Lu, P., Sun, X., Wang, Y., Zhu, Z.A.: Asymptotically optimal strategy-proof mechanisms for two-facility games. In: The 11th ACM Conference on Electronic Commerce (EC), pp. 315–324 (2010)
14. Lu, P., Wang, Y., Zhou, Y.: Tighter bounds for facility games. In: Leonardi, S. (ed.) WINE 2009. LNCS, vol. 5929, pp. 137–148. Springer, Heidelberg (2009)
15. Procaccia, A.D., Tennenholtz, M.: Approximate mechanism design without money. In: The 10th ACM conference on Electronic Commerce (EC), pp. 177–186 (2009)
16. Yapicioglu, H., Smith, A.E., Dozier, G.: Solving the semi-desirable facility location problem using bi-objective particle swarm. Eur. J. Oper. Res. (EJOR) **177**(2), 733–749 (2007)
17. Ye, D., Mei, L., Zhang, Y.: Strategy-proof mechanism for obnoxious facility location on a line. In: Xu, D., Du, D., Du, D. (eds.) COCOON 2015. LNCS, vol. 9198, pp. 45–56. Springer, Heidelberg (2015)
18. Zhang, Q., Li, M.: Strategy proof mechanism design for facility location games with weighted agents on a line. J. Comb. Optim. (JOCO) **28**(4), 756–773 (2014)
19. Zou, S., Li, M.: Facility location games with dual preference. In: The International Conference on Autonomous Agents and Multiagent Systems (AAMAS), pp. 615–623 (2015)

Realtime Channel Recommendation: Switch Smartly While Watching TV

Li Ning[1], Zhongying Zhao[2], Rong Zhou[1(✉)], Yong Zhang[1],
and Shengzhong Feng[1]

[1] Shenzhen Institutes of Advanced Technology,
Chinese Academy of Sciences, Shenzhen, China
{li.ning,rong.zhou,zhangyong,sz.feng}@siat.ac.cn
[2] Department of Information Science and Engineering,
Shandong University of Science and Technology, Qingdao, China
zzysuin@163.com

Abstract. Existing works on TV recommendation mostly focus on determining users preferences of TV shows. A realistic system should also consider the dynamics of the shows information. In this paper, we consider the *profit maximization problem* for real-time channel recommendation: given a specified user and a time window, a recommender algorithm is required to decide when and how to switch among n channels, each of which contains at most k live shows. The objective is to maximize the users overall profit, i.e., the total score via watching shows minus the total cost by switching among channels. For the offline version, an exact algorithm with the time complexity $O(kn^2)$ is proposed, a lower bound $\Omega(n \log n)$ of the time complexity is given for any exact algorithm. The online version is also studied. For both the non-restricted and the restricted variants, algorithms with running time $O(n \log n)$ and constant competitive ratios are presented respectively.

Keywords: Channel recommendation · Profit maximization · Online algorithms

1 Introduction

Recommender systems have been widely studied for many business environments [9], such as online bookstores, video web-services, and personalized advertising. TV shows, for which the users' preferences vary a lot, are also good sources for building recommender systems. In particular, as set-top-boxes become more and more popular, users of digital TV service can interactive with their TV sets in a much more convenient way. At the same time, modern TV services provide the users with much many choices of the channels and the shows. In many countries, a basic package for digital TV service contains tens of channels, which consequently causes users' additional efforts locating an interesting one. All these conveniences and challenges lead to a strong motivation for the research of TV recommender systems.

© Springer International Publishing Switzerland 2016
D. Zhu and S. Bereg (Eds.): FAW 2016, LNCS 9711, pp. 183–193, 2016.
DOI: 10.1007/978-3-319-39817-4_18

However, as summarized in [7], the study of TV recommendation is not so active, partly due to fact that distinct shows, which are the items to recommend, are usually available in specified time windows. This requires the study of TV recommendation to involve the dynamics of shows information in the model. Otherwise, the study can focus on how to derive users' preferences on those considered items.

Hence, it should be emphasized that the TV recommendation problem studied in this paper have the following characters which make it much different from the video recommendation problem and worth study.

- **Live Shows Coming in Channels.** The recommendation problem we considered is about live TV shows, which as mentioned are available in specified time windows. In particular, our algorithm recommends a series of channels (which host the live shows) given a user and a time window, and the "strategy" to switch among them. Note that in general, a channel presents exactly one show at each time. Hence, at a time point, recommending a channel is equivalent to recommend an according show. Furthermore, the goal of channel recommendation algorithms is to optimize users' overall experience while watching and switching among the channels.
- **Partial Evaluation of the Score on Watching Shows.** We assume for each pair of user and show is associated with a *score* (real number), which reflects the user's happiness on watching this show. Furthermore, when the user switches before a show ends, only part of the score, which is proportional to the watching time, is counted. It should be mentioned that the determination of the watching scores is not covered in this work. The most existing researches on the TV recommendation focus on this kind of determination methods, and any reasonable approaches among them can be harnessed as a pre-processing step to our work.
- **Switch Cost.** Besides the reduction of the watching scores, we also penalize the recommender system with a cost for switching among channels. This setting is harnessed due to the following reason: A switch among channels usually requires users' efforts. For example, to accept a recommended channel, it may need the user to do some operations (like pressing remote control buttons), or to break one's mind which is highly related to the show under watching. Hence, one may feel annoyed if there are too many switches. With such switch costs, the recommender algorithm should balance between recommending high-scored channels and making annoying switches to improve the user's overall experience.
- **Differences from Scheduling.** At the first glance at our model (formally introduced in Sect. 2), one may found it similar to the job-shop scheduling problem. However, there is a big difference between them. Even though we count the watching scores according to individual shows, the cost are charged only when there is a switch among the channels containing sequences of live shows. In the works about job-shop scheduling, the costs (if considered) are charged as long as there is a migration between different jobs. In particular, we have also considered the annoying cost, whose value is dynamically determined according to the remaining time when a switch happens before a show ends.

Related Works. In contrast to the active studies on recommender systems for movies, music, and books, there are much fewer works focusing on recommending TV shows. In this part, we briefly introduce some representative researches.

Even for those works on recommending TV shows, they put most of efforts on determining the attractiveness of a show to a user. For example, in 2003, Lee et al. [6] considered a multi-agent framework for recommending shows, in which a decision-tree-based approach was proposed to derive users' preferences about the shows. In [3], the authors proposed a hybrid method which combined content-based approach and item-based collaborative filtering approach. They also involved singular value decomposition to improve the prediction accuracy. In 2013, a framework to derive users' preferences on the shows has been proposed in [4].

In TV show recommendation problem, a "user" usually denotes a shared user account of a digital set-top-box. This situation causes more diverse user behaviors. By considering this situation, Goren-Bar et al. [5] studied the problem on recommending TV shows for multiple members.

In [2], Arias et al. have considered the problem to construct a news channel which fits the specified individual's interests, in mobile devices. Their model is based on the assumption that users watch the mobile TV programs in very short intervals, and hence the considered shows should be short accordingly.

In a recent paper [7], the authors considered the problem similar to the one we are considering in this paper. They aimed to find the "proper" time and the "good" shows to recommend, with the consideration of users' cost. Different from our settings, their cost model is designed with some assumption of "human memory", and the complicated modeling architecture makes the theoretical analysis even harder. Hence, in [7] the efficiency of the algorithm is measured by its performance on some real world data sets.

Although the problem considered in this paper has significant differences from the job-shop scheduling problem, we still get some heuristics from the scheduling strategies. By defining switches as *preemptives*, the job-shop scheduling problem is well studied in [1]. For more details on job-shop scheduling problems, one can refer to [8].

Our Contributions. In this paper, we study the problem of channel recommendation, in which the shows are live. When a user is watching TV at some time point, the available shows are on the air at the same time in the considered channels. For the (basic, offline) profit maximization problem (Sect. 3), we formally defined the rules of counting the watching score for a series of recommended channels, and the profit measurement of switch strategies. Considering n channels with live shows for recommendation, we proved that any algorithm that solves the profit maximization requires $\Omega(n \log n)$ time. On the other hand, we proposed an exact algorithm (Algorithm 2) of time complexity $O(n^2)$.

We also considered the problem of online profit maximization (Sect. 4), in which the information of coming shows are released as time goes on, and the switch cost is no longer a constant but proportional to the remaining time when a switch happens before a show ends. This setting is motivated by the fact that switching in the middle of the show may be much more annoying than switching close to the end. For both of these two online problems, we proved that in general, there is no algorithm with bounded competitive ratio, while under mild conditions, our proposed algorithms (respectively for two problems) run in $O(n \log n)$ time and achieve constant competitiveness.

2   Preliminaries

In this paper, we consider the *channels* as the items for recommendation. Formally speaking, a *show* is associated with a time period. In particular, for a show with index i, let $s_i \in \mathbb{R}$ and $f_i \in \mathbb{R}$ denote the beginning and the finishing time of $show_i$ respectively. For each $show_i$, there is an evaluation score $score_i \in \mathbb{R}$. A *channel* is a sequence of shows, in which there is no overlap between the associated time periods, and for any two consecutive shows, their time periods are adjacent. For example, let $show_i$ and $show_{i+1}$ are consecutively placed, then $s_{i+1} = f_i$. A *window* is a time period indicated by a beginning time, and a finishing time.

Definition 1 (Show Benefit with Respect to a Window). *Given a window w with beginning time $begin(w) \in \mathbb{R}$ and finishing time $end(w) \in \mathbb{R}$, the benefit of $show_i$ with respect to w, denoted by $p_w(show_i)$, is defined to be 0 if the time interval (s_i, f_i) has no overlap with window w (i.e. $s_i > end(w)$ or $f_i < begin(w)$); otherwise, $p_w(show_i)$ has the value*

$$score_i \cdot \frac{\min\{end(w), f_i\} - \max\{begin(w), s_i\}}{f_i - s_i}.$$

Intuitively, for $show_i$, its profit w.r.t. a time window w, is in proportion to the overlapping part of time periods (s_i, f_i) and $(begin(w), end(w)$. (See Fig. 1.)

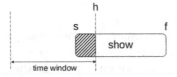

Fig. 1. The profit is $score \cdot (h - s)/(f - s)$.

Definition 2 (Channel Profit with Respect to a Window). *Consider a channel c consisting of shows $show_0, show_1, \ldots show_{k-1}$. The profit $p_w(c)$ of channel c w.r.t. a time window w is defined by the sum of the profit of all shows w.r.t. w:*

$$p_w(c) := \sum_{i=0}^{k-1} p_w(show_i)$$

Profit Maximization Problem. Consider a set $\mathcal{C}$ of n channels. A time window $w = (begin, end)$ is given, it is required to chop w into several time windows w_0, w_1, ..., and associate each time window with a channel. Denote the channel associated with w_i by $c(w_i)$, and the profit of $c(w_i)$ w.r.t. w_i by p^i. For a chopping (and associating) strategy involving m time windows w_0, w_1, ..., the *net profit* is defined by

$$\sum_{i=0}^{m-1} p^i - m \cdot \alpha$$

where $\alpha \geq 0$ is a constant parameter indicating the cost of a switch among channels.

The target of the *profit maximization problem* is to find the best chopping strategy that leads to the maximum total net profit. Note that once the chopping strategy is given, it is direct to associate the channel with maximum profit for each time window.

3   Offline Profit Maximization and Its Complexity

Given a set C of n channels, let S denotes the set of all starting time points s_i and finishing time points f_i. Without loss of generality, assume that all points in S are sorted in increasing order. That is, for $S := \{t_0, t_1, \ldots\}$, it holds that $t_i < t_j$ *iff* $i < j$.

Lemma 1. *Consider profit maximization during a time interval $[t_1, t_2]$, in which there is no show begins or ends. Assume there is a switch from channel c_1 to c_2 at time $t \in (t_1, t_2)$. Then the total score will not be decreased by changing the switch time to either t_1 or t_2.*

Proof. Let μ_1 and μ_2 denotes the unit scores of channel c_1 and channel c_2 over time $[t_1, t_2]$ (precisely speaking, the unit scores of the shows accordingly), respectively. the score gained over time $[t_1, t_2]$ is $\mu_1 \cdot (t - t_1) + \mu_2 \cdot (t_2 - t) - \alpha$, which is monotone as a function of t. This concludes the lemma.   □

Based on Lemma 1, we get Algorithm 1 in dynamic programming that solves the profit maximization problem in polynomial time. Recalling that S is sorted in increasing order, in Algorithm 1, define $p^i(c)$ to be the maximum profit over the time period starting from t_i and with the "first" channel selected to be $c \in C$, i.e. $[t_i, t_{i+1}]$ has been associated with channel c. We calculate $p^i(c)$ in a backward fashion. That is, conditioned on the fact that $p^j(c)$ are already known with $j > i$, the value of $p^i(c)$ can be calculated by considering the following two cases,

- **the case where switch at t_{i+1}:** the according optimal profit is $p_{[t_i, t_{i+1}]}(c) - \alpha + \max_{c' \neq c} p^{i+1}(c')$;
- **the case where don't switch at t_{i+1}:** the according optimal profit is $p_{[t_i, t_{i+1}]}(c) + p^{i+1}(c)$.

Algorithm 1. Backward dynamic programming.

1: set $p^{|S|-1}(c) = 1$ for all $c \in C$;
2: **for** $s = |S| - 2 \to 0$ **do**
3:    **for** $c \in C$ **do**
4:       $p^s(c) := p_{[t_s, t_{s+1}]}(c) + \max\{p^{s+1}(c), \max_{c' \neq c} p^{s+1}(c') - \alpha\}$;
5:    **end for**
6: **end for**
7: **print** $\max_c p^0(c)$;

Running Time of Algorithm 1. Considering there are n channels and k is the maximum number of shows contained in any chnnel from C, all $p_{[t_s, t_{s+1}]}(c)$ can be calculated in $O(kn^2)$ time. In order to quick access to $\max_{c'} p^t(c')$ for each time point s, we can build a maximum heap that supports $O(1)$ time query and single update in $O(\log m)$ time. Recall that there are at most kn shows, and hence there are at most kn heap updates in all. Consequently, we know Algorithm 1 runs in time of $O(kn^2 + kn \log(kn)) = O(kn^2)$, which is $O(n^2)$ when k is constant. Formally, we conclude Theorem 1.

Theorem 1. (Off-PM is Solvable in Polynomial Time). *Given n channels, where the maximum number of shows contained by a channel is constant, the chopping strategy which leads to the maximum total net profit can be found in $O(n^2)$ time.*

Lower Bound. One should not be surprised about the fact that the time needed for solving the profit maximization problem is at least linear in the number of channels.

Lemma 2. *There is an instance, for which in order to get the maximum total net profit, it needs $\Omega(n)$ to find the according chopping strategy.*

Proof. Consider n channels, each of which covers the time period from 0 to $k \cdot n$, by $2k$ shows. For the i-th channel (with $i \in \{1, \ldots, n\}$, its shows form k groups, each of which consists of two shows. In the j-th group with $j \in \{1, \ldots, k\}$, the first show starts at $(j-1) \cdot (n+1)$ and ends at $(j-1) \cdot (n+1) + i$, and has score 0; the second show starts at $(j-1) \cdot (n+1) + i$ and ends at $j \cdot (n+1)$, and has score $(j-1) \cdot n + i$. (See Fig. 2 for an example.)

For any constant α (e.g. $\alpha = 1$), it is easy to check in order to achieve the maximum total net profit, we need at least $(k-1) \cdot n$ time windows, which implies it needs $\Omega(n)$ time to find (and construct) the corresponding chopping strategy. $\square$

Theorem 2. (Number of Steps is Necessarily at the Logarithmic Order). *In the worst case, $\Omega(n \log n)$ steps is needed to solve an instance of the profit maximization problem with n channels.*

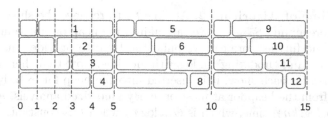

Fig. 2. An instance with $n = 4$ channels and $\alpha = 1$. To achieve the maximum total net profit, it needs $2n + n - 1 = 11$ time windows.

Proof. In the proof, we show that for an arbitrary instance of sorting n integers, with additional time of $O(n)$, it can be reduced to an instance of the profit maximization problem, and $O(n)$ is needed to "interpret" the answer.

Assume that we are given n integers $x_0, x_1, \ldots, x_{n-1}$ (without loss of generality, assume they are all positive, and the minimum is 1). Then for each integer x_i, we construct an channel c_i with two shows, where

1. the first show of c_i starts at 0 and ends at x_i, with evaluation score 0;
2. the second show of c_i starts at x_i and ends at $\max\{x_i\} + 1$, with evaluation score x_i.

For this instance, the cost of switching to another channel is 0, i.e., $\alpha = 0$. Now we get a profit maximization problem with n channels $c_0, c_1, \ldots, c_{n-1}$, within the time interval $(0, \max\{x_i\} + 1)$ with no switching cost.

Solving this profit maximization problem gives a series of windows $w_i = (s_i, f_i)$ with $i = 0, \ldots, n - 1$, and the associated channel $c(w_i)$s. Let x_i' denotes the integer corresponding to the channel associated with the i-th window w_i. Then it is clear that the sequence $\{x_0', x_1', \ldots, x_{n-1}'\}$ is a sorting sequence of the given n integers $x_0, x_1, \ldots, x_{n-1}$, in ascending order. □

4 Online Profit Maximization

In the online profit maximization, shows are released as time goes. Therefore, the backward dynamic programming does not work.

We introduce another algorithm that solves the offline profit maximization problem in greedy strategy when the switch is free (i.e. $\alpha = 0$). In Algorithm 2, the offline profit maximization problem can be solved in time of optimal order.

Algorithm 2. Greedy switch.

1: **for** $i = 0, 1, \ldots$ **do**
2:    set $S_{(t_i, t_{i+1})} := \{shows\ that\ overlaps\ with\ (t_i, t_{i+1})\}$;
3:    associate the channel containing the show with maximum unit score among $S_{(t_i, t_{i+1})}$;
4: **end for**
5: **print** every time windows with the associated channels.

Running Time of Algorithm 2. At time t_i, to retrieve the show with maximum unit score among $S_{(t_i, t_{i+1})}$, we can maintain a heap of unit values of shows in $S_{(t_i, t_{i+1})}$, and furthermore can we build the heap to provide "maximum element accessing" and "updating" in logarithmic time. Note that there are at most kn shows, then for each show, it is inserted into the heap for exactly once, and it is deleted from the heap for at most once. By amortized analysis, Algorithm 2 runs in $O(kn \log(kn))$ time, which is $O(n \log n)$ when k is constant.

4.1   Online Algorithms

Theorem 3. *Consider an instance of online profit maximization problem with n channels, each of which contains at most k shows, Algorithm 2 runs in $O(n \log n)$. And if $\alpha = 0$ (the switches are charged for free), the chopping strategy it returns leads to the maximum total net profit.*

Proof. When $\alpha = 0$, there is no cost to treat a time window as the concatenation of several shorter ones. Hence, Algorithm 2 is the optimal chopping strategy. □

It is easy to find out that the "free switch" condition is necessary for Algorithm 2 to achieve the optimal. More inspired, this algorithm has an ability that it can find the optimal channel with a underlying time interval $[t_i, t_{i+1}]$ with only the knowledge of the shows available currently. This is useful when the show forecast comes in stream. In other words, Algorithm 2 is online.

Consider an online algorithm **ALG** to solve this online profit maximization problem, and denote by **OPT** the optimal algorithm for the offline profit maximization problem. Note that an instance σ for the offline profit maximization problem can be easily transformed to an instance for the online profit maximization problem by releasing shows at the time they begin. Then for such an instance σ, we use σ_{off} and σ_{on} to denote its offline version and online version respectively. Let $\mathcal{E}_{OPT}^{\sigma_{off}}$ and $\mathcal{E}_{ALG}^{\sigma_{on}}$ be the according efficiencies for the solutions returned by **OPT** and **ALG**. Then the competitive ratio of **ALG** is defined as

$$\sup_{\sigma} \frac{\mathcal{E}_{OPT}^{\sigma_{off}}}{\mathcal{E}_{ALG}^{\sigma_{on}}}.$$

Our target is to design online algorithm with the competitive ratio as close to 1 as possible. According to Theorem 3, if the switch is charged for free, the Algorithm 2 achives competitive ratio 1. However, in general, there is no **ALG** with bounded competitive ratio. To illustrate this fact, consider following example.

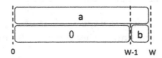

Fig. 3. General case to show the unbounded competitive ratio.

Consider the example instance shown in Fig. 3. Assume that the cost parameter α satisfies $\alpha > W(a + b)$, which implies that any algorithm starting with the upper channel has no motivation to switch. Note that at time 0, any online algorithm **ALG** has no idea about the relation between a and b. Consider two instances which are just different at b,

- instance 1: $b = W \cdot a$;
- instance 2: $b = 0$.

It can be verified that for any deterministic online algorithm **ALG**, it achieves at most a fraction of $1/W$ of the optimal, in either instance 1 or instance 2. When W is extremely large, it implies the competitive ratio of **ALG** is unbounded.

Furthermore, it should be noticed that involving randomness does not avoid the unbounded competitive ratio. Because of the lack of knowledge of upcoming shows, the probability that a randomized algorithm starts with a "perfect" channel (the one in the optimal solution) is of $O(1/n)$, which is arbitrarily small as n can be very large. Note that this is true no matter whether the selection of the first channel is uniform or not.

4.2   Online Profit Maximization with Annoying Cost

We have considered the online version of the profit maximization problem, in which the shows are released in stream. We will introduce another feature named *annoying cost* which also raises naturally in practice.

Definition 3 (Annoying Cost). *If there is a switch of channel c at time t (close the underlying window at time t and start a new window associated with a different channel), then the cost of such a switch annoying cost is defined to be $(f_i - t)/(f_i - s_i)$. Recall that s_i and f_i are the beginning time and the ending time of $show_i$, which is the one playing in c at time t.*

The idea of annoying cost is not new. In [7], the authors argued that the recommendation may suffer a penalty if it annoys people too much. Note that with the definition here, the annoying cost will be large if the current show is far from the end. At the opposite situation, if the current show is about to end, the switch is charged for a very low cost. With annoying cost, it is also sufficient to only consider the switch operations at the time points when some show begins or ends. Recall the proof of Lemma 1, with annoying cost, the total score during $[t_1, t_2]$ is now

$$\mu_1 \cdot (t - t_1) + \mu_2 \cdot (t_2 - t) - (f_1 - t)/(f_1 - s_1)$$

which is still monotone on t.

Online Profit Maximization with Annoying Cost. With the annoying cost, the efficiency of a sequence of time windows $W := w_0, w_1, \ldots$ (using channel assignment σ) w.r.t. a given set of channels C is defined to be

$$\mathcal{E}_C^\sigma := \sum_{w \in W} (p^w(c) - \alpha_w^c),$$

where α_w^c denotes the annoying cost of switching to channel $c \in C$ associated with window w, under the channel assignment σ. Then the profit maximization problem with annoying cost is: given a set of channels for which the shows are released online, find the window W as well as the channel assignment σ that achieves the optimal efficiency.

5   Conclusions and Future Work

In this paper, we studied the problem to optimize the channel switch strategy, so as to make personalized real-time channel recommendation. Different from the video recommendation, channel recommendation has to work without the control of shows' starting times and finishing times, i.e. the show under consideration is live.

As conclusions, we proved that in the offline case, the running time for any algorithm solving the profit maximization problem has the lower bound $\Omega(n \log n)$, where n is the number of channels; and, there exists a backward dynamic programming algorithm solving the problem in $O(n^2)$ time.

We also introduced two online versions of the profit maximization problem, and presented respectively two algorithms with constant competitive ratio, when the switch cost is not too large. Furthermore, the proposed online algorithms runs in $O(n \log n)$ time when the number of live shows in each channel is upper bounded by a constant.

In future works, it is interesting to explore the algorithms for offline version with $o(n^2)$ time. For the online version, it worths further studies on discovering other non-trivial situation with bounded competitive ratio.

Acknowledgments. This research is supported by NSFC (No. 61402461, 61303167, 61433012, U1435215), Open Project of Guangxi Key Laboratory of Trusted Software (No. KX201535), Scientific Research Foundation of Shandong University of Science and Technology for Recruited Talents (No. 2015RCJJ069), Shenzhen grant JCYJ20140509174140680, GJHS20130402135334984, and National High-tech R&D Program of China (863 Program) 2015AA050201.

References

1. Abdeddaïm, Y., Maler, O.: Preemptive job-shop scheduling using stopwatch automata. In: Katoen, J.-P., Stevens, P. (eds.) TACAS 2002. LNCS, vol. 2280, pp. 113–126. Springer, Heidelberg (2002)
2. Arias, J.J.P., Vilas, A.F., Redondo, R.P.D., Gil-Solla, A., Cabrer, M.R., Duque, J.G.: Making the most of tv on the move: my newschannel. Inf. Sci. **181**(4), 855–868 (2011)
3. Barragáns-Martínez, A.B., Costa-Montenegro, E., Burguillo-Rial, J.C., Rey-López, M., Mikic-Fonte, F.A., Peleteiro-Ramallo, A.: A hybrid content-based and item-based collaborative filtering approach to recommend TV programs enhanced with singular value decomposition. Inf. Sci. **180**(22), 4290–4311 (2010)

4. Chang, N., Irvan, M., Terano, T.: A TV program recommender framework. In: KES, pp. 561–570 (2013)
5. Goren-Bar, D., Glinansky, O.: Fit-recommending TV programs to family members. Comput. Graph. **28**(2), 149–156 (2004)
6. Lee, W.-P., Yang, T.-H.: Personalizing information appliances: a multi-agent framework for TV programme recommendations. Expert Syst. Appl. **25**(3), 331–341 (2003)
7. Oh, J., Kim, S., Kim, J., Yu, H.: When to recommend: a new issue on TV show recommendation. Inf. Sci. **280**, 261–274 (2014)
8. Pinedo, M.L.: Scheduling: Theory, Algorithms, and Systems, 3rd edn. Springer Publishing Company, Incorporated, New York (2008)
9. Ricci, F., Rokach, L., Shapira, B.: Introduction to recommender systems handbook. In: Ricci, F., Rokach, L., Shapira, B., Kantor, P.B. (eds.) Recommender Systems Handbook, pp. 1–35. Springer, US (2011)

Empirical Study of Phase Transition of Hamiltonian Cycle Problem in Random Graphs with Degrees Greater Than One

Wei Peng[1]([✉]), Dongxia Wang[2], and Xinwen Jiang[1]

[1] Science and Technology on Parallel and Distributed Processing Laboratory,
National University of Defense Technology, Changsha 410073, Hunan, China
`wpeng@nudt.edu.cn`, `xinwenjiang@sina.com`
[2] National Key Laboratory of Science and Technology on Information System
Security, Beijing Institute of System Engineering, Beijing, China
`dongxiawang@126.com`

Abstract. It is well-known that many NP-complete problems will undergo phase transitions along with the change of some problem-specific critical parameter values. It has been shown that the phase transition will occur at an average node degree $\log(n) + \log\log(n)$ for Hamiltonian cycle problem in random graphs with n nodes. In this paper, we prove that random graphs with such critical average node degrees tend to be hamiltonian graphs if their node degrees are greater than one. Using an improved backtracking algorithm with pruning operations, we try to find the areas where hard problem instances can be found with high probability. For random graphs with degrees greater than 1, the experimental results have demonstrated that hard cases can be found with high probability when graphs take lower average degrees, and the phase transition occurs at lower average degrees, too. Empirically, the phase transition between hamiltonicity and non-hamiltonicity occurs when the average degree is $1.1601 + 0.2418\log(n)$ for random graphs with degrees greater than one.

Keywords: Hamiltonian cycle problem · Phase transition · NP-completeness · Empirical study

1 Introduction

It has been found that the existence of hard instances of NP-complete problems is closely correlated with the phase transition phenomenon. Cheeseman et al. firstly studied the issue and they found that hard cases occur at a critical value of an 'order' parameter of the problem [1,2]. Later, the phase transition phenomena are well studied for some NP-complete or NP-hard problems, e.g.,

---

W. Peng—This work was supported in part by the National Natural ScienceFoundation of China under grant No. 61272010, 61271252, and the research project of National University of Defense Technology.

D. Zhu and S. Bereg (Eds.): FAW 2016, LNCS 9711, pp. 194–204, 2016.
DOI: 10.1007/978-3-319-39817-4_19

the k-satisfiability (k-SAT) problem [3–6], the Hamiltonian Cycle (HC) problem [7,8] and the Traveling Salesman Problem (TSP) [9].

It is believed that hard instances of a NP-complete problem normally locate in the phase transition area of the problem. For example, the ratio of clauses to variables, denoted as α, is an order parameter for k-SAT problems. For $k \geq 3$, there is a threshold value of α when a phase transition occurs from unsatisfiable state to satisfiable state. Correspondingly, it is relatively easy to find a satisfiable solution at low values of α, or show the formulae unsatisfiable at high values of α. It is most difficult to solve the problem at values of α near the threshold.

For Hamiltonian cycle problem, Cheeseman et al. have shown that the probability of a random graph containing a HC will change quickly from almost 0 to almost 1 for some critical values of average connectivity. The critical average connectivity, or average node degree, is $\log(n) + \log\log(n)$ from theoretic prediction, which is supported by Cheeseman's experimental results [1], and Vandegriend's experimental results [7]. However, by using an improved backtracking algorithm, Vandegriend et al. have shown that almost all tested random graphs of 100 to 1500 vertices (or nodes) are easily solved. Thus, it raises the question where the really hard HC problem instances are. Denote a node with degree x by *degree-x node*. A graph is non-hamiltonian if it has a degree-0 or degree-1 node. Intuitively, hard problem instances should be found in a smaller graph space, e.g., a graph space without degree-0 or degree-1 vertices.

In this paper, we studied the HC problem in random graphs with node degrees greater than 1. Firstly, we proved that random graphs with node degrees greater than 1 tend to be hamiltonian graphs at the phase transition point where average node degrees take $\log(n) + \log\log(n)$. Using a backtracking algorithm with pruning operations, we studied the phase transition area for these random graphs. We found that hard instances can appear more frequently and phase transitions occur at new points which are far from the old value of $\log(n) + \log\log(n)$. Using a non-linear regression method, we estimated that the phase transition between hamiltonicity and non-hamiltonicity occurs when the average degree is $1.1601 + 0.2418\log(n)$ for random graphs with degrees greater than one. Such results suggest that the problem space in the phase transition region need be examined more thoroughly with degree constraints.

The rest of the paper is organized as follows. Section 2 describes the related work. Section 3 presents the theoretical analysis of the studied problem and the experiment method. The experimental results and the corresponding discussions are given in Sect. 4 and the paper is concluded in Sect. 5.

2   Related Work

The phase transition behavior of k-SAT problem is well studied in previous work. It has been shown that the phase transition from unsatisfiable state to satisfiable state is characterized by an order parameter α, the ratio of clauses to variables [4–6]. The critical value of α is 1 for 2-SAT problems. However, it remains an open issue to locate the exact critical value of α when $k \geq 3$. Empirical results suggest that the critical value $\alpha_c \approx 4.2$ for 3-SAT problems.

Besides the phase transition of satisfiable state, phase transition will occur on the solution structure of k-SAT problems, too. Xu and Li have shown that the similarity of different satisfiable solutions will change abruptly at a critical value of α when $k \geq 5$ [6]. The set of literals which take true values in every satisfying truth assignment is called as the backbone of a k-SAT problem. The backbone size has been related to problem hardness [5,10]. Zhang has shown that the backbone of 3-SAT and MAX 3-SAT will experience phase transitions, too [11]. Specifically, the backbone of MAX 3-SAT with size 0.5 appears almost at the time when 3-SAT is satisfiable with probability 0.5. Mézard et al. have shown that there is an intermediate phase below the transition threshold α_c which accounts for the increasing of complexity in search algorithms [12]. Krzakala et al. have studied the distribution of solutions of the k-SAT problem and the q-coloring problem [13]. Their research work has demonstrated that the solution clusters will split along with the increasing of α and the ordered sequence of solution cluster size will converge to a Poisson-Dirichlet process.

Model RB is proposed as a prototypical random constraint satisfaction problem (CSP) with growing domain size [14]. Under the model, the critical values of phase transition can be obtained exactly. Zhao et al. have discovered that the solution space of Model RB is a connected cluster under a threshold value of the model parameter p. Beyond the threshold value, the set of solutions begins to split into many disconnected clusters, similar to the phenomenon of k-SAT problem [15].

As an important problem in graph theory, much analytic work has been done on the Hamiltonian cycle problem. With regard to the phase transition property, it is proved that a random graph with n vertices and $\frac{n}{2}(\log(n) + \log \log(n) + c)$ edges is hamiltonian with probability p_c tending to $\exp(-\exp(-c))$ as $n \to \infty$ [16]. Based on the theoretic result, a polynomial time algorithm HAM is proposed for the HC problem. Theoretically, the probability for the HAM algorithm to find a Hamiltonian cycle in a random graph with $\frac{n}{2}(\log(n) + \log \log(n) + c)$ edges is $\exp(-\exp(-c))$ as $n \to \infty$, too [17]. Thus, it has been proved that algorithms with polynomial expected running time exist for the HC problem. Other theoretic results about the Hamiltonian properties of random graphs can be found in [18].

Different researchers have examined the phase transition behavior of the Hamiltonian cycle problem empirically. It is experimentally verified that the phase transition for Hamiltonicity is very close to the phase transition for biconnectivity in random graphs with n vertices and m edges, denoted by $G(n, m)$ or $G_{n,m}$ graphs [19,20]. The critical value of average degree is about $\log(n)$ in their results, while Cheeseman's results have shown that the phase transition occurs when the average degree is about $\log(n) + \log \log(n)$. The latter is supported by Vandegriend and Culberson's experimental results again [7]. By Komlós and Szemerédi's theoretic results, phase transition occurs when the average degree is $c(\log(n) + \log \log(n))$ with $c = 1$ for the HC problem. Vandegriend and Culberson's experimental results have shown that c is between 1.08 and 1.10 for random graphs with $n \leq 1500$. For directed Hamiltonian Cycle problem, the critical edge number of phase transition is proved to be $cn(\log(n) + \log \log(n))$ with $c = 1$ [21] and it is verified experimentally that $c = 0.9$ for directed graphs with $n \leq 3162$ [8].

There are a number of theoretic results on the sufficient conditions for a graph to be hamiltonian. Although many conditions have been proved, a necessary and sufficient condition is still absent for the HC problem. See [22] for a survey on this issue.

3   Hamiltonian Cycles in Random Graphs with Degrees Greater Than One

For the HC problem, it is proved that there will be an algorithm with polynomial expected running time for the HC problem. So, our question is where the really hard problem instances are, with the fact in mind that the HC problem is NP-complete. Because a graph is not hamiltonian if one node degree is 0 or 1, we focus on the graphs with node degrees greater than 1 (or ≥ 2) in the paper. Note, all graphs are considered as undirected graphs implicitly in the paper. Let Γ_0 denote the set of graphs with n vertices and m edges and Γ_1 denote the subset of Γ_0 where all node degrees of each graph in Γ_1 are greater than one. A graph chosen randomly from Γ_0 is denoted by $G_{n,m}$, while a graph chosen randomly from Γ_1 is denoted by $G_{n,m}^{\delta>1}$.

The following issues are investigated:

- The critical point of phase transition between Hamiltonian and non-Hamiltonian for $G_{n,m}^{\delta>1}$, in term of average node degree;
- The critical value of average node degree when the probability of hard instances reaches its maximal point.

Firstly, based on the result of Komlós and Szemerédi, we have the following theorem.

Theorem 1. For a random $G_{n,m}^{\delta>1}$ graph G with n vertices and m edges, if

$$m = \frac{n}{2}log(n) + \frac{n}{2}loglog(n) + c_n n \qquad (1)$$

for a constant value c and some numerical sequence c_n where $c_n \to c$ or $c_n \to \infty$ when $n \to \infty$, then

$$\lim_{n \to \infty} Pr(G_{n,m}^{\delta>1} \ is \ hamiltonian) = 1. \qquad (2)$$

Proof: Consider the following events:

- A: $G_{n,m}^{\delta>1}$ is hamiltonian;
- B: $G_{n,m}$ is hamiltonian;
- C: the minimum degree of $G_{n,m}$ is at least 2.

Based on Bayesian formula, we have

$$Pr(A) = Pr(B|C) = \frac{Pr(BC)}{Pr(C)}. \tag{3}$$

Because the minimum degree of every hamiltonian graph is at least 2, so $Pr(BC) = Pr(B)$. Thus,

$$Pr(A) = \frac{Pr(B)}{Pr(C)} \tag{4}$$

Komlós and Szemerédi have proved that [16]

$$\lim_{n \to \infty} Pr(B) = \lim_{n \to \infty} Pr(C) = \begin{cases} 0, & c_n \to -\infty \\ e^{-e^{-2c}}, & c_n \to c \\ 1, & c_n \to \infty \end{cases} \tag{5}$$

so if $c_n \to c$ or $c_n \to \infty$ when $n \to \infty$, we have

$$\lim_{n \to \infty} Pr(A) = 1. \tag{6}$$

Next, we study the phase transition issues through experiments. The hardness of a problem instance depends on not only the properties of the instance, but also the algorithm used to tackle it. In this paper, we adopt the similar algorithm used by Vandegriend and Culberson [7]. The algorithm is an exact algorithm which uses backtracking to find a solution. The main difference is that we do not use the iterated restart technique in the algorithm. The algorithm includes two stages: initial pruning and recursive search.

(1) Initial pruning. The input graph is initially checked against some simple conditions:
   - *Ore's condition*: If $deg(x) + deg(y) \geq n$ for any nonadjacent vertex pair x and y in an undirected graph G, then G is hamiltonian;
   - *Cut-point condition*: If there is a cut-point in G, then G is non-hamiltonian;
   - *Fan's condition* [23]: If for any vertex pair x and y in a 2-connected graph G, $distance(x, y) = 2$ implies $max(deg(x), deg(y)) \geq (n/2)$, then G is hamiltonian.

   Then, the graph is pruned by checking vertices with only two adjacent vertices (called as *degree-2* vertices). The pruning rules include:
   - *rule 1*: If w and v are two neighbors of a degree-2 vertex u, then remove the vertex u and the edges $\langle w, u \rangle$, $\langle v, u \rangle$, add an edge $\langle w, v \rangle$ if it does not exist and mark it as a *forced edge*;
   - *rule 2*: If a vertex u is attached with two forced edges, then remove other edges attached with it, and apply the rule 1 on u to remove it;
   - *rule 3*: If a vertex u is incident to three or more forced edges in the graph G, then G is non-hamiltonian.

   After the pruning, cut-points are checked again to detect non-hamiltonian graphs.

(2) Recursive search. The search procedure uses backtracking to find a hamiltonian cycle in the input graph. During the search procedure, edges on the trial path are marked as *forced edges* and the pruning rules in the first stage are applied again to accelerate the search process.

A timeout threshold is set so that the search procedure can be terminated after a specified time. If the input graph can be determined to be hamiltonian or non-hamiltonian in the specified time, then the problem instance is classified as an easy case, or else it is considered as a hard one. With an appropriate setting of the timeout threshold, we can investigate the occurring probability of hard problem instances.

While there are problem instances which can not be solved within the specified time, we try to solve them using the SparseHAM algorithm [24], a variation of Bollobás's HAM algorithm [17]. The algorithm tries to extend a partial path from one of its end-points, or from one vertex in the middle of the partial path if the two end-points are connected. If the path can not be extended, then rotational transformation is used to get a new end-point of the path so that the path can be extended further.

We made some modifications to improve the original SparseHAM algorithm, including:

- If there is an edge marked as forced edge in the input graph, then the edge is used to create the initial path;
- Forced edges are detected and used with high priority if it is not contained in a path and avoided to be removed from the path;
- Middle results are saved so that backtracking can be performed if attempts on the current partial path are all failed.

Again, the modified SparseHAM algorithm will terminate with the limitation of the timeout threshold. It should be noted that the hamiltonian property of the input graph remains unknown if the SparseHAM algorithm fails to find a hamiltonian cycle.

4   Experiment Results

We consider random graphs with n vertices where n ranges from 100 to 6000. The $G_{n,m}$ graphs are studied firstly, then the $G_{n,m}^{\delta>1}$ graphs are tested. For graphs with n vertices, the edge number m increases with the prescribed average degree increasing from 2 to $2(\log(n) + \log\log(n) + 1)$. The edge number stops to increase if the ratio of hamiltonian graphs is larger than 99.9 %.

For each parameter setting, we generate 10^5 graphs and run the algorithm once on each graph. Then, the number of hard instances are summed to estimate the occurring probability of hard instances. The probability of hamiltonian graph is estimated using the ratio of hamiltonian graphs in all solved instances.

The timeout threshold will affect the decision of easy or hard instances. During the trial running, we find that most random problem instances can be solved at the microsecond level. So, we set the timeout threshold to be 2 s. Experiments were performed on a PC server with Intel Xeon CPU E5-2650 which has 8 cores and 2 threads per core.

4.1 Results with Backtracking Algorithm

Firstly, we investigate the dynamics of the hard instance ratio with the increasing of vertex number using the backtracking algorithm. The HAM algorithm is not executed at this stage. Figure 1 shows the variations of hard instance ratios for random graphs with 2000, 4000 and 6000 vertices. We compare the results of $G_{n,m}$ and $G_{n,m}^{\delta>1}$ random graphs.

Figure 1 shows that the curve of hard instance ratio increases with the increasing of average degree, and then decreases after reaching some critical points. It verifies that the "easy-hard-easy" pattern still exists in term of hard instance ratio for the Hamiltonian cycle problem. Compared to the results of $G_{n,m}$ graphs, the "increase-then-decrease" curves are quite distinct for $G_{n,m}^{\delta>1}$ graphs. Besides, the peaks come up at lower values of average degree for curves of $G_{n,m}^{\delta>1}$ graphs than $G_{n,m}$ graphs. Roughly, the critical average degree is about 12 for $G_{n,m}$ graphs, but is about 6.5 for $G_{n,m}^{\delta>1}$ graphs.

We take a close look at the curves of hard instance ratio when $n = 6000$ and compare them with the curves of hamiltonian graph ratios. Figure 2 shows that the peaks of curves of hard instance ratio are coupled with the phase transition of hamiltonian graph ratio. For $G_{n,m}$ graphs, the peak of curve of hard instance ratio appears when the average degree is 12.16. The ratio of hamiltonian graphs is close to 0.5 when the average degree is 11.6, while the theoretic prediction of phase transition is $\log(n) + \log\log(n) = 10.862782$. For $G_{n,m}^{\delta>1}$ graphs, the peak of curve of hard instance ratio appears at 6.6, while the ratio of hamiltonian graphs is close to 0.5 at 6.52. Both values are far from the phase transition point of $G_{n,m}$ graphs. According to Theorem 1, the probability of hamiltonian graphs approaches to 1 when the average degree is $\log(n) + \log\log(n)$ for $G_{n,m}^{\delta>1}$ graphs. It is verified by the experimental results.

Figure 3 shows the changing of critical degrees with the increasing of graph size. The critical degree of hard instances is the average degree when the ratio of hard instances reaches its maximal value. We show two curves of critical degree values when the ratio of hamiltonian graphs reaches 0.5 and 0.99, respectively. The curves of theoretic prediction values of phase transition are also plotted for comparison.

From the figure, for $G_{n,m}$ graphs, we can see that the critical degree of hard instances and the critical degree of hamiltonian graphs (0.5) come close to the theoretic values of phase transition. It conforms to the theoretic analysis results. However, for $G_{n,m}^{\delta>1}$ graphs, the critical degree of hard instances and the critical degree of hamiltonian graphs are much less than $\log(n) + \log\log(n)$. At present, the theoretic critical degree of phase transition for $G_{n,m}^{\delta>1}$ graphs remains unknown, which need be studied in the future. Another meaningful observation is that the critical degree of hard instances is very close to the critical degree of hamiltonian graphs (0.5). Thus, the conjecture still holds that most hard

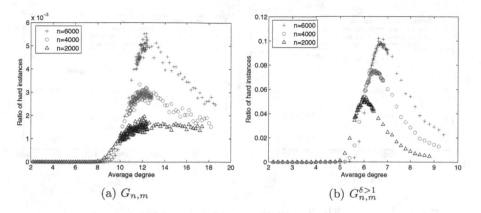

Fig. 1. Variations of ratios of hard instances

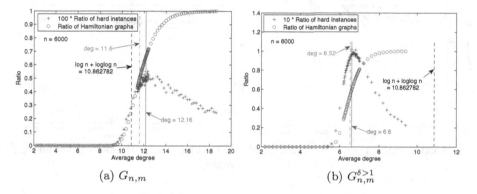

Fig. 2. Coupling of hard instance ratios and hamiltonian graph ratios

instances appear at the phase transition area where hamiltonian graphs and non-hamiltonian graphs go fifty-fifty.

Moreover, we apply non-linear regression analysis on the curves of $G_{n,m}^{\delta>1}$ graphs. We use the function $log(x)$ to fit the experimental results. For the curve of critical degrees of hamiltonian graphs where the hamiltonicity probability is 0.5, it can be approximated by $1.1601 + 0.2418\log(n)$. For the curve of critical degrees of hard instances, it can be fitted by $1.1269 + 0.2490\log(n)$. It seems that the critical degrees are mainly determined by the factor $\frac{1}{4}\log(n)$.

4.2   Results with Backtracking and SparseHAM Algorithm

Since the backtracking algorithm can not solve all random-generated graphs, a modified SparseHAM algorithm is used afterwards. For each $G_{n,m}^{\delta>1}$ graph which can not be solved by the backtracking algorithm in prescribed time, we apply the SparseHAM algorithm on it with the same timeout limitation. The results are plotted in Fig. 4.

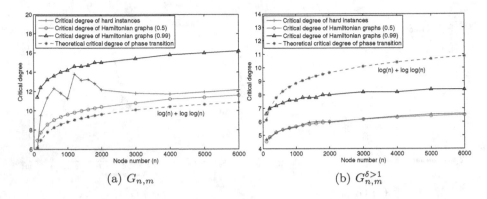

(a) $G_{n,m}$ (b) $G_{n,m}^{\delta>1}$

Fig. 3. Variations of critical degrees

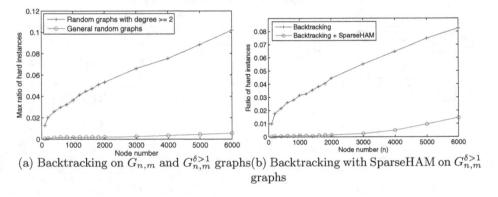

(a) Backtracking on $G_{n,m}$ and $G_{n,m}^{\delta>1}$ graphs (b) Backtracking with SparseHAM on $G_{n,m}^{\delta>1}$ graphs

Fig. 4. Comparison of ratios of hard instances

In Fig. 4a, it shows that the maximal ratio of hard instances increases with the increasing of graph size. The ratio of $G_{n,m}^{\delta>1}$ graphs increases faster than the ratio of $G_{n,m}$ graphs due to the smaller sampling space as node degrees are restricted to be greater than 1. The SparseHAM algorithm is applied on instances which can not be solved by the backtracking algorithm in specified time. From Fig. 4b, we can see that the maximal ratio of hard instances is greatly reduced. It verifies the effectiveness of the SparseHAM algorithm. However, the effectiveness of the SparseHAM algorithm is lowered with the increasing of graph size.

5   Conclusion

In this paper, we have studied the phase transition phenomenon of Hamiltonian cycle (HC) problem in random graphs with node degrees greater than 1 (called $G_{n,m}^{\delta>1}$ graphs). Although the phase transition occurs at the average connectivity $\log(n) + \log\log(n)$ for random graphs without node degree constraints, we found that the phase transition occurs at a much lower value of average connectivity if the node degree constraint is applied. Specifically, we proved theoretically that $G_{n,m}^{\delta>1}$ graphs tend to be hamiltonian if their average degrees take the values of $\log(n) + \log\log(n) + c_n$ and the numerical sequence c_n satisfies $\lim_{n\to\infty} c_n = c$ or $\lim_{n\to\infty} c_n = \infty$ where c is a constant value. Using a backtracking algorithm with pruning operations, we have shown experimentally that the phase transition area of $G_{n,m}^{\delta>1}$ graphs has been shifted to areas of lower average connectivity, compared to $G_{n,m}$ graphs without node degree constraints. Empirically, the critical average degree of phase transition can be estimated by $1.1601 + 0.2418 \log(n)$ for $G_{n,m}^{\delta>1}$ graphs.

The work presented here is preliminary and the problem space in the phase transition area need be studied more thoroughly for $G_{n,m}^{\delta>1}$ random graphs. An open theoretic problem is also left:

What is the exact point of phase transition for the Hamiltonian cycle problem in random graphs with node degrees greater than one?

References

1. Cheeseman, P., Kanefsky, B., Taylor, W.M.: Where the really hard problems are. In: Proceedings of 12th International Joint Conference on Artificial Intelligence (IJCAI 1991), Sydney, Australia, pp. 331–337 (1991)
2. Cheeseman, P., Kanefsky, B., Taylor, W.M.: Computational complexity and phase transitions. In: Workshop of Physics and Computation (PhysComp 1992), Dallas, Texas, USA, pp. 63–68 (1992)
3. Mitchell, D., Selman, B., Levesque, H.: Hard and easy distributions of SAT problem. In: Proceedings of the 10th Conference on Artificial Intelligence (AAAI 1992), pp. 459–465, San Jose, CA (1992)
4. Kirkpatrick, S., Selman, B.: Critical behavior in the satisfiability of random boolean expressions. Science **264**, 1297–1301 (1994)
5. Monasson, R., Zecchina, R., Kirkpatrick, S., Selman, B., Troyansky, L.: Determining computational complexity from characteristic 'phase transitions'. Nature **400**, 133–137 (1999)
6. Xu, K., Li, W.: Phase transition of the SAT problem. Sci. China Ser. E **29**(4), 354–360 (1999)
7. Vandegriend, B., Culberson, J.: The G(n, m) phase transition is not hard for the Hamiltonian cycle problem. J. Artif. Intell. Res. **9**, 219–245 (1998)
8. Jäger, G., Zhang, W.: An effective algorithm for and phase transitions of the directed Hamiltonian cycle problem. J. Artif. Intell. Res. **39**, 663–687 (2010)
9. Zhang, W.: Phase transitions and backbones of the asymmetric traveling salesman problem. J. Artif. Intell. Res. **21**, 471–497 (2004)
10. Parkes, A.J.: Clustering at the phase transition. In: Proceedings of the 14th National Conference on Artifial Intelligence and the 9th Conference on Innovative Applications of Artifial Intelligence, Rhode Island, USA pp. 340–345 (1997)
11. Zhang, W.: Phase transitions and backbones of 3-SAT and maximum 3-SAT. In: Walsh, T. (ed.) CP 2001. LNCS, vol. 2239, pp. 153–167. Springer, Heidelberg (2001)
12. Mézard, M., Parisi, G., Zecchina, R.: Analytic and algorithmic solution of random satisfiability problems. Science **297**, 812–815 (2002)

13. Krzakala, F., Montanari, A., Ricci-Tersenghi, F., Semerjian, G., Zdeborová, L.: Gibbs states and the set of solutions of random constraint satisfaction problems. Proc. Natl. Acad. Sci. U.S.A **104**(25), 10318–10323 (2007)
14. Xu, K., Li, W.: Exact phase transitions in random constraint satisfaction problems. J. Artif. Intell. Res. **12**, 93–103 (2000)
15. Zhao, C., Zhang, P., Zheng, Z., Xu, K.: Analytical and belief-propagation studies of random constraint satisfaction problems with growing domains. Phys. Rev. E **85**, 016106 (2012)
16. Komlós, J., Szemerédi, E.: Limit distribution for the existence of a Hamiltonian cycle in a random graph. Discrete Math. **43**, 55–63 (1983)
17. Bollobás, B., Fenner, T.I., Frieze, A.M.: An algorithm for finding Hamilton paths and cycles in random graphs. Combinatorica **7**(4), 327–341 (1987)
18. Brunet, G.: A Brief Introduction to Hamilton Cycles in Random Graphs. Department of Computer Science, University of Toronto, Toronto, Canada (2005)
19. Frank, J., Martel, C.: Phase transitions in the properties of random graphs. In: Workshop on Studying and Solving Really Hard Problems, First International Conference on Principles and Practice of Constraint Programming, Cassis, France, pp. 62–69 (1995)
20. Frank, J., Gent, I.P., Walsh, T.: Asymptotic and finite size parameters for phase transitions: Hamiltonian circuit as a case study. Inf. Process. Lett. **65**(5), 241–245 (1998)
21. McDiarmid, C.J.H.: Cluster percolation and random graphs. Math. Program. Stud. **13**, 17–25 (1980)
22. Li, H.: Generalizations of Dirac's theorem in Hamiltonian graph theory - A survey. Discrete Math. **313**, 2034–2053 (2013)
23. Fan, G.: New sufficient conditions for cycles in graphs. J. Comb. Theor. Ser. B **37**, 221–227 (1984)
24. Frieze, A.M.: Finding Hamilton cycles in sparse random graphs. J. Comb. Theor. Ser. B **44**, 230–250 (1988)

Can a Breakpoint Graph be Decomposed into None Other Than 2-Cycles?

Lianrong Pu and Haitao Jiang$^{(\boxtimes)}$

School of Computer Science and Technology, Shandong University, Jinan, China
htjiang@sdu.edu.cn

Abstract. Breakpoint graph is a key data structure to study genome rearrangements. The problem of Breakpoint Graph Decomposition (BGD), which asks for a largest collection of edge-disjoint cycles in a breakpoint graph, is a crucial step in computing rearrangement distances between genomes. This problem for genomes of unsigned genes is proved NP-hard, and the best known approximation ratio is 1.4193+ϵ [1]. In this paper, we present a polynomial time algorithm to detect whether a breakpoint graph can be decomposed into none other than 2-cycles. Our algorithm can be used to detect if there exists a sorting scenario between two genomes without reusing any breakpoints.

Keywords: Breakpoint graph · Genome rearrangement · Cycle decomposition

1 Introduction

In 1993, Bafna and Pevzner first introduce an important data structure, *breakpoint graph*, to study the problem of sorting by reversals. A reversal inverts a subsequence in the genome. Given two genomes, the problem of computing reversal distance between them asks to find a shortest sequence of reversals that transform one genome into the other. This distance is often used as a measure of the evolutionary distance between genomes. Usually we use a permutation π to represent one genome, and the identity permutation to represent the other genome, thus we call this problem as *sorting by reversals*. The problem of sorting unsigned genomes by reversals was first studied by Kececioglu and Sankoff [2], and later proved to be NP-hard through the reduction from the problem of Breakpoint Graph Decomposition (BGD) by Caprara [3].

The problem of Breakpoint Graph Decomposition (BGD) is a crucial step in computing reversal distances as well as other rearrangement distances. For example, the reversal distance is given by the number of elements in the permutation plus one, minus the number of cycles, plus the number of hurdles, and plus one if a fortress is present [4]. The number of elements in the permutation is a constant number given by the input genomes, and the presence of hurdles and fortresses can be safely ignored in most cases [5–7]. Thus the reversal distance relies on the number of cycles in BGD. Similarly, computing the translocation

© Springer International Publishing Switzerland 2016
D. Zhu and S. Bereg (Eds.): FAW 2016, LNCS 9711, pp. 205–214, 2016.
DOI: 10.1007/978-3-319-39817-4_20

distances [8,9] and double-cut-and-join [10] distance is also related to a solution for BGD. Moreover, Breakpoint Graph Decomposition (BGD) also has applications in ortholog assignment in comparative genomics [11,12] and approximation of exemplar breakpoint distance in the presence of duplicated genes [13].

The BGD problem for genomes of unsigned genes is proved NP-hard [3], and the best known approximation ratio is $1.4193+\epsilon$ [1]. Lin and Jiang transform the instance of Breakpoint Graph Decomposition into an instance of the 6-Set Packing problem — because each subset intersects with at most 6 other subsets, this problem admits an approximation ratio of $1.4193+\epsilon$, based on a suitable modification on the collection of edge-disjoint length-2 cycles [1].

In this paper, we study the BGD problem in terms of edge-disjoint length-2 cycles. We present a polynomial time algorithm to detect whether a breakpoint graph can be decomposed into none other than 2-cycles. A cycle decomposition with none other than 2-cycles will be called as a *2-cycle decomposition* of the breakpoint graph in the following. More specifically, we first simplified the breakpoint graph and construct a *matching graph* for the simplified breakpoint graph. Then we further simplified the matching graph for which we will propose a polynomial time algorithm to detect whether it admits an *excellent match*. We will return a 2-cycle decomposition for the breakpoint graph if the simplified matching graph admits an excellent match. Our algorithm also can be used to test whether there exists a sorting scenario between two genomes without reusing any breakpoints (e.g., under the infinite sites model of genome evolution [14]) and have potential applications in deriving better approximation algorithms for computing rearrangement distance and improving ortholog assignment in comparative genomics.

2   Preliminaries

Let $\pi = (\pi_1, ..., \pi_n)$ be a permutation of $\{1, ..., n\}$, we can construct the *breakpoint graph*, $G(\pi)$, between π and the identity permutation as follows:

(1) add elements $\pi_0 = 0$ and $\pi_{n+1} = n + 1$ to π,
(2) construct a corresponding vertex for each element in π,
(3) join two vertices i and j by a black edge if i and j are in consecutive positions in π and $|i - j| \neq 1$, and join i and j by a gray edge if they are not in consecutive positions in π and $|i - j| = 1$.

Note that there is no cycle in $G(\pi)$ only containing edges of same color. A cycle in $G(\pi)$ is called *alternating* if the colors of every two consecutive edges in this cycle is different. In the following, when we mention cycles in a breakpoint graph, they are alternating cycles. The length of a cycle is the number of black edges (or gray edges) in this cycle. A cycle with length i is denoted by i-cycle.

A key technique used in the approximation algorithms for computing rearrangement distances is *Breakpoint Graph Decomposition (BGD)*. Each vertex in $G(\pi)$ is either isolated, or incident to one black edge and one gray edge, or incident to two black edges and two gray edges. If we split each vertex in $G(\pi)$

that incident to two black edges and two gray edges into two new vertices, and make each new vertex incident to one black edge and one gray edge respectively. Then we will get a new graph consisting of edge-disjoint cycles. This new graph corresponds to a *cycle decomposition* of $G(\pi)$. If a cycle decomposition of $G(\pi)$ contains none other than 2-cycles, we will call such a cycle decomposition as a *2-cycle decomposition* of $G(\pi)$. In this paper, we will detect if there exists a 2-cycle decomposition of $G(\pi)$ in polynomial time.

In order to detect a 2-cycle decomposition of $G(\pi)$, we will construct a new graph, denoted by *matching graph* $F(\pi)$, from $G(\pi)$. The matching graph $F(\pi)$ is constructed as follows:

(1) construct a vertex for each black edge in $G(\pi)$,
(2) connect two vertices u and v of $F(\pi)$ by an edge if the two black edges corresponding to u and v are contained in a 2-cycle in $G(\pi)$.

Here each edge in $F(\pi)$ corresponds to a 2-cycle in $G(\pi)$. Jiang *et al.* [1] showed that the degree of a vertex in $F(\pi)$ is at most 3 by the following lemma.

Lemma 1. *Every edge in a breakpoint graph belongs to at most 3 different 2-cycles.*

If there exists a perfect match M of $F(\pi)$ and all the 2-cycles corresponding to edges in M are edge-disjoint, then these 2-cycles provides a 2-cycle decomposition of $G(\pi)$. We call such a perfect match as an *excellent match* of $F(\pi)$.

3   Simplification on the Breakpoint Graph and the Matching Graph

In this section, we will apply a rule to simplify the breakpoint graph $G(\pi)$ and derive a simplified graph $G'(\pi)$. Then we construct a matching graph $F(\pi)$ from $G'(\pi)$. Afterwards, we further simplify the matching graph $F(\pi)$ to derive $F'(\pi)$. In the following, we will prove that there exists an excellent match of $F'(\pi)$ if and only if there exists a 2-cycle decomposition of $G(\pi)$.

Before presenting the rules to simplify the graphs, we first introduce a lemma that will be used later. For a 2-cycle u, if the four edges of u are also covered by two edge-disjoint 2-cycles v_0 and v_1, then we say u is *covered* by v_0 and v_1.

Lemma 2. *Assume that (1) $G_2(\pi)$ is a 2-cycle decomposition of $G(\pi)$, (2) 2-cycle u is not contained in $G_2(\pi)$, (3) u is covered by two cycles in $G_2(\pi)$. Then there exists another 2-cycle decomposition of $G(\pi)$ in which u is included.*

Proof. Assume that u is covered by two 2-cycles v_0 and v_1 in $G_2(\pi)$. If we view the cycles as edge sets, then $u \subset v_0 \bigcup v_1$. Let $u_1 = (v_0 \bigcup v_1) - u$, it is easy to see that u_1 is a 2-cycle too. Since the edge set of 2-cycles v_0 and v_1 is completely same with which of u and u_1, we can replace v_0 and v_1 with u and u_1 without changing other 2-cycles in $G_2(\pi)$. Thus we get another 2-cycle decomposition of $G(\pi)$ in which u is included.   □

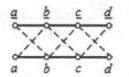 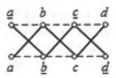

Fig. 1. A double tangle induced by vertices $\{a, b, c, d, \underline{a}, \underline{b}, \underline{c}, \underline{d}\}$, the solid lines represent black edges and the dotted lines represent gray edges, note two possible representations.

Now we will show how to simplify breakpoint graph $G(\pi)$ first. For four consecutive vertices $\{a, b, c, d\}$ in $G(\pi)$, if (1) the adjacent vertices of b and c are four consecutive vertices, say $\{\underline{a}, \underline{b}, \underline{c}, \underline{d}\}$, and the adjacent vertices of b are not consecutive as well as c, (2) the adjacent vertices of $\underline{b}$ and $\underline{c}$ are $\{a, b, c, d\}$, and the adjacent vertices of $\underline{b}$ are not consecutive as well as $\underline{c}$, then the subgraph induced by $\{a, b, c, d, \underline{a}, \underline{b}, \underline{c}, \underline{d}\}$ is defined as a *double tangle* (as shown in Fig. 1). If $G(\pi)$ contains a double tangle, we will remove it from $G(\pi)$ to derive a simplified graph $G'(\pi)$. Note that removing a subgraph from $G(\pi)$ includes removing all the edges in this subgraph from $G(\pi)$ as well as all the vertices with degree 0. Clearly, $G'(\pi)$ is also a breakpoint graph. In order to show that the above simplification is valid, we prove the following lemma,

Lemma 3. *There exists a 2-cycle decomposition of breakpoint graph $G(\pi)$, if only if there exists a 2-cycle decomposition of breakpoint graph $G'(\pi)$, where $G'(\pi)$ is derived from $G(\pi)$ by removing all the double tangles.*

Proof. Assume that $G(\pi)$ contains a double tangle as shown in Fig. 1. Let $v_0 = (a, \underline{b}, \underline{a}, b)$, $v_1 = (b, \underline{c}, \underline{b}, c)$, $v_2 = (c, \underline{d}, \underline{c}, d)$ be 2-cycles in $G(\pi)$.

(1) If $\leftarrow$: Assume there exists a 2-cycle decomposition of $G'(\pi)$, denoted by $G'_2(\pi)$. It is easy to prove that by adding v_0, v_1 and v_2 to $G'_2(\pi)$, we will obtain a 2-cycle decomposition of $G(\pi)$.

(2) Only if $\rightarrow$: Assume there exists a 2-cycle decomposition of $G(\pi)$, denoted by $G_2(\pi)$. If v_0, v_1 and v_2 are included in $G_2(\pi)$, then it is easy to see that $G_2(\pi) - v_0 - v_1 - v_2$ is a 2-cycle decomposition of $G'(\pi)$. Otherwise, we can always transform $G_2(\pi)$ to get another 2-cycle decomposition of $G(\pi)$ in which v_0, v_1 and v_2 are included.

Note that, a 2-cycle in $G_2(\pi)$ containing edge $(\underline{b}, \underline{c})$ must be $(a, b, \underline{c}, \underline{b})$, $(b, c, \underline{b}, \underline{c})$ or $(\underline{b}, \underline{c}, d, c)$. It cannot be cycle $(a, \underline{b}, \underline{c}, d)$, otherwise we will have a cycle (a, b, c, d) with only black edges, a contradiction to the property of breakpoint graph.

(2.1) If the 2-cycle in $G_2(\pi)$ containing edge $(\underline{b}, \underline{c})$ is $(a, b, \underline{c}, \underline{b})$, then $(\underline{a}, \underline{b}, c, b)$ must be another 2-cycle in $G_2(\pi)$ since it is the only 2-cycle containing edge $(\underline{a}, \underline{b})$ and edge-disjoint with cycle $(a, b, \underline{c}, \underline{b})$. Then we can replace 2-cycles $(a, b, \underline{c}, \underline{b})$ and $(\underline{a}, \underline{b}, c, b)$ in $G_2(\pi)$ with v_0 and v_1. If v_2 is contained in $G_2(\pi)$, then we get a 2-cycle decomposition of $G(\pi)$ containing v_0, v_1

and v_2. Otherwise, v_2 must be covered by two 2-cycles in $G_2(\pi)$, thus it can be introduced into $G_2(\pi)$ by Lemma 2.

(2.2) If the 2-cycle in $G_2(\pi)$ containing edge $(\underline{b}, \underline{c})$ is $(b, c, \underline{b}, \underline{c})$, namely v_1 is contained in $G_2(\pi)$. Then v_0 and v_2 are either contained in $G_2(\pi)$ or covered by two 2-cycles in $G_2(\pi)$ respectively. If they are contained in $G_2(\pi)$, then $G_2(\pi)$ a 2-cycle decomposition of $G(\pi)$ containing v_0, v_1 and v_2. Otherwise, v_0 and v_2 can be introduced into $G_2(\pi)$ by Lemma 2.

(2.3) If the 2-cycle in $G_2(\pi)$ containing edge $(\underline{b}, \underline{c})$ is $(\underline{b}, \underline{c}, d, c)$, we have the similar argument from case (2.1).

In summary, if $G(\pi)$ a 2-cycle decomposition, we can always transform $G_2(\pi)$ to get another 2-cycle decomposition of $G(\pi)$ in which v_0, v_1 and v_2 are included. □

Let's formalize the above process to simplify the breakpoint graph as *double tangle selection rule*:

Double Tangle Selection Rule: If $G(\pi)$ contains a double tangle (as shown in Fig. 1), remove it from $G(\pi)$.

After applying double tangle selection rule, we get a new graph denoted by $G'(\pi)$. We then construct a matching graph $F(\pi)$ from $G'(\pi)$ as introduced in the preliminaries. In the following, we further simplify the matching graph $F(\pi)$ into $F'(\pi)$, for which we design a polynomial algorithm to detect an excellent match in the next section.

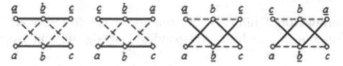

Fig. 2. A tangle induced by vertices $\{a, b, c, \underline{a}, \underline{b}, \underline{c}\}$, the solid lines represent black edges and the dotted lines represent gray edges, note four possible representations.

For three consecutive vertices $\{a, b, c\}$ and $\{\underline{a}, \underline{b}, \underline{c}\}$, if b is incident to $\underline{a}$ and $\underline{c}$ while $\underline{b}$ is incident to a and c, then we define the subgraph induced by vertices $\{a, b, c, \underline{a}, \underline{b}, \underline{c}\}$ as a *tangle* (as shown in Fig. 2). Let $v_0 = (a, \underline{b}, \underline{a}, b)$, $v_1 = (b, \underline{c}, \underline{b}, c)$ be two edge-disjoint 2-cycles in the tangle. If there exists a tangle in breakpoint graph $G'(\pi)$, then there must exist two edges corresponding to v_0 and v_1 in $F(\pi)$. We can always delete these two edges in $F(\pi)$ to get a simpler graph $F'(\pi)$. By the following lemma, we will find that $F'(\pi)$ is a valid simplification of $F(\pi)$.

Lemma 4. *There exists an excellent match of $F(\pi)$, if and only if there exists an excellent match of $F'(\pi)$, where $F'(\pi)$ is derived from $F(\pi)$ by deleting two edges corresponding to two edge-disjoint 2-cycles in a tangle of $G'(\pi)$.*

Proof. Assume that $G'(\pi)$ contains a tangle as shown in Fig. 2. Let $v_0 = (a, \underline{b}, \underline{a}, b)$, $v_1 = (b, \underline{c}, \underline{b}, c)$, $u_0 = (a, b, \underline{c}, \underline{b})$, $u_1 = (b, c, \underline{b}, \underline{a})$ be 2-cycles in $G'(\pi)$. We also use v_0, v_1, u_0 and u_1 to denote the corresponding edges in $F(\pi)$. Without loss of generality, we delete edges v_0 and v_1 from $F(\pi)$.

(1) If ←: Assume there exist an excellent match of $F'(\pi)$, denoted by $M'(\pi)$. Note that the vertex set of $F(\pi)$ is completely same with which of $F'(\pi)$. Thus $M'(\pi)$ is also an excellent match of $F(\pi)$.

(2) Only if →: Assume there exist an excellent match of $F(\pi)$, denoted by $M(\pi)$. If edges v_0 and v_1 are not contained in $M(\pi)$, apparently $M(\pi)$ is also an excellent match of $F'(\pi)$. Otherwise, we can always transform $M(\pi)$ to get another excellent match of $F(\pi)$ in which v_0 and v_1 are excluded.

(2.1) If both edges v_0 and v_1 are contained in $M(\pi)$, then we can replace edges v_0 and v_1 in $M(\pi)$ with edges u_0 and u_1 to get another excellent match of $F(\pi)$ in which v_0 and v_1 are excluded.

(2.2) If just one of edges v_0 and v_1 is contained in $M(\pi)$, without loss of generality, we can assume v_0 is contained in $M(\pi)$ while v_1 is not. Let $G'_2(\pi)$ denote the 2-cycle decomposition of $G'(\pi)$ corresponding to $M(\pi)$, then cycle v_0 is contained in $G'_2(\pi)$ while cycle v_1 is not. However, cycle v_1 must be covered by two cycles in $G'_2(\pi)$. By Lemma 2, cycle v_1 can be introduced to $G'_2(\pi)$. Then we obtain a corresponding excellent match of $F(\pi)$ containing edges v_0 and v_1. Similar argument from case (2.1) can be applied.

In conclusion, if $F(\pi)$ admits an excellent match, we can always transform $M(\pi)$ to get another excellent match of $F(\pi)$ in which v_0 and v_1 are excluded.   □

Let's formalize the above process to simplify the matching graph $F(\pi)$ as *tangle elimination rule*:

Tangle Elimination Rule: If $G'(\pi)$ contains a tangle (as shown in Fig. 2), delete two edges in $F(\pi)$ which correspond to two edge-disjoint 2-cycles in the tangle.

It is easy to see that the double tangles and tangles in $G(\pi)$ can be found in polynomial time, thus we can complete the simplification process in this section polynomially. Since we have proved that there exists an excellent match of $F'(\pi)$ if and only if there exists a 2-cycle decomposition of $G(\pi)$. In the next section we will propose a polynomial algorithm to detect whether $F'(\pi)$ admits an excellent match or not.

4   Our Algorithm

In this section, we will try to find an excellent match in matching graph $F'(\pi)$ if it exists. Assume edges (or vertices) contained in an excellent match are called *matching edge* (or *matching vertices*). Note that if there exists a vertex in $F'(\pi)$ with degree 0, then $F'(\pi)$ doesn't admit an excellent match and our algorithm returns false. Thus, in the following we always assume that the degree of a vertex in $F'(\pi)$ is at least 1.

If there exists a vertex with degree 1 in $F'(\pi)$, say v_1. Let v_2 be the only adjacent vertex of v_1. In order to get an excellent match in $F'(\pi)$, edge (v_1, v_2) must be chosen as a matching edge. Otherwise, v_1 can not be a matching vertex.

Note that when one edge is chosen as a matching edge, the two vertices incident to this edge will be deleted from $F'(\pi)$ as well as the edges incident to these two vertices.

When there are no vertices in $F'(\pi)$ with degree 1 any more, by Lemma 1, the degree of vertices left in $F'(\pi)$ is 2 or 3. However, by the following lemma, we will show that there are no vertices with degree 3 left in $F'(\pi)$.

Lemma 5. *If $F'(\pi)$ contains a vertex with degree 3, then it must contain at least one vertex with degree less than 2.*

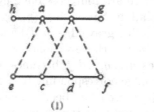 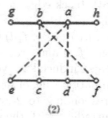

Fig. 3. (1) and (2) are two presentations of a black edge belonging to three different 2-cycles.

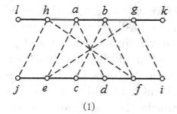 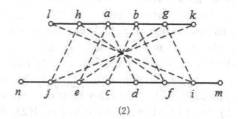

Fig. 4. (1) the breakpoint graph after making black edges (b, g) and (a, h) belong to two different 2-cycles. (2) the breakpoint graph after making black edges (f, i) and (e, j) belong to two different 2-cycles.

Proof. Let u be a vertex in $F'(\pi)$ with degree 3, and (a, b) be the corresponding black edge in $G'(\pi)$ that belongs to 3 different 2-cycles. Then $G'(\pi)$ must contain an induced subgraph as shown in Fig. 3(1) or (2). Note that Fig. 3(1) and (2) are symmetric. Thus, without loss of generality, we will prove the case shown in Fig. 3(1), where black edge (a, b) belongs to 2-cycles (a, b, c, d), (a, b, c, e) and (a, b, f, d). Now we prove it by contradictory. Assume all vertices in $F'(\pi)$ have degree at least 2, *i.e.*, all black edges in $G'(\pi)$ belong to at least two different 2−cycles.

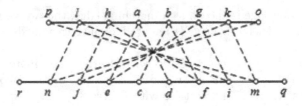

Fig. 5. The breakpoint graph after making black edges (i, m) and (j, n) belong to two different 2-cycles.

Let's start with black edge (b, g). There are four possible 2-cycles that may contain black edge (b, g): (g, b, c, e) (by adding gray edge (g, e)), (g, b, c, d) (by adding gray edge (g, d)), (g, b, f, d) (by adding gray edge (g, d)) and (g, b, f, i) (by adding black edge (f, i) and gray edge (g, i)). However, once we add gray edge (g, d), the subgraph induced by vertices a, b, g, c, d, f becomes a tangle. After using tangle elimination rule, either the edge corresponding to 2-cycle (a, b, f, d) or the edge corresponding to 2-cycle (a, b, c, d) has been removed from $F'(\pi)$ already. This contradicts with the assumption that the degree of u is 3 in $F'(\pi)$. Thus, we can not add gray edge (g, d) to Fig. 3(1). The only way to make (b, g) belong to two 2-cycles is adding gray edges (g, e), (g, i) and black edge (f, i) to obtain 2-cycles (g, b, c, e) and (g, b, f, i). For black edge (a, h), according to same argument gray edges (h, f), (h, j) and black edge (e, j) are added to obtain 2-cycles (h, a, d, f) and (h, a, e, j) (as shown in Fig. 4(1)).

Now let's switch to black edge (f, i). It is already contained in 2-cycle (f, i, g, b). To find one more 2-cycle containing it, we will need to add gray edge (i, l) to get 2-cycle (i, f, h, l). For black edge (e, j), according to same argument, gray edge (j, k) will be added to get 2-cycle (j, e, g, k). Since vertices i and j are incident to two gray edges respectively now, we need to add two new black edges (i, m) and (j, n) as shown in Fig. 4(2). In order to make black edges (i, m) and (j, n) belong to two 2-cycles respectively, we will need to add new gray edges (m, k), (m, p) and (n, l), (n, o) to Fig. 4(2). Afterwards, there will have new black edges (m, q) and (n, r) need to be included in two 2-cycles respectively (as shown in Fig. 5). In order to make sure each black edge belongs to at least two different 2-cycles, this iterative process will not stop and the corresponding breakpoint graph grows into infinite size, a contradiction. Thus $F'(\pi)$ must contain at least one vertex with degree less than 2. □

By Lemma 5, all the vertices left in $F'(\pi)$ are with degree 2. $F'(\pi)$ must be a set of edge disjoint cycles. If there exists an odd cycle in $F'(\pi)$, $F'(\pi)$ does not have an excellent match. The algorithm returns false. Otherwise, all cycles left in $F'(\pi)$ are even cycles, and the algorithm obtains a perfect match by choosing edges alternatively. If such a perfect match is an excellent match, then our algorithm returns the corresponding 2-cycles otherwise it returns false.

We summarize our polynomial algorithm to detect an excellent match of matching graph $F'(\pi)$ in Fig. 6.

Algorithm *to detect excellent match of $F'(\pi)$*

Input: A matching graph $F'(\pi)$.

Output: Return $M(\pi)$ as an excellent match of $F'(\pi)$ if $F'(\pi)$ has an excellent match
　　　　otherwise return false.

1　For each vertex with degree 1 in $F'(\pi)$, say v_1, do

2　　$M(\pi) \leftarrow (v_1, v_2)$, where v_2 is the only one adjacent vertex of v_1.

3　　remove vertices v_1 and v_2 from $F'(\pi)$ as well as the edges incident to them.

4　End for

5　If there exist a odd cycle left in $F'(\pi)$, return false.

6　For each even cycle left in $F'(\pi)$, say $(v_1, v_2, ..., v_{2m})$, do

7　　if cycles corresponding to edge set $\{(v_1, v_2), (v_3, v_4), ..., (v_{2m-1}, v_{2m})\}$ are edge-disjoint,

8　　　$M(\pi) \leftarrow \{(v_1, v_2), (v_3, v_4), ..., (v_{2m-1}, v_{2m})\}$.

9　　　remove this cycle from $F'(\pi)$

10　　else if cycles corresponding to edge set $\{(v_2, v_3), (v_4, v_5), ..., (v_{2m}, v_1)\}$ are edge-disjoint,

11　　　$M(\pi) \leftarrow \{(v_2, v_3), (v_4, v_5), ..., (v_{2m}, v_1)\}$.

12　　　remove this cycle from $F'(\pi)$

13　　else return false.

14　End for

15　return $M(\pi)$ as an excellent match of $F'(\pi)$

Fig. 6. The polynomial algorithm to detect an excellent match of matching graph $F'(\pi)$, where $M(\pi)$ represents the excellent match of $F'(\pi)$ if it exists.

Theorem 6. *If there exists an excellent match of $F'(\pi)$, by Lemma 3 and the construction of matching graph, we can reconstruct a 2-cycle decomposition of $G(\pi)$ in polynomial time. Otherwise, $G(\pi)$ doesn't admit a 2-cycle decomposition.*

5　Conclusion

In this paper, we have proposed a polynomial-time algorithm to detect whether a breakpoint graph admit a 2-cycle decomposition or not. The 2-cycle decomposition of a breakpoint graph can be used to sort permutations by reversals without the reuse of breakpoints (e.g., under the infinite sites model of genome evolution [14]). The cycle decomposition of breakpoint graph is also a crucial step in other sorting problems such as sorting by translocations [8,9], or sorting by DCJs [10], and we will extend our results to study these problems in the future. Although detecting a 2-cycle decomposition of a breakpoint graph is in P, the complexity of finding the cycle decomposition with maximum 2-cycles is still unknown, for which we are also interested to explore.

References

1. Lin, G., Jiang, T.: A further improved approximation algorithm for breakpoint graph decomposition. J. Comb. Optim. **8**(2), 183–194 (2004)
2. Kececioglu, J., Sankoff, D.: Exact and approximation algorithms for the inversion distance between two chromosomes. Comb. Pattern Matching Fourth Ann. Symp. **684**, 87–105 (1993)

3. Caprara, A.: Sorting by reversals is difficult. In: Proceedings of the First Annual International Conference on Computational Molecular Biology (1997)
4. Hannenhalli, S., Pevzner, P.A.: Transforming cabbage into turnip (polynomial algorithm for sorting signed permutations by reversals). In: Proceedings of 27th Annual ACM Symposium on Theory of Comput (STOC) (1995)
5. Caprara, A.: On the tightness of the alternating-cycle lower bound for sorting by reversals. J. Comb. Optim. **3**(2), 149–182 (1999)
6. Swenson, K.M., Lin, Y., Rajan, V., Moret, B.M.E.: Hurdles and sorting by inversions: combinatorial, statistical, and experimental results. J. Comput. Biol. **16**(10), 1339–1351 (2009)
7. Swenson, K.M., Lin, Y., Rajan, V., Moret, B.M.E.: Hurdles hardly have to be heeded. In: Nelson, C.E., Vialette, S. (eds.) RECOMB-CG 2008. LNCS (LNBI), vol. 5267, pp. 241–251. Springer, Heidelberg (2008)
8. Hannenhalli, S.: Polynomial-time algorithm for computing translocation distance between genomes. Proc. Sixth Ann. Symp. Comb. Pattern Matching (CPM) **937**, 162–176 (1995)
9. Kececioglu, J., Ravi, R.: Of mice and men: algorithms for evolutionary distances between genomes with translocation. In: Proceedings Sixth Annual ACM-SIAM Symposium, Discrete Algorithms (SODA) (1995)
10. Jiang, H., Zhu, B., Zhu, D.: Algorithms for sorting unsigned linear genomes by the DCJ operations. Bioinformatics **27**(3), 311–316 (2010)
11. Chen, X., Zheng, J., Fu, Z., Nan, P., Zhong, Y., Lonardi, S., Jiang, T.: Assignment of orthologous genes via genome rearrangement. ACM/IEEE Trans. Comput. Bio. Bioinf. **2**(4), 302–315 (2005)
12. Shao, M., Lin, Y., Moret, B.: An exact algorithm to compute the DCJ distance for genomes with duplicate genes. In: Sharan, R. (ed.) RECOMB 2014. LNCS, vol. 8394, pp. 280–292. Springer, Heidelberg (2014)
13. Chen, Z., Fu, B., Zhu, B.: The approximability of the exemplar breakpoint distance problem. In: Cheng, S.-W., Poon, C.K. (eds.) AAIM 2006. LNCS, vol. 4041, pp. 291–302. Springer, Heidelberg (2006)
14. Ma, J., Ratan, A., Raney, B.J., Suh, B.B., Miller, W., Haussler, D.: The infinite sites model of genome evolution. Proc. Nat. Acad. Sci. **105**(38), 14254–14261 (2008)

A Much Faster Branch-and-Bound Algorithm for Finding a Maximum Clique

Etsuji Tomita[1]([⊠]), Kohei Yoshida[1], Takuro Hatta[1], Atsuki Nagao[1], Hiro Ito[1,2], and Mitsuo Wakatsuki[1]

[1] The University of Electro-Communications, Chofu, Tokyo 182-8585, Japan
tomita@ice.uec.ac.jp
[2] CREST, JST, Chiyoda, Tokyo 102-0076, Japan

Abstract. We present improvements to a branch-and-bound maximum-clique-finding algorithm MCS (WALCOM 2010, LNCS 5942, pp. 191–203) that was shown to be fast. First, we employ an efficient approximation algorithm for finding a maximum clique. Second, we make use of appropriate sorting of vertices only near the root of the search tree. Third, we employ a lightened approximate coloring mainly near the leaves of the search tree. A new algorithm obtained from MCS with the above improvements is named MCT. It is shown that MCT is much faster than MCS by extensive computational experiments. In particular, MCT is shown to be faster than MCS for gen400_p0.9_75 and gen400_p0.9_65 by over 328,000 and 77,000 times, respectively.

1 Introduction

We define a *clique* as a complete subgraph in which all pairs of vertices are adjacent to each other. Algorithms for finding a maximum clique (e.g., [18]) in a given graph have received much attention especially recently, since they have many applications. There has been much theoretical and experimental work on this problem [3,20]. In particular, while finding a maximum clique is a typical NP-hard problem, considerable progress has been made for solving this problem *in practice*. Furthermore, much faster algorithms are required in order to solve many practical problems. Along this line, Tomita et al. developed a series of branch-and-bound algorithms MCQ [16], MCR [17], and MCS [18] among others that run fast in practice. It was shown that MCS is relatively fast for many instances tested.

In this paper, we present improvements to MCS in order to make it much faster. First, we turn back to our original MCS [14] that employs an approximation algorithm for the maximum clique problem in order to obtain an initial lower bound on the size of a maximum clique. We choose here another approximation algorithm called *k-opt local search* [7] that runs quite fast. Second, we sort vertices as in MCR [17] and MCS [18] only appropriately near the root of the search tree. This technique is based on our successful earlier result [8]. Third, we employ lightened approximate coloring mainly near the leaves of the search

© Springer International Publishing Switzerland 2016
D. Zhu and S. Bereg (Eds.): FAW 2016, LNCS 9711, pp. 215–226, 2016.
DOI: 10.1007/978-3-319-39817-4_21

tree [8]. A new algorithm obtained from MCS with the above improvements is named MCT. It is shown that MCT is much faster than MCS by extensive computational experiments.

2   Definitions and Notation

We are concerned with a simple undirected graph $G = (V, E)$ with a finite set V of vertices and a finite set E of edges. The set V of vertices is considered to be *ordered*, and the i-th element in it is denoted by $V[i]$. A pair of vertices v and w are said to be adjacent if $(v, w) \in E$. For a vertex $v \in V$, let $\Gamma(v)$ be the set of all vertices that are adjacent to v in $G = (V, E)$. We call $|\Gamma(v)|$ the degree of v. For a subset $R \subseteq V$ of vertices, $G(R) = (R, E \cap (R \times R))$ is an *induced subgraph*. An induced subgraph $G(Q)$ is said to be a *clique* if $(v, w) \in E$ for all $v, w \in Q \subseteq V$, with $v \neq w$. In this case, we may simply say that Q is a clique. A largest clique in a graph is called a *maximum clique*, and the number of vertices in a maximum clique in $G(R)$ is denoted by $\omega(R)$.

3   Maximum Clique Algorithm MCS

3.1   Search Tree

The preceding branch-and-bound algorithm MCS [18] begins with a small clique and continues by finding larger and larger cliques. More precisely, we maintain global variables Q and Q_{max}, where Q consists of the vertices of the current clique and Q_{max} consists of the vertices of the largest clique found so far. Let $R \subseteq V$ consist of vertices (*candidates*) that can be added to Q. We begin the algorithm by letting $Q := \emptyset$, $Q_{max} := \emptyset$, and $R := V$ (the set of all vertices). We select a certain vertex p from R, add it to Q ($Q := Q \cup \{p\}$), and then compute $R_p := R \cap \Gamma(p)$ as the new set of *candidate* vertices. Such a procedure is represented by a *search tree*, where the root is V and, whenever $R_p := R \cap \Gamma(p)$ is applied then R_p is a child of R. The edge between R and $R_p := R \cap \Gamma(p)$ is called a *branch*.

3.2   Approximate Coloring: Numbering

In order to prune unnecessary searching, we used *greedy approximate coloring* or *Numbering* of the vertices in MCS. That is, each $p \in R$ is *sequentially* assigned a minimum possible positive integer value $No[p]$, called the *Number* or *Color* of p, such that $No[p] \neq No[r]$ if $(p, r) \in E$. Consequently, we have that $\omega(R) \leq Max\{No[p] | p \in R\}$. Hence, if $|Q| + Max\{No[p] | p \in R\} \leq |Q_{max}|$ holds, we need not continue the search for R.

After *Numbers* (*Colors*) are assigned to all vertices in R, we sort the vertices in nondecreasing order with respect to their *Numbers*. Vertices are expanded for searching from the rightmost to the leftmost on this R. Let $Max\{No[r] | r \in R\} = maxno$ and $C_i = \{r \in R | No[r] = i\}$, where $i = 1, 2, \ldots, maxno$.

3.3   Re-NUMBER

Because of the *bounding condition* mentioned above, if $No[r] \leq |Q_{max}| - |Q|$, then it is not necessary to search from vertex r. When we encounter a vertex p with $No[p] > |Q_{max}| - |Q|$, we attempt to change its *Number* by **Procedure Re-NUMBER** described in Fig. 1, where No_p denotes the original value of $No[p]$ and $No_{th} := |Q_{max}| - |Q|$ stands for $No_{threshold}$. Try to find a vertex q in $\Gamma(p)$ such that $No[q] = k_1 \leq No_{th} - 1$, with $|C_{k_1}| = 1$. If such q is found, then try to find NUMBER k_2 such that no vertex in $\Gamma(q)$ has *Number* k_2. If such number k_2 is found, then exchange the NUMBERs of q and p so that $No[q] = k_2$ and $No[p] = k_1$. When the vertex q with NUMBER k_2 is found in Fig. 1, $No[p]$ is changed from No_p to k_1 ($\leq No_{th} - 1$); thus, *it is no longer necessary to search from p*.

Procedure. Re-NUMBER was first proposed in MCS [14] and is shown to be quite effective [14, 18, 19].

procedure Re-NUMBER$(p, No_p,$
$\qquad\qquad No, C_1, C_2, ..., C_{maxno})$
begin
$\quad No_{th} := |Q_{max}| - |Q|;$
$\quad$**for** $k_1 := 1$ **to** $No_{th} - 1$ **do**
$\quad\quad$**if** $|C_{k_1} \cap \Gamma(p)| = 1$ **then**
$\quad\quad\quad q :=$ the element in $(C_{k_1} \cap \Gamma(p))$;
$\quad\quad\quad$**for** $k_2 := k_1 + 1$ **to** No_{th} **do**
$\quad\quad\quad$**if** $C_{k_2} \cap \Gamma(q) = \emptyset$ **then**
$\quad\quad\quad\quad$ {Exchange the *Numbers*
$\qquad\qquad\qquad\qquad$ of p and q.}
$\quad\quad\quad\quad C_{No_p} := C_{No_p} - \{p\};$
$\quad\quad\quad\quad C_{k_1} := (C_{k_1} - \{q\}) \cup \{p\};$
$\quad\quad\quad\quad No[p] := k_1;$
$\quad\quad\quad\quad C_{k_2} := C_{k_2} \cup \{q\};$
$\quad\quad\quad\quad No[q] := k_2;$
$\quad\quad\quad\quad$**return**
$\quad\quad\quad$**fi od fi**
$\quad$**od**
end { of Re-NUMBER}

Fig. 1. Procedure Re-NUMBER

3.4   EXTENDED INITIAL SORT-NUMBER

At the beginning of MCR and MCS, vertices are sorted in nondecreasing order from the rightmost to the leftmost mainly with respect to their degrees [17, 18]. In addition, vertices are assigned initial *Numbers*. More precisely, the steps from {SORT} to just above EXPAND(V, No) in Fig. 4 (Algorithm MCR) in [17] is named *EXTENDED INITIAL SORT-NUMBER* to V. Note that *global variable* Q_{max} can be updated by "**then** $Q_{max} := R_{min}$" at the final stage of Fig. 4 (Algorithm MCR) in [17].

Here, MCS introduced another new *adjunct ordered set* V_a of vertices in order to preserve the order of the vertices sorted by EXTENDED INITIAL SORT-NUMBER. Approximate coloring is carried out in the order of V_a from the left to the right. Lastly, we reconstruct the adjacency matrix in MCS just after EXTENDED INITIAL SORT-NUMBER. This is to establish a more effective use of the cache memory.

The algorithm obtained as above is named MCS [18, 19].

4   Improved Algorithms

4.1   Effective Use of an Approximate Solution

When the algorithm MCS was first proposed in [14], the first part of MCS consisted of a procedure for finding an approximately maximum clique of the given graph. Its approximation algorithm named init-lb [14] is a local search algorithm based on our previous work [15]. It finds a near-maximum clique in a very short time, and the result is used as an initial lower bound of the size of a maximum clique. It demonstrated the effectiveness of an approximate solution for finding an exactly maximum clique. More precisely, when a sufficiently large near-maximum clique Q'_{max} is found, we let $Q_{max} := Q'_{max}$ at the beginning of MCS. Then $No_{th} := |Q_{max}| - |Q|$ becomes large and the bounding condition becomes more effective.

The final version of MCS presented in [18,19] excluded a procedure for finding an approximately maximum clique. This is because it is important to examine the performance of the main body of MCS [18] itself independently of an approximation algorithm. Batsyn et al. [1] and Maslov et al. [12] also demonstrated the effectiveness of an approximate solution, independently.

We have many approximation algorithms for finding a maximum clique [20], while finding a good approximate solution for the maximum clique problem is considered to be very hard [21]. The most important problem is a proper choice of the trade-off between the quality of the approximate solution and the time required to obtain it. We now turn back to our original MCS in [14] and choose another approximation algorithm called *k-opt local search* [7]. It does not necessarily give the best quality solution, but it runs quite fast and it is easy to control the above trade-off. The *k-opt local search* repeats a number of local searches from different vertices of the given graph. In this repetition, we select a vertex with the largest degree one by one from the sorted vertices with respect to their degrees by EXTENDED INITIAL SORT-NUMBER. When the number of repetitions becomes large, the quality of the solution increases but with increased running time.

In order to give a proper compromise between the high quality of the solution and the time required to obtain it for the given graph $G = (V, E)$ with $n = |V|, m = |E|$, and $dens = 2m/n(n-1)$ (*density*), we have chosen the number *rep* of *repetitions* as follows by preliminary experiments:

$$rep = \min\{20n^{1/2} \times dens^3, \ n\} \quad for \ \ n \geq 1.$$

Hereafter, a procedure for finding an approximate maximum clique of the given graph $G = (V, E)$ under the above condition is named KLS(V, Q'_{max}) and its solution is given to Q'_{max}.

The new MCS that is composed of a combination of the KLS procedure and MCS in [18] as above is named MCS$_1$.

4.2   EXTENDED INITIAL SORT-NUMBER Near the Root of the Search Tree

It is shown that both search space and overall running time are reduced when vertices are sorted in a nondecreasing order with respect to their degrees prior to application of a branch-and-bound depth-first search for finding a maximum clique [4, 5, 15, 16]. All of the preceding algorithms MCQ, MCR, and MCS employ such sorting of vertices at the root level (*depth* = 0) of the search trees. It is also made clear that if the vertices are sorted as above and followed by *Numbering* at every depth of the search tree then the resulting search space becomes more reduced but with much more overhead of time [8].

Therefore, it becomes important to choose a good trade-off between the reduction of the search space and the time to realize it. For an earlier algorithm MCLIQ [15] that is a predecessor of MCQ, we proposed a technique to solve the above trade-off and reduced the overall running time successfully in the way as follows [8]:

(i)   At the first stage near the root of the search tree, we apply sorting of vertices followed by *Numbering*. ([8])

(ii) In the second stage of the search tree, we apply *Numbering* without new sorting of vertices. (Just as in [15])

(iii) In the third stage of the search tree near the leaves, we expand vertices by only inheriting the order of vertices and the previous NUMBERs. (Just as in [5]).

The above techniques are considered to be promising for any algorithm for finding a maximum clique if we control these three stages appropriately. So, we apply the techniques of [8] to MCS. Here, we make full use of the adjunct ordered set V_a of vertices in MCS [18] in which vertices are sorted in nondecreasing order with respect to their degrees from the rightmost (end) to the leftmost (front) by EXTENDED INITIAL SORT-NUMBER in [18]. In addition, we avoid the set R of vertices in MCS [18] so that we are free from the task of reconstructing such R in which vertices are sorted with respect to their NUMBERs. From now on, we rename V_a as R, for simplicity. So, be careful that the set R in this paper corresponds to V_a, and not to R in MCS [18].

Hereafter, the NUMBERing procedure combined with Re-NUMBER is named NUMBER-R and is shown in Fig. 2. This is exactly the first half of the **procedure** Re-NUMBER-SORT in Fig. 2 of MCS [18].

A slightly stronger procedure Re-NUMBER1 is defined as the one obtained from **procedure** Re-NUMBER by replacing "**for** $k_2 := k_1 + 1$ **to** No_{th} **do**" by "**for** $k_2 := 1$ **to** $k_1 - 1$ **and** $k_1 + 1$ **to** No_{th} **do**". Another slightly modified **procedure** NUMBER-R+(R, No) is defined as the one obtained from **procedure** NUMBER-R(R, No) by replacing "**if** $(k > No_{th})$ **and** $(k = maxno)$ **then**" by "**if** $(k > No_{th})$ **then**" and "Re-NUMBER-R" by "Re-NUMBER1" in NUMBER-R(R, No). That is, the condition for applying Re-NUMBER is relaxed in **procedure** NUMBER-R+(R, No).

procedure NUMBER-R(R, No)
begin
  {NUMBER}
  $maxno := 0$;
  $C_1 := \emptyset$;
  for $i := 1$ **to** $|R|$ **do**
    { Conventional greedy
      approximate coloring }
    $p := R[i]$;
    $k := 1$;
    while $C_k \cap \Gamma(p) \neq \emptyset$
      do $k := k + 1$ **od**
    if $k > maxno$ **then**
      $maxno := k$;
      $C_{maxno} := \emptyset$
    fi
    $C_k := C_k \cup \{p\}$;
    $No[p] := k$;

    { - Re-NUMBER starts - }
    $No_{th} := |Q_{max}| - |Q|$;
    if $(k > No_{th})$ **and**
      $(k = maxno)$ **then**
      Re-NUMBER($p, k, No,$
        $C_1, C_2, ..., C_{maxno}$) ;
      if $C_{maxno} = \emptyset$ **then**
        $maxno := maxno - 1$
      fi
    fi
    { - Re-NUMBER ends - }

  od
end { of NUMBER-R }

Fig. 2. Procedure NUMBER-R

procedure NUMBER-RL($R, No, newNo$)
begin
  $No_{th} := |Q_{max}| - |Q|$;
  for $i := 1$ **to** $|R|$ **do**
    $C_i := \emptyset$;
  od
  $maxno := 1$;
  for $i := 1$ **to** $|R|$ **do**
    if $No[R[i]] \leq No_{th}$ **then**
      $k := No[R[i]]$;
      if $k > maxno$ **then** $maxno := k$ **fi**
      $C_k := C_k \cup \{R[i]\}$; $newNo[R[i]] := k$;
    fi
  od
  for $i := 1$ **to** $|R|$ **do**
    if $No[R[i]] > No_{th}$ **then**
      $p := R[i]$; $k := 1$;
      while $C_k \cap \Gamma(p) \neq \emptyset$
        do $k := k + 1$ **od**
      if $k > maxno$ **then**
        $maxno := k$;
      fi
      $C_k := C_k \cup \{p\}$;
      $newNo[p] := k$;
      if $(k > No_{th})$ **then**
        Re-NUMBER1($p, k, No,$
          $C_1, C_2, ..., C_{maxno}$) ;
        if $C_{maxno} = \emptyset$ **then**
          $maxno := maxno - 1$
      fi
    fi
  od
end { of NUMBER-RL}

Fig. 3. Procedure NUMBER-RL

At the first stage near and including the root of the search tree, we sort a set of vertices by EXTENDED INITIAL SORT-NUMBER to R followed by Numbering by NUMBER-R+(R, No). The procedure is shown in Fig. 4 with "$Th_1 = 0.4, Th_2 = 0$" instead of "$Th_1 = 0.4, Th_2 = 0.1$" at {Switches}. It is experimentally confirmed that NUMBER-R+(R, No) is better than NUMBER-R(R, No), since NUMBER-R+(R, No) is applied only a few times with better results but with more overhead than NUMBER-R(R, No).

This task of preprocessing (of sorting vertices followed by NUMBER-R) is time-consuming. So, as stated at the beginning of Sect. 4.2, it is important to change this first stage to the second stage at an appropriate switching depth that is near the root of the search tree. First, for a vertex $p \in R$ at a certain depth of the search tree, consider $newR := R_p = R \cap \Gamma(p)$ that is a child of R. If the ratio $|\{v|No[v] > No_{th}\}|/|newR|$ becomes large, it is considered that much more preprocessing becomes appropriate. In addition, when $dens$ (density) of the graph becomes larger it generally requires more time for finding a maximum clique and then much more number of preprocessing becomes appropriate. As a result, we consider the following value:

$$T = \frac{|\{v|No[v] > No_{th}\}|}{|newR|} \times dens.$$

From preliminary experiments, we have chosen that if $T \geq 0.4$ then we continue the same procedure described for the first stage. Otherwise, we switch the stage to the second stage. Thus, we let $Th_1 := 0.4$ in Fig. 4. The new procedure obtained from Fig. 4 by replacing "$Th_1 := 0.4, Th_2 := 0.1$" by "$Th_1 := 0.4, Th_2 := 0$" at {Switches} is named MCS$_2$. Here, we control the $stage = 1$ so that it never returns back to $stage = 1$ after it changed to the second or the third $stage(\neq 1)$. Konc and Janežič [9] also improved MCQ [16] successfully in a similar way as in [8], independently.

4.3   Lightened Numbering Mainly Near the Leaves of the Search Tree

Mainly near the leaves of the search tree, the ratio $|\{v|No[v] > No_{th}\}|/|newR|$ tends to be small. In this third stage, it is preferable to lighten the task of preprocessing before expansion of vertices. So, we only inherit the order of vertices from that in their parent depth, as in the second stage. In addition, we inherit the assigned NUMBERs from those assigned to their parents only if their NUMBERs are less than or equal to No_{th}. If we inherit all the assigned NUMBERs from those assigned to their parents as in [5] the resulting bounding condition becomes too weak. In order to remedy this weakness, if the inherited NUMBERs from those assigned to their parents are greater than No_{th} then we give them new NUMBERs. For vertices whose inherited NUMBERs from their parents are greater than No_{th} we newly give them NUMBERs by sequential numbering combined with Re-Numbering. For this Re-Numbering we adopt stronger Re-NUMBER1 instead of Re-NUMBER since Re-Numbering is required not so many times in this stage. The resulting procedure in this stage named **procedure** NUMBER-RL is shown in Fig. 3.

From preliminary experiments, we have chosen to turn to the new $stage = 3$ if the previously given value $T = (|\{v|No[v] > No_{th}\}|/|newR|) \times dens$ is less than 0.1. Then we let $Th_2 := 0.1$ in Fig. 4. The **procedure** NUMBER-RL is weaker than the previous **procedure** NUMBER-R for obtaining strong bounding condition, but it requires less overhead than the previous one. However, if the given graph is too dense then **procedure** NUMBER-RL becomes too weak and the number of branches of the search tree grows quite large. So, we choose to go to new $stage = 3$ only if $dens \leq 0.95$. In addition, a simpler algorithm is generally better than sophisticated algorithms for sparse graphs.

procedure MCT($G = (V, E)$)
begin
  global $Q := \emptyset$;
  global $Q_{max} := \emptyset$;
  global $dens := 2|E|/|V|(|V| - 1)$;
    {$density$}
  if $dens \leq 0.1$ **then**
    MCS($\overline{G} = (V, E)$));
  else
    $Th_1 := 0.4$; $Th_2 := 0.1$;
      {Switches}
    Apply *EXTENDED INITIAL*
        SORT-NUMBER to V;
      {Q_{max} can be updated.}
    Reconstruct the adjacency
        matrix as described in [18];
    KLS(V, Q'_{max});
    if $Q_{max} < Q'_{max}$ **then**
    $Q_{max} := Q'_{max}$ **fi**
    NUMBER-R+(V, No);
      $stage := 1$;
    EXPAND ($V, No, stage, Th_1, Th_2$);
  fi
  output Q_{max} {Maximum clique}
end { of MCT}

Fig. 4. Procedure MCT

procedure EXPAND($R, No, stage, Th_1, Th_2$)
begin
  for $i := |R|$ **downto** 1 **do**
    $p := R[i]$;
    if ($stage = 1$ **and** $|Q| + \max_{v \in R}\{No[v]\} > |Q_{max}|$)
    or ($stage \neq 1$ **and** $|Q| + No[p] > |Q_{max}|$) **then**
      $Q := Q \cup \{p\}$;
      $newR := R \cap \Gamma(p)$;   {preserving the order}
      if $newR \neq \emptyset$ **then**
        $No_{th} := |Q_{max}| - |Q|$;
        $T := \frac{|\{v|No[v]>No_{th}\}|}{|newR|} \times dens$;
        if $stage = 1$ **and** $Th_1 \leq T$ **then**
          Apply *EXTENDED INITIAL*
            SORT-NUMBER to R;
          NUMBER-R+($newR, newNo$);
          {The initial value of $newNo$ has no significance.}
          $newstage := 1$;
        else if $dens > 0.95$ **or** $Th_2 \leq T$ **then**
          NUMBER-R($newR, newNo$);
          $newstage := 2$;
        else
          NUMBER-RL($newR, No, newNo$);
          $newstage := 3$;
        fi
        EXPAND($newR, newNo, newstage, Th_1, Th_2$)
      else if $|Q| > |Q_{max}|$ **then** $Q_{max} := Q$ **fi**
      $Q := Q - \{p\}$;
      $R := R - \{p\}$;   {preserving the order}
    fi
  od
end { of EXPAND }

Fig. 5. Procedure EXPAND

So, if $dens \leq 0.1$ we choose simpler algorithm MCS [18] without relying on any new technique introduced in this paper (Fig. 4).

The resulting algorithm obtained by taking the total techniques in Sects. 4.1, 4.2 and 4.3 to improve MCS [18] is named MCT (The 'T' is for 'Total'.) and is shown in Fig. 4.

5   Computational Experiments

In order to demonstrate the effectiveness of the techniques given in the previous section, we carried out computational experiments. All the algorithms were implemented in C language. The computer had an Intel core i7-4790 CPU of 3.6 GHz clock with 8 GB of RAM and 8 MB of cache memory. It worked on a Linux operating system with a compiler gcc -O3. The *dfmax running time for DIMACS benchmark instances* [6] for r300.5, r400.5 and r500.5 are 0.14, 0.90 and 3.44 seconds, respectively.

5.1   Stepwise Improvement

Table 1 shows stepwise improvement from MCS to MCT for selected graphs chosen from the next Table 2.

Table 1. Comparison of MCS, MCS$_1$, MCS$_2$ and MCT

Graph	Times [sec]				Branches [$\times 10^{-6}$]			
	MCS	MCS$_1$	MCS$_2$	MCT	MCS	MCS$_1$	MCS$_2$	MCT
brock400_1	288	256	182	116	89	77	52	55
brock800_4	1,768	1,751	1,256	819	381	380	258	270
C250.9	1,171	926	774	404	255	197	154	186
gen400_p0.9_55	22,536	1,651	1,970	167	2,895	181	210	61
gen400_p0.9_65	57,385	5.73	6.07	0.74	7,628	0.33	0.34	0.13
gen400_p0.9_75	108,298	1.38	1.38	0.33	17,153	0.05	0.05	0.02
p_hat700-3	900	456	438	216	88	43	40	54
p_hat1000-2	85	47	46	29	13	6.6	6.3	10
p_hat1500-2	6,299	2,964	2,832	1,560	560	253	234	400
san400_0.7_1	0.26	0.06	0.06	0.06	22,771	200	0	0
frb-30-15-2	1,048	691	773	116	229	135	148	61

(1) Improvement from MCS to **MCS$_1$** by an approximate solution in Sect. 4.1: The improvement is particularly quite effective for the gen graph family. MCS$_1$ is faster than MCS for gen400_p0.9_75 and gen400_p0.9_65 by more than 78,000 and 10,000 times, respectively. This technique is effective for almost all graphs but with few exceptions as for the MANN graph family.

(2) Improvement from MCS$_1$ to **MCS$_2$** by EXTENDED INITIAL SORT-NUMBER in Sect. 4.2: This technique is effective mainly for the brock graph family by around 1.4 times. For some graphs such as the gen and frb graph families, the effect is negative.

(3) Improvement from MCS$_2$ to **MCT** by Lightened Numbering in Sect. 4.3: This technique is effective for almost all graphs in reducing computing time in spite of increased numbers of branches in general. MCT is faster than MCS$_2$ for gen400_p0.9_55 and gen400_p0.9_65 by more than 11 and 8 times, respectively, where their numbers of branches are also reduced.

5.2   Overall Improvement

Table 2 shows the result of the overall improvement from MCS to MCT in computing time for the benchmark graphs where the columns *sol* and *time* below KLS show the solution and the computing time by KLS, respectively. The benchmark graphs include brock - DSJ graphs in DIMACS [6] and the frb family in BHOSLIB [2]. They also include random graphs of r200.8 - r20000.1 where r*n.p* stands for a random graph with the number of vertices = n and the edge probability = p. The averages are taken over 10 random graphs except for r200.9 and r200.95 whose averages are taken over 100 random graphs. The state-of-the-art result of BBMCX (MCX *for short*) [13] by Segundo et al. is included. Here, its computing time is calibrated on the established way in the Second DIMACS Implementation Challenge [6], where our computer is calculated to be 1.30 times

Table 2. CPU time [sec] for benchmark graphs

Graph Name	n	dens	ω	KLS sol	KLS time	MCS [18]	MCT	MCX [13]	MaxC [11]	I&M [12]	BG14 [1]
brock200_1	200	0.75	21	21	0.01	0.36	0.23	**0.18**	0.34	4.41	2.51
brock400_1	400	0.75	27	25	0.08	288	**116**	150	205	188	302
brock400_2	400	0.75	29	24	0.08	124	**52**	68	96	94	132
brock400_3	400	0.75	31	24	0.08	195	**86**	120	160	145	211
brock400_4	400	0.75	33	25	0.08	103	**46**	68	100	72	87
brock800_1	800	0.65	23	21	0.22	4,122	**1,950**	2,690	4,560	4,000	4,220
brock800_2	800	0.65	24	21	0.22	3,683	**1,630**	2,420	4,000	3,460	3,780
brock800_3	800	0.65	25	21	0.22	2,540	**1,110**	1,590	2,510	2,360	2,650
brock800_4	800	0.65	26	20	0.22	1,768	**819**	1,100	1,850	1,680	1,870
C250.9	250	0.90	44	44	0.08	1,171	404	713	**268**		
C2000.5	2000	0.50	16	15	0.59	33,899	**21,027**				
gen200_p0.9_44	200	0.90	44	44	0.05	0.174	**0.076**	0.155	0.115	1.68	
gen200_p0.9_55	200	0.90	55	55	0.06	0.458	**0.068**	0.312	0.142	2.43	0.917
gen400_p0.9_55	400	0.90	55	53	0.25	22,536	**167**	19,400		46,500	2,960
gen400_p0.9_65	400	0.90	65	65	0.26	57,385	**0.74**	66,100	36,700	2,130	19
gen400_p0.9_75	400	0.90	75	75	0.28	108,298	**0.33**	47,200	9,980	83.5	7.8
MANN_a27	378	0.99	126	126	0.81	0.26	1.05	0.18	**0.16**	1.30	
MANN_a45	1035	0.99	345	344	21.5	53.4	75.5	32.0	22.7	**17.3**	55.1
p_hat300-3	300	0.74	36	36	0.06	0.99	**0.28**	0.66	1.16	6.72	3.62
p_hat500-3	500	0.75	50	50	0.22	57.1	**17.4**	33.3	39.6	50.3	59.5
p_hat700-3	700	0.75	62	62	0.46	900	**216**	680	879	552	767
p_hat1000-2	1000	0.49	46	46	0.23	85	**29**	73	101	204	113
p_hat1000-3	1000	0.74	68	68	1.00	305,146	**38,800**				
p_hat1500-1	1500	0.25	12	11	0.03	1.8	**1.4**	2.0	10	478	422
p_hat1500-2	1500	0.51	65	65	0.73	6,299	**1,560**	3,850	8,030	5,350	5,430
san1000	1000	0.50	15	10	0.06	1.02	**0.21**	0.68	0.72	449	158
san200_0.7_1	200	0.70	30	30	0.01	**0.0037**	0.0133	0.0115	0.0092	7.62	
san200_0.9_1	200	0.90	70	70	0.07	0.0848	0.0727	0.0385	**0.0131**	1.35	
san400_0.7_1	400	0.70	40	40	0.06	0.26	**0.06**	0.14	0.13	15.80	6.69
san400_0.7_2	400	0.70	30	30	0.05	0.0589	**0.0519**	0.0923	0.0638	19.3	
san400_0.7_3	400	0.70	22	18	0.05	0.665	**0.273**	0.391	0.433	26.9	11.6
sanr200_0.7	200	0.70	18	18	0.01	0.15	0.11	**0.079**	0.17	5.05	1.03
sanr200_0.9	200	0.90	42	42	0.06	15.3	4.67	7.38	**4.21**	4.62	10.2
sanr400_0.5	400	0.50	13	13	0.01	0.351	0.274	**0.186**	0.688	34.9	17.6
sanr400_0.7	400	0.70	21	21	0.06	77.3	**40.7**	44.5	81.2	86.2	81.4
DSJC500.5	500	0.50	13	13	0.02	1.53	1.20	**0.81**	2.84		
DSJC1000.5	1000	0.50	15	15	0.12	141	**93**	102	265		
keller5	776	0.75	27	27	0.34	82,421	10,000	30,300	**4,980**	5,780	82,500
frb30-15-1	450	0.82	30	28	0.15	740	**156**	1,029	560		
frb30-15-2	450	0.82	30	30	0.15	1,048	**116**	672	758		
frb30-15-3	450	0.82	30	28	0.15	670	**124**	350	477		
frb30-15-4	450	0.82	30	28	0.15	2,248	**535**	1,157	955		
frb30-15-5	450	0.82	30	28	0.15	972	**156**	801	705		
r200.8	200	0.8	24-27	24-27	0.028	1.66	**0.78**	0.95	1.08		
r200.9	200	0.9	39-43	39-43	0.060	27.0	10.7	14.8	**6.2**		
r200.95	200	0.95	58-66	58-66	0.098	21.1	10.3	30.2	**2.5**		
r500.6	500	0.6	17-18	16-17	0.056	18	11	**10**	22		
r500.7	500	0.7	22-23	21-22	0.101	723	**340**	423	564		
r1000.4	1000	0.4	12	11	0.045	5.99	5.14	**4.52**	14.5		
r1000.5	1000	0.5	15-16	14-15	0.122	134	**92**	103	231		
r5000.1	5000	0.1	7	5-6	0.149	**1.17**	**1.17**	1.19	68		
r5000.2	5000	0.2	9-10	7-8	0.21	45	**39**	68	78		
r5000.3	5000	0.3	12	10-11	0.52	2,283	**1,875**				
r10000.1	10000	0.1	7	5-6	0.58	**14**	**14**	20	684		
r10000.2	10000	0.2	10	8-9	0.87	1,303	**1,139**				
r15000.1	15000	0.1	8	6	1.30	**62**	**62**	114	2,749		
r20000.1	20000	0.1	8	6-7	2.31	**234**	**234**				

faster than that in [13]. The calibrated computing time of MaxCLQ (MaxC *for short*) [10,11] by Li and Quan is also included from [13]. The calibrated computing time by ILS&MCS (I&M *for short*) [12] and BG14 [1] are added on the assumption that the performance of each MCS is the same, for reference, too. The boldface entries indicate the fastest time in the row.

The result shows that MCT is faster than MCS for graphs gen400_p0.9_75, gen400_p0.9_65, gen400_p0.9_55, frb-30-15-2, keller5, p_hat1000-3, gen200_p0.9_55, frb-30-15-5 and frb-30-15-3 by over 328,000, 77,000, 134, 9.0, 8.2, 7.8, 6.7, 6.2 and 5.4 times, respectively. MCT is faster than MCS for graphs san1000, frb-30-15-1, san400_0.7_1, frb-30-15-4, p_hat700-3 and p_hat1500-2 by over 4 times, and for graphs p_hat300-3, p_hat500-3 and sanr200_0.9 by over 3 times. In Table 2, MCT is faster than MCS by more than 2 times for the other 16 graphs including r200.9, r200.8, r500.7 and r200.95. Except for few special graphs as in MANN family and for easy graphs that can be solved in a very short time, MCT is faster than MCS for almost all graphs in the instances tested.

MCT is faster than the other algorithms in Table 2 for many instances. Note that MaxCLQ (MaxC) is fast for dense graphs.

In conclusion, MCT has achieved significant improvement over MCS, that is, MCT is much faster than MCS.

Acknowledgements. We express our sincere gratitude to the referees, E. Harley, and T. Toda for their useful comments and help. This research was supported in part by MEXT&JSPS KAKENHI Grants, JST CREST grant and Kayamori Foundation grant.

References

1. Batsyn, M., Goldengorin, B., Maslov, E., Pardalos, P.M.: Improvements to MCS algorithm for the maximum clique problem. J. Comb. Optim. **27**, 397–416 (2014)
2. http://www.nlsde.buaa.edu.cn/~kexu/benchmarks/graph-benchmarks.htm
3. Bomze, I.M., Budinich, M., Pardalos, P.M., Pelillo, M.: The maximum clique problem. In: Du, D.-Z., Pardalos, P.M. (eds.) Handbook of Combinatorial Optimization, Supplement, vol. A, pp. 1–74. Kluwer Academic Publishers, Boston (1999)
4. Carraghan, R., Pardalos, P.M.: An exact algorithm for the maximum clique problem. Oper. Res. Lett. **9**, 375–382 (1990)
5. Fujii, T., Tomita, E.: On efficient algorithms for finding a maximum clique, Technical report of IECE, AL81-113, 25–34 (1982)
6. Johnson, D.S., Trick, M.A. (eds.): Cliques, Coloring, and Satisfiability. DIMACS Series in DMTCS, vol. 26. American Mathematical Society, Boston (1996)
7. Katayama, K., Hamamoto, A., Narihisa, H.: An effective local search for the maximum clique problem. Inf. Process. Lett. **95**, 503–511 (2005)
8. Kohata, Y., Nishijima, T., Tomita, E., Fujihashi, C., Takahashi, H.: Efficient algorithms for finding a maximum clique, Technical report of IEICE, COMP89-113, 1–8 (1990)
9. Konc, J., Janežič, D.: An improved branch and bound algorithm for the maximum clique problem. MATCH Commun. Math. Comput. Chem. **58**, 569–590 (2007)
10. Li, C.M., Quan, Z.: An efficient branch-and-bound algorithm based on MaxSAT for the maximum clique problem. In: AAAI Conference on AI, pp. 128–133 (2010)

11. Li, C.M., Quan, Z.: Combining graph structure exploitation and propositional reasoning for the maximum clique problem. In: Proceedings of the IEEE ICTAI, pp. 344–351 (2010)
12. Maslov, E., Batsyn, M., Pardalos, P.M.: Speeding up branch and bound algorithms for solving the maximum clique problem. J. Glob. Optim. **59**, 1–21 (2014)
13. Segundo, P.S., Nikolaev, A., Batsyn, M.: Infra-chromatic bound for exact maximum clique search. Comput. Oper. Res. **64**, 293–303 (2015)
14. Sutani, Y., Higashi, T., Tomita, E., Takahashi, S., Nakatani, H.: A faster branch-and-bound algorithm for finding a maximum clique, Technical report of IPSJ, 2006-AL-108, 79–86 (2006)
15. Tomita, E., Kohata, Y., Takahashi, H.: A simple algorithm for finding a maximum clique, Technical report of the Univ. of Electro-Commun., UEC-TR-C5(1) (1988)
16. Tomita, E., Seki, T.: An efficient branch-and-bound algorithm for finding a maximum clique. In: Calude, C.S., Dinneen, M.J., Vajnovszki, V. (eds.) DMTCS 2003. LNCS, vol. 2731, pp. 278–289. Springer, Heidelberg (2003)
17. Tomita, E., Kameda, T.: An efficient branch-and-bound algorithm for finding a maximum clique with computational experiments. J. Glob. Optim. **37**, 95–111 (2007)
18. Tomita, E., Sutani, Y., Higashi, T., Takahashi, S., Wakatsuki, M.: A simple and faster branch-and-bound algorithm for finding a maximum clique. In: Rahman, M.S., Fujita, S. (eds.) WALCOM 2010. LNCS, vol. 5942, pp. 191–203. Springer, Heidelberg (2010)
19. Tomita, E., Sutani, Y., Higashi, T., Wakatsuki, M.: A simple and faster branch-and-bound algorithm for finding a maximum clique with computational experiments. IEICE Trans. Inf. Syst. **E96–D**, 1286–1298 (2013)
20. Wu, Q., Hao, J.K.: A review on algorithms for maximum clique problems. Eur. J. Oper. Res. **242**, 693–709 (2015)
21. Zuckerman, D.: Linear degree extractors and the inapproximability of max clique and chromatic number. In: Proceedings of the STOC 2006, pp. 681–690 (2006)

A PTAS for the Multiple Parallel Identical Multi-stage Flow-Shops to Minimize the Makespan

Weitian Tong[1,3], Eiji Miyano[2], Randy Goebel[3], and Guohui Lin[3(✉)]

[1] Department of Computer Sciences,
Georgia Southern University, Georgia, Statesboro 30460, USA
wtong@georgiasouthern.edu
[2] Department of Systems Design and Informatics,
Kyushu Institute of Technology, Iizuka, Fukuoka 820-8502, Japan
miyano@ces.kyutech.ac.jp
[3] Department of Computing Science,
University of Alberta, Edmonton, Alberta T6G 2E8, Canada
{rgoebel,guohui}@ualberta.ca

Abstract. In the *parallel k-stage flow-shops* problem, we are given m identical k-stage flow-shops and a set of jobs. Each job can be processed by any one of the flow-shops but switching between flow-shops is not allowed. The objective is to minimize the makespan, which is the finishing time of the last job. This problem generalizes the classical parallel identical machine scheduling (where $k = 1$) and the classical flow-shop scheduling (where $m = 1$) problems, and thus it is NP-hard. We present a polynomial-time approximation scheme for the problem, when m and k are fixed constants. The key technique is to enumerate over schedules for *big* jobs, solve a linear programming for *small* jobs, and add the fractional small jobs at the end. Such a technique has been used in the design of similar approximation schemes.

Keywords: Multiprocessor scheduling · Flow-shop scheduling · Makespan · Linear program · Polynomial-time approximation scheme

1 Introduction

In the *parallel k-stage flow-shop* problem, we are given m parallel identical k-stage flow-shops $F_1, F_2, \ldots, F_m$ and a set of n jobs $\mathcal{J} = \{J_1, J_2, \ldots, J_n\}$. These k-stage flow-shops are the *classic* flow-shops, each contains exactly one machine at every stage, *i.e.*, k sequential machines. Every job has k operations, and it can be assigned to exactly one of the m flow-shops for processing; once it is assigned to the flow-shop, its k operations are then respectively processed on the k sequential machines in the flow-shop. The goal is to minimize the makespan, which is the completion time of the last job. We denote the problem for simplicity as (m, k)-PFS. Let $M_{\ell,1}, M_{\ell,2}, \ldots, M_{\ell,k}$ denote the k sequential

© Springer International Publishing Switzerland 2016
D. Zhu and S. Bereg (Eds.): FAW 2016, LNCS 9711, pp. 227–237, 2016.
DOI: 10.1007/978-3-319-39817-4_22

machines in the flow-shop F_ℓ, for every ℓ. The job J_i is represented as a k-tuple $(p_{i,1}, p_{i,2}, \ldots, p_{i,k})$, where $p_{i,j}$ is the processing time for the j-th operation, that is, J_i needs to be processed non-preemptively on the j-th machine in the flow-shop to which the job is assigned. For all i, j, $p_{i,j}$ is a non-negative real number.

It is clear to see that, when $m = 1$, the (m, k)-PFS problem is the classic *flow-shop scheduling* [5] (a k-stage flow-shop); when $k = 1$, the (m, k)-PFS problem is the classic *multiprocessor scheduling* [5] (m parallel identical machines). When the two-stage flow-shops are involved, *i.e.*, $k = 2$, the $(m, 2)$-PFS problem has been previously studied in [4,13,23,26]. Here we first review the complexity and the approximation algorithms for the flow-shop scheduling and the multiprocessor scheduling problems.

For the k-stage flow-shop problem, it is known that when $k = 2$ or 3, there exists an optimal schedule that is a *permutation schedule*, in which the jobs are processed on all the k machines in the same order; but when $k \geq 4$, it is shown [3] that there may exist no optimal schedule that is a permutation schedule. Johnson [17] presented an $O(n \log n)$-time algorithm for the two-stage flow-shop problem, where n is the number of jobs; the k-stage flow-shop problem becomes *strongly* NP-hard when $k \geq 3$ [6]. After several efforts [2,6,7,17], Hall [12] designed a polynomial-time approximation scheme (PTAS) for the k-stage flow-shop problem, for any fixed constant $k \geq 3$. Due to the strong NP-hardness, such a PTAS is the best possible unless P = NP. When k is a part of the input (*i.e.*, an arbitrary integer), Williamson *et al.* [25] showed that the flow-shop scheduling cannot be approximated within 1.25; nevertheless, it remains unknown whether this case is APX-complete, that is, whether the problem admits a constant ratio approximation algorithm.

For the m-parallel identical machine scheduling problem, it is NP-hard when $m \geq 2$ [5]. When m is a fixed integer, the problem admits a pseudo-polynomial time exact algorithm [5], and Sahni [21] showed that this exact algorithm can be used to construct a *fully* PTAS (FPTAS); when m is a part of the input, the problem becomes strongly NP-hard, but still admits a PTAS by Hochbaum and Shmoys [14]. The *list-scheduling* algorithm by Graham [8] is a $(2 - 1/m)$-approximation, for arbitrary m.

Besides the (m, k)-PFS problem, another generalization of the flow-shop scheduling and the multiprocessor scheduling is the so-called *hybrid k-stage flow-shop* problem [19,20]. A hybrid k-stage flow-shop is a *flexible* flow-shop, that contains $m_j \geq 1$ parallel identical machines in the j-th stage, for $j = 1, 2, \ldots, k$. The problem is abbreviated as $(m_1, m_2, \ldots, m_k)$-HFS. A job J_i is again represented as a k-tuple $(p_{i,1}, p_{i,2}, \ldots, p_{i,k})$, where $p_{i,j}$ is the processing time for the j-th operation, which can be processed non-preemptively on any one of the m_j machines in the j-th stage. The objective of the $(m_1, m_2, \ldots, m_k)$-HFS problem is also to minimize the makespan. One clearly sees that when $m_1 = m_2 = \ldots = m_k = 1$, the problem reduces to the classic k-stage flow-shop problem; when $k = 1$, the problem reduces to the classic m-parallel identical machine scheduling problem.

The literature on the hybrid k-stage flow-shop problem $(m_1, m_2, \ldots, m_k)$-HFS is also rich [19,20], especially on the hybrid two-stage flow-shop problem (m_1, m_2)-HFS. First, $(1, 1)$-HFS is the classic two-stage flow-shop problem which can be optimally solved in $O(n \log n)$ time [17], where n is the number of jobs. When $\max\{m_1, m_2\} \geq 2$, Hoogeveen et al. [15] showed that the (m_1, m_2)-HFS problem is strongly NP-hard. The special cases $(m_1, 1)$-HFS and $(1, m_2)$-HFS have attracted many researchers' attention [1,9–11]; the interested reader might refer to [24] for a survey on the hybrid two-stage flow-shop problem with a single machine in one stage.

For the general hybrid k-stage flow-shop problem $(m_1, m_2, \ldots, m_k)$-HFS, when all the m_1, m_2, ..., m_k are fixed integers, Hall [12] claimed that the PTAS for the classic k-stage flow-shop problem can be extended to a PTAS for the $(m_1, m_2, \ldots, m_k)$-HFS problem. Later, Schuurman and Woeginger [22] presented a PTAS for the hybrid two-stage flow-shop problem (m_1, m_2)-HFS, even when the numbers of machines m_1 and m_2 in the two stages are a part of the input. Jansen and Sviridenko [16] generalized this result to the hybrid k-stage flow-shop problem $(m_1, m_2, \ldots, m_k)$-HFS, where k is a fixed integer while $m_1, m_2, \ldots, m_k$ can be a part of the input. Due to the inapproximability of the classic k-stage flow-shop problem, when k is arbitrary, the $(m_1, m_2, \ldots, m_k)$-HFS problem cannot be approximated within 1.25 unless P = NP [25]. In addition, there are plenty of heuristic algorithms in the literature for the general hybrid k-stage flow-shop problem, and the interested readers can refer to the survey by Ruiz et al. [20].

Compared to the rich literature on the hybrid k-stage flow-shop problem, the parallel k-stage flow-shop problem is much less studied. In fact, the general (m, k)-PFS problem is almost untouched, except only the two-stage flow-shops are involved [4,13,23,26]. He et al. [13] first studied the m parallel identical two-stage flow-shop problem $(m, 2)$-PFS, motivated by an application from the glass industry. In their work, the $(m, 2)$-PFS problem is formulated as a mixed-integer programming and an efficient heuristic is proposed [13]. Vairaktarakis and Elhafsi [23] also studied the $(m, 2)$-PFS problem, in order to investigate the hybrid k-stage flow-shop problem. Among other results, Vairaktarakis and Elhafsi [23] presented an $O(nP^3)$-time dynamic programming algorithm for solving the NP-hard $(2, 2)$-PFS problem optimally, where n is the number of jobs and P is the sum of all processing times. That is, the $(2, 2)$-PFS problem can be solved exactly in pseudo-polynomial time.

The NP-hardness of $(2, 2)$-PFS implies that the general $(m, 2)$-PFS problem is NP-hard, when either m is a part of the input (arbitrary) or m is a fixed integer greater than one. Zhang et al. [26] studied on how to approximate the $(m, 2)$-PFS problem, more precisely only for the special case where $m = 2$ or 3. They designed a 3/2-approximation algorithm when $m = 2$ and a 12/7-approximation algorithm when $m = 3$ [26]. Both algorithms are variations of Johnson's algorithm and the main idea is first to sort all the jobs using Johnson's algorithm into a sequence, then to cut this sequence into two (three, respectively) parts for the two (three, respectively) two-stage flow-shops in order to minimize the makespan. Recently, Dong *et al.* [4] extended the dynamic programming algorithm for the $(2,2)$-PFS problem to solve the $(m,2)$-PFS problem, for any fixed $m \geq 2$, in $O(nmP^{2m+1})$-time and $O(P^{2m})$-space. They then designed an FPTAS for the $(m,2)$-PFS problem out of this exact pseudo-polynomial time algorithm.

In this paper, we present a PTAS for the (m,k)-PFS problem when m and k are fixed integers. Our PTAS borrows some design ideas from the PTAS for the classic k-stage flow-shop problem by Hall [12]. The key technique is to enumerate over schedules for *big* jobs, then to solve a linear programming for *small* jobs to obtain the assignments for most of them in each schedule, followed by adding the fractional small jobs at the end. Such a technique has been used in the design of similar approximation schemes.

2   A PTAS for the (m, k)-PFS Problem

In the sequel, a *schedule* for an instance of the (m, k)-PFS problem is an assignment of non-negative starting times to all the operations of the given jobs, each on one of the m flow-shops, and a *feasible* schedule is one in which the assignment meets the processing restrictions: (1) each job can have at most one of its operations undergoing processing at any point in time, (2) each operation of a job must be processed on a machine non-preemptively for the specified length of time, and (3) each machine can process at most one operation at any point in time. We use π^* to denote an optimal schedule and its makespan is denoted by OPT.

For ease of presentation, we let $P_i = \sum_{j=1}^{k} p_{ij}$ denote the total processing time of the job J_i over all k machines, and assume without loss of generality that $P_1 \geq P_2 \geq \ldots \geq P_n$; we also let $Q_j = \sum_{i=1}^{n} p_{ij}$ denote the total processing time of all the jobs in the j-th stage machines. Define $P = \sum_{i=1}^{n} P_i = \sum_{j=1}^{k} Q_j$. The following lemma bounds OPT, the proof of which is omitted due to space limit.

Lemma 1. *We have the following upper and lower bounds on OPT:*

$$\max\left\{ \frac{P}{mk}, P_1 \right\} \leq OPT \leq \frac{P}{m} + P_1.$$

We normalize the job processing time by dividing each p_{ij} by the quantity $2 \cdot \max\{P/m, P_1\}$, for all i, j. This way, we have

$$\frac{1}{2k} \leq \text{OPT} \leq 1. \tag{1}$$

Note that from the proof of Lemma 1 we also have $C^\pi \leq 1$, where π is the schedule produced by the list scheduling algorithm and C^π denotes its makespan. We aim to find a better schedule than π and therefore, in the sequel, we consider only those feasible schedules having a makespan less than or equal to 1.

We use $[n]$ to denote the set $\{1, 2, \ldots, n\}$, for every integer $n \geq 1$. For some real number $\gamma \in (0, 1)$, which will be determined later (in Eq. (4)), we partition the job set $\mathcal{J}$ into two subsets of *big* jobs and *small* jobs, as follows.

$$\mathcal{B} = \{J_i \mid \exists j \in [k], \; p_{ij} \geq \gamma\}, \text{ and } \mathcal{S} = \{J_i \mid \forall j \in [k], \; p_{ij} < \gamma\}. \tag{2}$$

The next lemma states that there are not too many big jobs, the proof of which is omitted due to space limit.

Lemma 2. *There are at most $\frac{mk}{\gamma}$ big jobs.*

At the high-level, the basic idea in our PTAS is as follows. First we compute the *configuration* for each feasible schedule (having a makespan ≤ 1), and the feasible schedules having the same configuration form into a group; that is, all feasible schedules are partitioned into groups by their configurations. Then for each group, we use its configuration to construct a feasible schedule such that its makespan is very close to the minimum makespan of the schedules in the group. Lastly, we return the constructed schedule with the minimum makespan over all the groups.

2.1   Configuration

Recall that π^* denotes an optimal schedule and its makespan is OPT, which is lower and upper bounded in Eq. (1). Recall also that the makespan of all the feasible schedules considered is at most 1. We will determine the parameter γ later (in Eq. (4)), which depends on the worst-case approximation ratio we want to achieve.

Let $\delta \in (0, 1)$ be a multiple of γ (again this multiple will be determined later, in Eq. (4)), and such that $\mu = 1/\delta$ is an integer. We call an interval of length δ a δ-interval. (In our discussion, these intervals are half open.) The time interval $[0, 1)$ is partitioned into μ consecutive δ-intervals; and we let I_t denote the t-th δ-interval $[(t - 1)\delta, t\delta)$, for each $t \in [\mu]$.

Given a feasible schedule π (with makespan ≤ 1), for each job J_i, we define its *assignment* as $X_i = (\ell, s_1, s_2, \ldots, s_k)$, where ℓ is the index of the flow-shop to which the job J_i is assigned in the schedule π, and s_j records the index of the δ-interval in which the j-th operation is started. That is, the machine $M_{\ell,j}$ starts processing the job J_i in the δ-interval $[(s_j - 1)\delta, s_j\delta)$. Let $X_{\mathcal{B}} = (X_i)_{J_i \in \mathcal{B}}$ and $X_{\mathcal{S}} = (X_i)_{J_i \in \mathcal{S}}$.

In the schedule π, for each δ-interval I_t, $t \in [\mu]$ and each machine $M_{\ell,j}$, $(\ell, j) \in [m] \times [k]$, we define $L_{t,\ell,j}$ to be the *workload* of small jobs, which is the total time inside the interval I_t the machine $M_{\ell,j}$ spends for processing small jobs. Furthermore, we always round $L_{t,\ell,j}$ up to the nearest multiple of γ. Let $L = (L_{t,\ell,j})_{(t,\ell,j) \in [\mu] \times [m] \times [k]}$.

Then $(X_{\mathcal{B}}, L)$ is defined as the *configuration* of the schedule π, or we say that the schedule π is *associated* with the configuration $(X_{\mathcal{B}}, L)$. It is important to note that the configuration does not have any information about the assignments of small jobs. Clearly, every feasible schedule is associated with exactly one configuration; the feasible schedules associated with the same configuration form a group. The following Lemma 3 states that there are not too many distinct configurations, of which the proof is omitted due to space limit. Let $\mathcal{C}$ be the collection of all configurations.

Lemma 3. *There are at most* $(m\mu^k)^{mk/\gamma}(\delta/\gamma + 1)^{mk\mu}$ *distinct configurations.*

2.2 The PTAS

We want to construct a feasible schedule for every configuration in $\mathcal{C}$, such that the makespan of the constructed schedule is very close to the minimum makespan among all the feasible schedules associated with the same configuration. For simplicity, we fix a configuration and assume that the optimal schedule π^* is associated with this configuration. That is, among all the feasible schedules associated with this configuration, the minimum makespan is OPT.

We describe an algorithm called SLIDE-I (see Fig. 1) that constructs a feasible schedule when the assignments of all the jobs of $\mathcal{J}$ are known, that is, more information than the configuration. Using the assignments, the algorithm first collects for each machine $M_{\ell,j}$ the set of operations it needs to start in the interval I_t; let $\mathcal{O}_{t,\ell,j}$ denote this set of operations, for every $(t, \ell, j) \in [\mu] \times [m] \times [k]$. Next, the machine $M_{\ell,j}$ processes the operations of $\mathcal{O}_{t,\ell,j}$ in a non-decreasing order of processing time (in fact, any order suffices as long as all operations can be started in the interval I_t), denoted as $\overrightarrow{\mathcal{O}}_{t,\ell,j}$ (in Lemma 4 we prove that all these operations can be started in the interval I_t, in particular in the non-decreasing order of processing time); thus the sub-schedule on $M_{\ell,j}$ is $\langle \overrightarrow{\mathcal{O}}_{1,\ell,j}, \overrightarrow{\mathcal{O}}_{2,\ell,j}, \ldots, \overrightarrow{\mathcal{O}}_{\mu,\ell,j} \rangle$. Lastly, the machine $M_{\ell,j}$ *delays* the processing by $2(j-1)\delta$ time.

Algorithm SLIDE-I:

Input: m parallel identical k-stage flow-shops, $\mathcal{J}$ with known assignments, γ, δ;
Output: A feasible schedule π (with makespan at most OPT $+ 2(k-1)\delta$).

Step 1. For each machine $M_{\ell,j}$ and each interval I_t:
    1.1. let $\mathcal{O}_{t,\ell,j}$ be the operation set with starting time in I_t on $M_{\ell,j}$;
    1.2. schedule the operations of $\mathcal{O}_{t,\ell,j}$ in non-decreasing processing time;
    1.2. let $\pi_{t,\ell,j}$ denote the sub-schedule for $\mathcal{O}_{t,\ell,j}$;
Step 2. For each machine $M_{\ell,j}$:
    2.1. concatenate $\pi_{t,\ell,j}$ in increasing t;
    2.2. let $\pi_{\ell,j}$ denote the sub-schedule on $M_{\ell,j}$;
Step 3. For each machine $M_{\ell,j}$:
    3.1. delay the sub-schedule $\pi_{\ell,j}$ by $2(j-1)\delta$ time;
Step 4. Return the final whole schedule denoted as π.

Fig. 1. A high-level description of SLIDE-I.

Lemma 4. *If in the configuration the assignments for all the jobs of $\mathcal{J}$ are known, then the algorithm* SLIDE-I *produces a feasible schedule with makespan at most* $OPT + 2(k-1)\delta$.

Proof. The proof is omitted due to space limit.   □

Unfortunately, given a configuration, we do not have the assignment information about the small jobs, but only the small job workload for each machine inside each δ-interval. We next try to obtain from the configuration the assignment information of "most" small jobs. To this purpose, we construct a linear program (LP) with the decision variables $y_{i,X}$, each for a small job and an assignment. That is, $y_{i,X} = 1$ if and only if the small job J_i has an assignment X in the given configuration. Recall that we use $\mathcal{S}$ to denote the set of small jobs and that there are at most $m\mu^k$ different assignments for each job.

(LP)

$$\sum_X y_{i,X} = 1, \quad \forall J_i \in \mathcal{S};$$

$$\sum_{J_i \in \mathcal{S}, X=(\ell, s_1, \dots, s_j=t, \dots, s_k)} p_{ij} y_{i,X} \leq L_{t,\ell,j}, \forall (t, \ell, j) \in [\mu] \times [m] \times [k];$$

$$y \geq 0.$$

In this LP, every small job J_i must have an assignment, and the workload of the small jobs on the machine $M_{\ell,j}$ inside the interval I_t must be less than or equal to $L_{t,\ell,j}$, due to rounding. Clearly, there are only $|\mathcal{S}| + km\mu$ constraints and therefore the number of variables $|\mathcal{S}|m\mu^k$ is considerably larger. It follows that a basic feasible solution to this LP has at most $|\mathcal{S}| + km\mu$ positive values. Note that for every small job J_i, if there is an X such that $y_{i,X}$ is a positive fractional value, then there must be another distinct X' such that $y_{i,X'}$ is a positive fractional value too. Suppose the total number of positive fractional values in the basic feasible solution is N. Let $\mathcal{S}_1$ denote the subset of small jobs for each of which there is an associated variable having value 1, that is, from the basic solution we know the assignment for each small job of $\mathcal{S}_1$; and let $\mathcal{S}_2 = \mathcal{S} - \mathcal{S}_1$ denote the subset of small jobs for each of which there are some (equivalently, at least two) associated variables having fractional values. It follows that $|\mathcal{S}_2| \leq N/2$, and thus $|\mathcal{S}_1| \geq |\mathcal{S}| - N/2$. Therefore, the total number of positive values in the basic solution is at least $|\mathcal{S}| - N/2 + N = |\mathcal{S}| + N/2$. From $|\mathcal{S}| + N/2 \leq |\mathcal{S}| + km\mu$ we have $N \leq 2km\mu$, and thus we conclude that

$$|\mathcal{S}_2| \leq \frac{N}{2} \leq km\mu. \tag{3}$$

We summarize the above result from the LP in the following lemma.

Lemma 5. *Given a configuration where the assignments for all the jobs of $\mathcal{B}$ are known, the assignments for most, but no more than $km\mu$, of jobs of $\mathcal{S}$ can be obtained by solving the constructed LP.*

Now we are ready to describe the second algorithm called SLIDE-II (see Fig. 2). In the first step, the algorithm uses the given configuration to a linear program LP as stated in the above, and obtains a basic solution to the LP. In the second step, the algorithm retrieves the assignments for the small jobs of S_1, and calls the algorithm SLIDE-I on the job set $B \cup S_1$ since it has the assignments for all the big jobs from the given configuration. Let π denote the achieved schedule. Lastly, the algorithm appends the small jobs of S_2 to the end of the schedule π, arbitrarily but each of the m flow-shops is assigned $|S_2|/m$ small jobs. (When $|S_2|/m$ is not integral, some flow-shops are assigned $\lceil |S_2|/m \rceil$ small jobs of S_2, while the others are assigned $\lfloor |S_2|/m \rfloor$ small jobs.)

Algorithm SLIDE-II:

Input: m parallel identical k-stage flow-shops, $\mathcal{J}$, a configuration, γ, δ;
Output: A feasible schedule π.

Step 1. Construct a linear program using the configuration and solve it;
    1.1. obtain the job subset S_1 with known assignments, and S_2;
Step 2. Run the algorithm SLIDE-I on the job subset $B \cup S_1$;
    2.1. obtain a partial schedule π;
Step 3. Append the jobs of S_2 to the end of the schedule π;
    3.1. each flow-shop is assigned with $|S_2|/m$ small jobs of S_2;
Step 4. Return the final whole schedule still denoted as π.

Fig. 2. A high-level description of SLIDE-II.

Lemma 6. *Given the configuration, the algorithm SLIDE-II produces a feasible solution with makespan at most $OPT + 2(k-1)(\delta + \gamma) + \mu\gamma + (k\mu + k - 1)\gamma$.*

Proof. The proof is omitted due to space limit. □

Our final algorithm, called SLIDE-III, for the (m,k)-PFS problem runs the algorithm SLIDE-II on every configuration to achieve a schedule, and returns the best schedule among them, *i.e.*, the one with the minimum makespan.

Theorem 1. *The algorithm SLIDE-III can be designed into a PTAS for the (m,k)-PFS problem.*

Proof. For any $\epsilon \in (0,1)$, we show how to set up the values for the parameters δ and γ such that the makespan of the schedule returned by the algorithm SLIDE-III is within $(1+\epsilon)OPT$. Recall that the job processing times have been normalized to ensure that Eq. (1) holds for OPT. Recall also that δ is a multiple of γ. For ease of presentation (to avoid the use of ceiling function) we assume $\epsilon = \frac{1}{T}$ for some positive integer T. Let

$$\delta = \frac{\epsilon}{8k(k-1)}, \text{ and } \gamma = \frac{\epsilon^2}{64(k+1)k^2(k-1)}. \tag{4}$$

From Lemma 2, the number of big jobs is at most mk/γ, which is polynomial in $m, k, \frac{1}{\epsilon}$. Moreover, when m and k are fixed constants and ϵ is given (and thus a constant as well), mk/γ is a constant. Similarly, from Lemma 3, the number of distinct configurations is at most $(m\mu^k)^{mk/\gamma}(\delta/\gamma + 1)^{mk\mu}$, which is a constant when m, k, ϵ are fixed constants. That is, the algorithm SLIDE-III makes only a constant number of calls to the algorithm SLIDE-II.

Inside the execution of the algorithm SLIDE-II, the constructed linear program LP contains $|\mathcal{S}| + km\mu$ constraints and $|\mathcal{S}|m\mu^k$ variables. That is, the size of the LP is polynomial when m, k, ϵ are fixed. Since a linear program can be solved in polynomial time, for example by the interior point method [18], and the running time of the algorithm SLIDE-I is polynomial in the number of jobs which have known assignments, the running time of the algorithm SLIDE-II is polynomial in the number of jobs. In summary, the algorithm SLIDE-III is polynomial in n, the number of jobs, when m, k, ϵ are fixed constants.

For the performance ratio, from Lemma 6 we only need to measure the additive error term against OPT. By $\mu = 1/\delta$ and Eq. (4), we have the following:

$$
\begin{aligned}
&2(k-1)(\delta+\gamma) + \mu\gamma + (k\mu + k - 1)\gamma \\
&= 2(k-1)\delta + (3(k-1) + (k+1)\mu)\gamma \\
&= 2(k-1)\delta + \left(3(k-1) + \frac{k+1}{\delta}\right)\gamma \\
&= \frac{\epsilon}{4k} + \left(3(k-1) + \frac{8(k+1)k(k-1)}{\epsilon}\right)\frac{\epsilon^2}{64(k+1)k^2(k-1)} \\
&= \frac{\epsilon}{4k} + \frac{3\epsilon^2}{64(k+1)k^2} + \frac{\epsilon}{8k} \\
&= \left(\frac{3}{4} + \frac{3\epsilon}{32(k+1)k}\right)\frac{1}{2k}\epsilon \\
&< \frac{1}{2k}\epsilon.
\end{aligned}
$$

It follows that the makespan of the schedule produced by the algorithm SLIDE-III is less than $\text{OPT} + \frac{1}{2k}\epsilon < (1+\epsilon)\text{OPT}$, by Eq. (1). This proves the theorem. $\square$

3   Conclusions

We presented a polynomial-time approximation scheme (PTAS) for the (m, k)-PFS problem, in which there are m parallel identical k-stage flow-shops. Our PTAS requires both m and k to be fixed integers. Since the classic k-stage flow-shop problem is strongly NP-hard for a fixed $k \geq 3$, our PTAS seems the best possible unless P = NP. The APX-hardness of the classic k-stage flow-shop problem when k is a part of the input implies the APX-hardness of the (m, k)-PFS problem when k is a part of the input. An open problem is to investigate the (in-)approximability of the (m, k)-PFS problem when m is a part of the input while k is a constant.

Acknowledgments. Tong was supported by the FY16 Startup Funding from the Georgia Southern University and an Alberta Innovates Technology Futures (AITF) Graduate Student Scholarship. Miyano is supported by the Grants-in-Aid for Scientific Research of Japan (KAKENHI), Grant Number 26330017. Goebel is supported by the AITF and the Natural Sciences and Engineering Research Council of Canada (NSERC). Lin is partially supported by the NSERC and his work was mostly done during his sabbatical leave at the Kyushu Institute of Technology, Iizuka Campus.

References

1. Chen, B.: Analysis of classes of heuristics for scheduling a two-stage flow shop with parallel machines at one stage. J. Oper. Res. Soc. **46**, 234–244 (1995)
2. Chen, B., Glass, C.A., Potts, C.N., Strusevich, V.A.: A new heuristic for three-machine flow shop scheduling. Oper. Res. **44**, 891–898 (1996)
3. Conway, R.W., Maxwell, W.L., Miller, L.W.: Theory of Scheduling. Addison-Wesley, Reading (1967)
4. Dong, J., Tong, W., Luo, T., Wang, X., Hu, J., Xu, Y., Lin, G.: An FPTAS for the parallel two-machine flowshop problem. Theor. Comput. Sci. (2016, in press)
5. Garey, M.R., Johnson, D.S.: Computers and Intractability: A Guide to the Theory of NP-Completeness. W. H. Freeman & Co., New York (1979)
6. Garey, M.R., Johnson, D.S., Sethi, R.: The complexity of flowshop and jobshop scheduling. Math. Oper. Res. **1**, 117–129 (1976)
7. Gonzalez, T., Sahni, S.: Flowshop and jobshop schedules: complexity and approximation. Oper. Res. **26**, 36–52 (1978)
8. Graham, R.L.: Bounds for certain multiprocessing anomalies. Bell Syst. Tech. J. **45**, 1563–1581 (1966)
9. Gupta, J.N.D.: Two-stage, hybrid flowshop scheduling problem. J. Oper. Res. Soc. **39**, 359–364 (1988)
10. Gupta, J.N.D., Hariri, A.M.A., Potts, C.N.: Scheduling a two-stage hybrid flow shop with parallel machines at the first stage. Ann. Oper. Res. **69**, 171–191 (1997)
11. Gupta, J.N.D., Tunc, E.A.: Schedules for a two-stage hybrid flowshop with parallel machines at the second stage. Int. J. Prod. Res. **29**, 1489–1502 (1991)
12. Hall, L.A.: Approximability of flow shop scheduling. Math. Program. **82**, 175–190 (1998)
13. He, D.W., Kusiak, A., Artiba, A.: A scheduling problem in glass manufacturing. IIE Trans. **28**, 129–139 (1996)
14. Hochbaum, D.S., Shmoys, D.B.: Using dual approximation algorithms for scheduling problems theoretical and practical results. J. ACM **34**, 144–162 (1987)
15. Hoogeveen, J.A., Lenstra, J.K., Veltman, B.: Preemptive scheduling in a two-stage multiprocessor flow shop is NP-hard. Eur. J. Oper. Res. **89**, 172–175 (1996)
16. Jansen, K., Sviridenko, M.I.: Polynomial time approximation schemes for the multiprocessor open and flow shop scheduling problem. In: Reichel, H., Tison, S. (eds.) STACS 2000. LNCS, vol. 1770, p. 455. Springer, Heidelberg (2000)
17. Johnson, S.M.: Optimal two- and three-stage production schedules with setup times included. Naval Res. Logistics Q. **1**, 61–68 (1954)
18. Karmarkar, N.: A new polynomial-time algorithm for linear programming. Combinatorica **4**, 373–395 (1984)
19. Lee, C.-Y., Vairaktarakis, G.L.: Minimizing makespan in hybrid flowshops. Oper. Res. Lett. **16**, 149–158 (1994)

20. Ruiz, R., Vázquez-Rodríguez, J.A.: The hybrid flow shop scheduling problem. Eur. J. Oper. Res. **205**, 1–18 (2010)
21. Sahni, S.K.: Algorithms for scheduling independent tasks. J. ACM **23**, 116–127 (1976)
22. Schuurman, P., Woeginger, G.J.: A polynomial time approximation scheme for the two-stage multiprocessor flow shop problem. Theor. Comput. Sci. **237**, 105–122 (2000)
23. Vairaktarakis, G., Elhafsi, M.: The use of flowlines to simplify routing complexity in two-stage flowshops. IIE Trans. **32**, 687–699 (2000)
24. Wang, H.: Flexible flow shop scheduling: optimum, heuristics and artificial intelligence solutions. Expert Syst. **22**, 78–85 (2005)
25. Williamson, D.P., Hall, L.A., Hoogeveen, J.A., Hurkens, C.A.J., Lenstra, J.K., Sevastjanov, S.V.: Short shop schedules. Oper. Res. **45**, 288–294 (1997)
26. Zhang, X., van de Velde, S.: Approximation algorithms for the parallel flow shop problem. Eur. J. Oper. Res. **216**, 544–552 (2012)

Kernelization of Two Path Searching Problems on Split Graphs

Yongjie Yang[1], Yash Raj Shrestha[2], Wenjun Li[3], and Jiong Guo[4(✉)]

[1] Universität des Saarlandes, Saarbrücken, Germany
`yyongjie@mmci.uni-saarland.de`
[2] Department of Management, Technology and Economics,
ETH Zürich, Zürich, Switzerland
`yshrestha@ethz.ch`
[3] Hunan Provincial Key Laboratory of Intelligent Processing of Big Data
on Transportation, School of Computer and Communication Engineering,
Changsha University of Science and Technology, Changsha, China
`liwenjun@csu.edu.cn`
[4] School of Computer Science and Technology,
Shandong University, Shandong, China
`jguo@sdu.edu.cn`

Abstract. In the k-VERTEX-DISJOINT PATHS problem, we are given a graph G and k terminal pairs of vertices, and are asked whether there is a set of k vertex-disjoint paths linking these terminal pairs, respectively. In the k-PATH problem, we are given a graph and are asked whether there is a path of length k. It is known that both problems are NP-hard even in split graphs, which are the graphs whose vertices can be partitioned into a clique and an independent set. We study kernelization for the two problems in split graphs. In particular, we derive a $4k$ vertex-kernel for the k-VERTEX-DISJOINT PATHS problem and a $\frac{3}{2}k^2 + \frac{1}{2}k$ vertex-kernel for the k-PATH problem.

1 Introduction

We study two path searching problems in split graphs, the so-called k-VERTEX-DISJOINT PATHS problem (k-VDP for short) and the k-PATH problem. In the k-PATH problem, we are given a graph G and are asked whether there is a path of length k. In the k-VDP problem, we are given a graph G and k vertex terminal pairs $\{s_1, t_1\}, ..., \{s_k, t_k\}$. The question is whether there are k vertex-disjoint paths $P_1, ..., P_k$ such that P_i, $1 \leq i \leq k$, is a path linking the vertices s_i and t_i.

A *split graph* is a graph whose vertices can be partitioned into a clique and an independent set. Split graphs form a significant graph class and have been extensively studied in the literature due to its wide applications [8,9,14,18]. It is known that both the k-VDP problem and the k-PATH problem are NP-hard in split graphs [8,15]. In this paper, we study kernelization algorithms for these two

---

This work is supported by the National Natural Science Foundation of China under Grants (61502054).

D. Zhu and S. Bereg (Eds.): FAW 2016, LNCS 9711, pp. 238–249, 2016.
DOI: 10.1007/978-3-319-39817-4_23

problems, and derive a $4k$ vertex-kernel for the k-VDP problem and a $\frac{3}{2}k^2 + \frac{1}{2}k$ vertex-kernel for the k-PATH problem.

Notation. All graphs considered in this paper are finite, simple and undirected. A graph G with *vertex set* V and *edge set* E is denoted by $G = (V, E)$. For simplicity, we also use $V(G)$ and $E(G)$ to denote the vertex set and the edge set of the graph G, respectively. An *edge* between two vertices v and u is denoted by (v, u) or (u, v). For a vertex v in G, we write $N_G(v)$ to denote its *neighborhood*, that is, $N_G(v) = \{u \in G \mid (u, v) \in E\}$. The *degree* of a vertex v is $|N_G(v)|$. For $A \subseteq V(G)$, let $N_G(A) = \cup_{u \in A} N_G(u) \setminus A$. We drop the subindex G if it is clear from the context. A *path* is a vertex sequence $(v_1, v_2, ..., v_t)$ such that (1) $v_i \neq v_j$ for every integers $1 \leq i \neq j \leq t$; and (2) $(v_i, v_{i+1}) \in E$ for every integer $1 \leq i \leq t - 1$. The *length* of a path is the number of edges in the path. A vertex subset C (resp. I) is a *clique* (resp. *independent set*) if there is an (resp. no) edge between every two vertices in C (resp. I). A *matching* M is a set of edges such that no two edges in M share a common vertex. A vertex v is *saturated* by a matching M if v is in an edge in M. A *perfect matching* is a matching that saturates all vertices.

A *split graph* is a graph whose vertices can be partitioned into a clique C and an independent set I, either of which may be empty; such a partition (C, I) is called a *split partition*. A split graph G with split partition (C, I) and edge set E is denoted by $G = (C \cup I, E)$. Notice that in general, a split graph can have more than one split partition.

A *parameterized problem* is a subset $Q \subseteq \Sigma^* \times \mathbb{N}$ for some finite alphabet Σ, where the second part is called the *parameter*. A *kernelization algorithm* (or simply, *kernelization*) for a parameterized problem Q is an algorithm that transforms each instance (x, k) of Q in time $(|x| + k)^{O(1)}$ into an instance (x', k') such that (1) $(x, k) \in Q$ if and only if $(x', k') \in Q$; (2) $k' \leq f(k)$ for some computable function f; and (3) $|x'| \leq g(k)$ for some computable function g. The new instance (x', k') is called a *kernel* of the problem, and $g(k)$ is the *size* of the kernel. If g is a polynomial function on k, we say the problem has a *polynomial kernel*. In the context of graph problems, many research papers use the term "vertex-kernel size" to refer to the kernel size counted as the number of vertices in the kernel, see, e.g., [2, 10, 17, 21, 22]. In this paper, we adopt this term.

2   Disjoint Paths

k-VERTEX-DISJOINT PATHS (k-VDP)

Input: A graph G and a collection $S = \{\{s_1, t_1\}, ..., \{s_k, t_k\}\}$ of k pairs of vertices of G, where no two of $\{s_1, ..., s_k, t_1, ..., t_k\}$ are identical.

Parameter: k.

Question: Is there a set $P = \{P_1, P_2, ..., P_k\}$ of k pairwise vertex-disjoint paths such that P_i is a path from s_i to t_i?

In the above definition, the $2k$ vertices $s_1, ..., s_k, t_1, ..., t_k$ are called *terminal vertices* and each pair $\{s_i, t_i\} \in S$ is called a *terminal pair*. The k-VDP problem is a fundamental graph problem with applications in a wide range of areas, including VLSI layout, transportation networks, network reliability and virtual circuit routing in high-speed networks or Internet [7,19]. Unfortunately, the k-VDP problem is NP-hard, even when restricted to special graphs such as planar graphs and interval graphs [12,13,16]. On the positive side, Robertson and Seymour [20] showed that the k-VDP problem is FPT in general graphs. However, assuming $NP \not\subseteq coNP/poly$, Bodlaender et al. [5] proved that the k-VDP problem does not have polynomial kernels. Given the negative result concerning the kernel lower bound in general graphs, Heggernes et al. [8] studied the k-VDP problem in split graphs. In fact, they studied a generalization of the k-VDP problem where the terminal vertices are not necessarily distinct, that is, $s_i = s_j$ and $t_i = t_j$ are possible for different i and j, and they define two paths "vertex disjoint" if they are not identical and the internal vertices in the two paths are distinct. To start with, they showed that this problem is NP-hard in split graphs. Their NP-hardness proof directly applies to the k-VDP problem in split graphs since in the reduction no two terminal vertices are identical (see the proof of Theorem 3 in [8] for further details). Then, they derived a $4k^2$ vertex-kernel for the problem in split graphs. However, applying their kernelization to the k-VDP problem in split graphs does not reduce the square kernel to a linear kernel. Their study of the generalization of the k-VDP problem is theoretically motivated by the close relationship between the problem of finding disjoint paths and the problem of finding topological minors in graphs, as stated in [8]. However, there exist many real-world applications where the terminal vertices are distinct [3,7,11].

In this section, we study the kernelization of the k-VDP problem in split graphs. In particular, we derive a $4k$ vertex-kernel for the problem. To this end, we introduce some reduction rules. Let $G = (C \cup I, E)$ be a split graph, and $S = \{\{s_1, t_1\}, \{s_2, t_2\}, ..., \{s_k, t_k\}\}$ be a collection of terminal pairs. Let $V(S) = \{s_1, ..., s_k, t_1, ..., t_k\}$ be the set of all terminal vertices.

Rule 1. If there is a non-terminal vertex $v \in I$, then remove v from the graph G.

Rule 2. If there is a terminal pair $\{s_i, t_i\} \in S$ such that $(s_i, t_i) \in E$, then remove both s_i and t_i from the graph, remove the terminal pair $\{s_i, t_i\}$ from S, and decrease the parameter k by one.

For ease of exposition, after each application of Rule 2 on a terminal pair $\{s_i, t_i\} \in S$, we denote the terminal pair $\{s_j, t_j\}$ for every $j > i$ by $\{s_{j-1}, t_{j-1}\}$, so that the indices of all terminal pairs are still consecutive.

Rule 3. If there are two terminal vertices $v, u \in V(S)$ such that $\{v, u\} \notin S$, $|\{v, u\} \cap I| = 1$ and $(u, v) \in E$, then remove the edge (v, u) from the graph G. Moreover, if there is a vertex $w \in C$ such that $I \cup \{w\}$ induces an independent set, remove the vertex w from C to I.

A reduction rule is *sound* if each application of the reduction rule does not affect the answer of the instance.

Lemma 1. *Rules 1–3 are sound.*

Proof. We prove the soundness of Rules 1–3 one by one.

Rule 1. Let $v \in I$ be a non-terminal vertex. Let G' be the graph obtained from G by applying Rule 1 on the vertex v. Obviously, G' is still a split graph. Moreover, if G' has k vertex-disjoint paths linking the terminal pairs in S, then so does G. It remains to prove the opposite direction. Let P be a set of k vertex-disjoint paths linking the k terminal pairs in S in G. If v is not in any path in P, then the paths in P still link the terminal pairs in S in G'; we are done. Now, suppose that v is in some path $P_i \in P$ linking the terminal vertices s_i and t_i. Let a and b be the two neighbors of v in the path P_i. Since G is a split graph and $v \in I$, it holds that $a, b \in C$. Thus, $(a, b) \in E$. Therefore, deleting v from P_i still results in a path linking s_i and t_i.

Rule 2. Let $\{s_i, t_i\} \in S$ be a terminal pair such that $(s_i, t_i) \in E$. Let G' be the graph obtained from G by applying Rule 2 on $\{s_i, t_i\}$. Moreover, let $S' = S \setminus \{\{s_i, t_i\}\}$. Obviously, G' is a split graph. It is clear that if G' has a set P' of $k - 1$ vertex-disjoint paths linking the terminal pairs in S', then $P' \cup \{(s_i, t_i)\}$ is a set of k vertex-disjoint paths linking the terminal pairs in S. It remains to prove the other direction. Let P be a set of k vertex-disjoint paths linking the terminal pairs in S in G. Moreover, let $P_i \in P$ be the path linking the terminal pair $\{s_i, t_i\}$. Clearly, no path in $P \setminus \{P_i\}$ contains s_i or t_i. Therefore, $P \setminus \{P_i\}$ is a set of $k - 1$ vertex-disjoint paths linking the terminal pairs in S' in G'.

Rule 3. Let v and u be two terminal vertices such that $\{v, u\} \notin S$, $|\{v, u\} \cap I| = 1$ and $(v, u) \in E$. Let $\{s_i, t_i\}$ and $\{s_j, t_j\}$ be the two terminal pairs involving v and u, respectively, that is, $v \in \{s_i, t_i\}$ and $u \in \{s_j, t_j\}$. Let G' be the graph obtained from G by applying Rule 3 on v and u. It is clear that G' is still a split graph. Moreover, if G' has a set P' of k vertex-disjoint paths linking the terminal pairs in S, then the paths in P' still link the terminal pairs in S in G. It remains to prove the other direction. Let $P = \{P_1, ..., P_k\}$ be a set of k vertex-disjoint paths linking the terminal pairs in S in G, where $P_i, P_j \in P$ are the paths that link the terminal pairs $\{s_i, t_i\}$ and $\{s_j, t_j\}$, respectively. Since paths in P are pairwise vertex disjoint, no path in $P \setminus \{P_i\}$ contains v, and no path in $P \setminus \{P_j\}$ contains u. Thus, the edge (v, u) is not included in any path in P. It is clear now that P is also a set of k vertex-disjoint paths that link the terminal pairs in S in G'.   □

Now we study a useful property. We say an instance is *reduced* by a set of reduction rules if no reduction rule in the set applies to the instance.

Lemma 2. *Let $(G = (C \cup I, E), S)$ be a reduced instance of the k-VDP problem by Rules 1–3. Then, $I = V(S)$.*

Proof. For the sake of contradiction, assume that $V(S) \neq I$. Then, there is a non-terminal vertex $w \in I$, or a terminal vertex $u \in C$. In the former case, Rule 1 applies, contradicting that $(G = (C \cup I, E), S)$ is a reduced instance. Therefore, it holds that $I \subseteq V(S)$. In the following, we shall show that in the latter case, either Rule 2 or Rule 3 applies. Let v be the vertex that forms a terminal pair with u. If $(u,v) \in E$, then Rule 2 applies; otherwise, if there is a vertex $x \in I$ such that $(u,x) \in E$, then the first part of Rule 3 applies; finally, if there is no edge between u and any vertex in I, then $I \cup \{u\}$ induces an independent set, then, the second part of Rule 3 applies.   □

Now we introduce another reduction rule that copes with a specific structure in the graph, the so-called crown decomposition which has been proved useful in many kernelization algorithms [1,2,6].

A *crown decomposition* of a graph G is a partition (A, H, R) of $V(G)$ such that (1) A is an independent set; (2) $H = N(A)$; and (3) the edges between A and H contain a matching in which all vertices in H are saturated.

The following lemma is due to [6].

Lemma 3. *Given an independent set B of a graph G such that $|N(B)| < |B|$, a crown decomposition (A, H, R) of G such that $A \subseteq B$, along with a matching between A and H that saturates H, can be found in polynomial time.*

Let $(G = (C \cup I, E), S)$ be a reduced instance of the k-VDP problem by Rules 1–3. We study a crown rule to further reduce the instance. To this end, we create an auxiliary graph G' which is obtained from G by removing all edges between vertices in C. Clearly, G' is a bipartite graph and $N_{G'}(C) \subseteq I$.

Crown Rule. If $|C| > |I|$, find a crown decomposition (A, H, R) of G' such that $A \subseteq C$ in polynomial time according to Lemma 3. Let M be a matching consisting of edges between A and H that saturates all vertices in H. Then, remove all vertices in A from the graph G that are not saturated by M.

Lemma 4. *Crown Rule is sound.*

Proof. Let $(G = (C \cup I, E), S)$ and G' be defined as above. Suppose that $|C| > I$ so that Crown Rule is applicable. Let (A, H, R) and M be the crown decomposition and the matching as stated in Crown Rule, respectively. Since $A \subseteq C$ and $N_{G'}(C) \subseteq I$, according to the definition of crown decomposition, we know that $H \subseteq I$. For a vertex $v \in H$, let $M(v)$ be the vertex in A that is matched with v in M. Moreover, let $M(H) = \{M(v) \mid v \in H\}$. Let $\bar{G}$ be the reduced graph by Crown Rule, obtained from G by removing all vertices in $A \setminus M(H)$. Clearly, $\bar{G}$ is still a split graph with the split partition $(\bar{C}, I)$, where $\bar{C} = C \setminus (A \setminus M(H))$. Moreover, due to Lemma 2, there is no terminal vertex in C. Thus, no terminal vertex is removed by Crown Rule. It is easy to see that if $\bar{G}$ has a set P of k vertex-disjoint paths linking the terminal pairs in S, then so does G. It remains to prove the opposite direction.

Let $P = \{P_1, P_2, ..., P_k\}$ be a set of k vertex-disjoint paths in G where P_i is a path linking the terminal pair $\{s_i, t_i\} \in S$. Since C is a clique in G, we assume that each path $P_i \in P$ contains at most one edge in $G[C]$ (if some path P_i contains more than one edge in $G[C]$, we can derive a new solution by removing all those edges in $G[C]$ from P_i except an one (v, u) such that $\{(v, s_i)(u, t_i)\} \subset E$). Therefore, each path P_i takes either the form (s_i, v, u, t_i) or the form (s_i, w, t_i), where u, v, w are vertices in C. The following procedure shows how to derive a solution P' for $\bar{G}$ from P by replacing some paths in P, so that no path in P' includes a vertex in $A \setminus M(H)$. By and large, it first copies each path $P_i \in P$ into P', and then replaces the intermediate vertices in the copy of P_i with vertices in $A \setminus M(H)$ when s_i or t_i are from H.

```
1  P' = ∅;
2  forall the path Pᵢ ∈ P do
3  |    let P'ᵢ = Pᵢ;
4  |    if P'ᵢ = (sᵢ, v, u, tᵢ) then
5  |    |    if sᵢ ∈ H then
6  |    |    |    replace v with M(sᵢ) in P'ᵢ;
7  |    |    if tᵢ ∈ H then
8  |    |    |    replace u with M(tᵢ) in P'ᵢ;
9  |    if P'ᵢ = (sᵢ, w, tᵢ) and sᵢ, tᵢ ∈ H then
10 |    |    let P'ᵢ = (sᵢ, M(sᵢ), M(tᵢ), tᵢ);
11 |    P' = P' ∪ {P'ᵢ};
```

Now we prove that P' is a solution for $\bar{G}$. It suffices to prove the following claims.

Claim 1. No element in P' contains a vertex in $A \setminus M(H)$.
Claim 2. Assume Claim 1, every element $P'_i \in P'$ is a path in $\bar{G}$.
Claim 3. Assume Claim 2, no two paths in P' share a common vertex.

In the following, we prove the claims one by one.

Proof of Claim 1. Let P'_i be an element in P'. According to the above procedure, P'_i is obtained from P_i by replacing the intermediate vertices with vertices in $M(H)$. In particular, if P_i contains four vertices, say $P_i = (s_i, v, u, t_i)$, then (a1) $P'_i = (s_i, M(s_i), u, t_i)$ if $s_i \in H, t_i \notin H$; (a2) $P'_i = (s_i, M(s_i), M(t_i), t_i)$ if $s_i, t_i \in H$; (a3) $P'_i = (s_i, v, M(t_i), t_i)$ if $s_i \notin H, t_i \in H$; and (a4) $P'_i = P_i = (s_i, v, u, t_i)$ if $s_i, t_i \notin H$ (see Lines 3–8). If P_i contains three vertices, say $P_i = (s_i, w, t_i)$, then (b1) $P'_i = (s_i, M(s_i), M(t_i), t_i)$ if $s_i, t_i \in H$; and (b2) $P'_i = P_i = (s_i, w, t_i)$; otherwise. Observe that if $s_i \in H$ (resp. $t_i \in H$), then $M(s_i) \in M(H)$ (resp. $M(t_i) \in M(H)$. Moreover, recall that $s_i, t_i \in I$. Since $A \subseteq C$, it holds that $s_i, t_i \notin A$. Furthermore, since $H = N_{G'}(A)$, if $s_i \notin H$ (resp. $t_i \notin H$), according to the construction of the graph G', $v, w \notin A$ (resp. $u, w \notin A$) in all cases they occur. In conclusion, P'_i contains no vertex in $A \setminus M(H)$ in all possibilities.

Proof of Claim 2. Let P'_i be an element in P'. If $P'_i = P_i$, then P'_i is clearly a path in $\bar{G}$ since P'_i does not contain any vertex in $A \setminus M(H)$. It remains to prove the claim for the case $P'_i \neq P_i$. See the above proof for Claim 1 for all possibilities of P'_i such that $P'_i \neq P_i$. Since $M(s_i), M(t_i), u, v$ are all from the clique $\bar{C}$ in $\bar{G}$ in all cases they occur, and $\{(s_i, M(s_i)), (t_i, M(t_i))\} \subseteq M \subseteq E(\bar{G})$, P'_i is a path in all possibilities.

Proof of Claim 3. According to the above procedure, each path $P'_i \in P'$ is obtained from the path $P_i \in P$ by replacing the intermediate vertices with some vertices in $M(H)$. Since no vertex belongs to two distinct paths in P, if there is a vertex that belongs to two different paths in P', the vertex must be from $M(H)$. Suppose that $a \in M(H)$ is a vertex that belongs to two different paths P_i' and P_j'. Then, at least one of the following holds: $P_i' \neq P_i$ or $P_j' \neq P_j$. Due to symmetry, suppose that $P_i' \neq P_i$. Due to the above procedure, P_i' contains four vertices. Without loss of generality, assume that $P_i' = (s_i, a, b, t_i)$ (the proof for the case that a is the neighbor of t_i in the path P_i', that is, $P_i' = (s_i, b, a, t_i)$ can be proved similarly by exchanging all occurrences of s_i and t_i in the following arguments). We first show that $a = M(s_i)$. If $|P_i| = 3$, then P_i' is obtained from P_i by replacing the intermediate vertex with $M(s_i)$ and $M(t_i)$ such that $M(s_i)$ is the neighbor of s_i in P_i' (see Lines 9–10). Thus, $a = M(s_i)$ in this case. Otherwise, since $a \in M(H) \subseteq A$, $s_i \in I$ and $H = N_{G'}(A)$, we know that $s_i \in H$. Then, due to Lines 4–8, $a = M(s_i)$. We proceed with the proof by considering all cases of P_j'. We shall show that all cases lead to some contradiction.

Case 1. $|P_j'| = 3$. In this case, $P_j' = P_j = (s_j, a, t_j)$. Since $a \in M(H) \subseteq A$, $s_j, t_j \in I$, and $H = N_{G'}(A)$, it holds that $s_j, t_j \in H$. However, according to Lines 9–10 in the above procedure, $P_j' = (s_j, M(s_j), M(t_j), t_j)$; a contradiction.

Case 2. $|P_j'| = 4$. In this case, if $|P_j| = 3$, then according to the above procedure, P_j' is obtained from P_j by replacing the intermediate vertex by $M(s_j)$ and $M(t_j)$ such that $M(s_j)$ is the neighbor of s_j and $M(t_j)$ is the neighbor of t_j in the path P_j' (see Lines 9–10). Thus, either $a = M(s_j)$ or $a = M(t_j)$. However, since $a = M(s_i)$ and all terminal vertices are distinct, both cases contradict with the fact that M is a matching. Assume now that $|P_j| = 4$. It is clear that a is either the neighbor of s_j or the neighbor of t_j in the path P_j'. Similar to the proof for Case 1, we can conclude that $s_j \in H$ in the former case and $t_j \in H$ in the latter case. Then, according to Lines 4–8, in the former case $a = M(s_j)$ and in the latter case $a = M(t_j)$. Again, since $a = M(s_i)$ and all terminal vertices are distinct, both cases contradict with the fact that M is a matching.   □

Theorem 1. *The k-VDP problem in split graphs admits a $4k$ vertex-kernel.*

Proof. Given an instance of the k-VDP problem, we use Rules 1–3 to reduce the instance until none of them is applicable. Let $(G = (C \cup I, E), k)$ be the resulting instance. Due to Lemma 2, I contains exactly the terminal vertices. Thus, it holds that $|I| = 2k$. Then, we apply Crown Rule if $|C| > |I|$. After this, $|C| \leq |I|$. Thus, the reduced graph G has at most $|C| + |I| \leq 4k$ vertices.

Due to Lemmas 1 and 4, the original instance is a yes-instance if and only if the reduced instance by Rules 1–3 and Crown Rule is a yes-instance. Since each application of a reduction rule reduces the graph by either at least one vertex or at least one edge, and takes polynomial time, the kernelization takes polynomial time.   □

3 k-Path

The k-PATH problem is defined as follows.
k-PATH

Input: A graph G.

Parameter: A positive integer k.

Question: Is there a path of length k in G?

The k-PATH problem is a well-known NP-hard problem. From parameterized complexity perspective, it is FPT in general graphs [4]. However, a polynomial-size kernel seems unlikely unless $PH = \sum_3^P$ [4]. This argument holds even when the problem is restricted in several special graph classes such as planar graphs [10]. In this section, we study kernelization of the k-PATH problem in split graphs. In particular, we derive a square vertex-kernel for the problem in this setting. Since the HAMILTONIAN PATH problem, a special case of the k-PATH problem with k being the number of vertices in the given graph, remains NP-hard in split graphs [15], so does the k-PATH problem.

We first study a reduction rule. Let $G = (C \cup I, E)$ be the given split graph with the split partition (C, I). We assume that $k > 3$ (if $k \le 3$, we solve the problem in polynomial time).

Rule 1. If there are two degree-1 vertices $v, u \in I$ who have the same neighbor in C, remove arbitrarily one of them from the graph G.

The correctness of the above reduction rule is based on the observation that if $k > 3$, any k-path contains at most one of v and u. In the following, we study a crown decomposition based reduction rule. We need the definition of tricrown decomposition and a useful property (a similar concept named "double crown decomposition" was studied in [17].) Roughly, the tricrown decomposition differs from the crown decomposition in that the independent set A is at least triple size of its neighborhood H. Moreover, there are three matchings consisting of edges between A and H whose vertices in A are disjoint. Moreover, every matching saturates H. The formal definition is as follows.

A *tricrown decomposition* of a graph G is a partition (A, H, R) of $V(G)$ such that

1. A is an independent set;
2. $H = N_G(A)$; and
3. there is a partition (A_1, A_2, A_3, A_4) of A such that there is a matching M_i between A_i and H which saturates all vertices in H, for every $i \in \{1, 2, 3\}$.

The following lemma is useful.

Lemma 5. *Given an independent set I of a graph G, if $|I| > 3|N(I)|$, then there is a tricrown decomposition (A, H, R) where $A \subseteq I$. Moreover, such a tricrown decomposition along with the three matchings as sated in the above definition can be found in polynomial time.*

The proof of the above lemma is similar to the proof of Lemma 9 in [17] which states that if $|I| > 2|N(I)|$, a double crown decomposition can be found in polynomial time. To exploit tricrown decomposition, we need to create an auxiliary graph. Let $(G = (C \cup I, E), k)$ be a reduced instance by Rule 1. Let (E_1, E_2) be a partition of E where E_1 is the set of edges with both ends in C, and

E_2 the set of edges between C and I. Moreover, let (J, I') be a partition of I such that I' is the set of all degree-1 vertices in I, and J is the set of the remaining vertices. We create an auxiliary bipartite graph G' with vertices $E_1 \cup J$. For ease of notation, for an edge $(u, v) \in E_1$, we denote by $e_{v,u}$ its corresponding vertex in G'. Both J and E_1 form independent sets in G'. The edges of G' are as follows: there is an edge between a vertex $e_{u,v} \in E_1$ and a vertex $w \in J$, if w is adjacent to both v and u in G. The crown rule is as follows.

Crown Rule. If $|J| > 3|E_1|$, then find a tricrown decomposition (A, H, R) of G' such that $A \subseteq J$ in polynomial time according to Lemma 3. Let (A_1, A_2, A_3, A_4) be a partition as stated in the definition of tricrown decomposition, then remove all vertices in A_4 from the graph G.

In the following, we prove the soundness of the above reduction rule.

Lemma 6. *Crown Rule for the k-PATH problem is sound.*

Proof. Suppose that $|J| > 3|E_1|$ so that Crown Rule is applicable. Let $\bar{G}$ be the reduced graph obtained from G by applying the above reduction rule. Let (A, H, R) be the tricrown decomposition of the graph G' as discussed above, and (A_1, A_2, A_3, A_4) be a partition of A as stated in Crown Rule. Moreover, let $V(H) - \{v \in G \mid \exists u \in G \text{ such that } c_{v,u} \in H\}$. Clearly, $\bar{G}$ is a split graph with a split partition $(C, A_1 \cup A_2 \cup A_3 \cup I')$. Clearly, if $\bar{G}$ has a k-path, so does G. It remains to prove the other direction.

Suppose that $P = (v_1, v_2, ..., v_k)$ is a k-path of G. Every vertex v_i with $1 < i < k$ in the path P is called an *inner vertex*. If P contains no vertex in A_4, P is a k-path of $\bar{G}$; we are done. Otherwise, we construct a k-path for $\bar{G}$ as follows. Let M_t, $t = 1, 2, 3$, be the matching between H and A_t that saturates H. For a vertex $e_{v,u} \in H$ and an integer $t \in \{1, 2, 3\}$, let $M_t(e_{v,u})$ be the vertex matched with $e_{v,u}$ in the matching M_t. Clearly, $M_t(e_{v,u}) \in A_t$ for every $t \in \{1, 2, 3\}$. Let v_i be an inner vertex. If $v_i \in A$, then both of its neighbors v_{i-1} and v_{i+1} in the path P are from C; and thus, $(v_{i-1}, v_{i+1}) \in E(G)$. Then, according to the construction of the graph G', $e_{v_{i-1}, v_{i+1}}$ is a vertex in G'. We replace all such inner vertices v_i that are in A with vertices only in $A_1 \cup A_2 \cup A_3$ as follows. First, for each inner vertex v_i, if $v_i \in A$, then we replace v_i in P with any one of the vertices in

$$\{M_1(e_{v_{i-1}, v_{i+1}}), M_2(e_{v_{i-1}, v_{i+1}}), M_3(e_{v_{i-1}, v_{i+1}})\} \setminus \{v_1, v_k\}.$$

According to the construction of G', we know that for every $t \in \{1, 2, 3\}$, $(M_t(e_{v_{i-1}, v_{i+1}}), v_{i-1}) \in E(G)$ and $(M_t(e_{v_{i-1}, v_{i+1}}), v_{i+1}) \in E(G)$. Moreover, since $M_t(e_{v_{i-1}, v_{i+1}}), v_{i-1}, v_{i+1} \in V(\bar{G})$, M_t is a matching between H and A_t for every $t \in \{1, 2, 3\}$, and no two inner vertices have the same two neighbors in the path P, it holds that after the above replacements the path P without the first and the last vertices v_1 and v_k is a $(k-2)$-path of the graph $\bar{G}$.

We may need further replace the vertex v_1 (resp. v_k) in the path P if $v_1 \in A_4$ (resp. $v_k \in A_4$). We first claim that if $v_1 \in A_4$ (resp. $v_k \in A_4$), then $v_2 \in V(H)$ (resp. $v_{k-1} \in V(H)$): if $v_1 \in A_4 \subseteq J$ (resp. $v_k \in A_4 \subseteq J$), then v_1 (resp. v_k)

has at least two neighbors u, w in C. Then, according to the construction of G', $e_{u,w} \in H$. Moreover, there is an edge between $e_{u,w}$ and v_1 (resp. v_k) in G'. If $v_2 \in \{u, w\}$ (resp. $v_{k-1} \in \{u, w\}$), we are done. Otherwise, since $v_1 \in J$ (resp. $v_k \in J$), it holds that $v_2 \in C$ (resp. $v_{k-1} \in C$). Then, $(v_2, u) \in E(G)$ (resp. $(u, v_{k-1}) \in E(G)$). According to the construction of G', it holds that $e_{v_2,u} \in H$ (resp. $e_{v_{k-1},u} \in H$). Thus, $v_2 \in V(H)$ (resp. $v_{k-1} \in V(H)$). This completes the proof of the claim.

Now we continue illustrating how to replace the vertex v_1 (resp. v_k) if $v_1 \in A_4$ (resp. $v_k \in A_4$). If there is a vertex $x \in C$ which is not on the path P, we replace v_1 (resp. v_k) in P with x. Since $V(H) \subseteq C$, $(x, v_2) \in E(\bar{G})$ (resp. $(x, v_{k-1}) \in E(\bar{G})$). Otherwise, according to the above claim, there is a vertex v_i in the path P where $i > 2$ (resp. $i < k - 1$) such that $e_{v_2,v_i} \in H$ (resp. $e_{v_{k-1},v_i} \in H$). Let $\mathcal{M}(e_{v_2,v_i}) = \{M_1(e_{v_2,v_i}), M_2(e_{v_2,v_i}), M_3(e_{v_2,v_i})\}$ (resp. $\mathcal{M}(e_{v_{k-1},v_i}) = \{M_1(e_{v_{k-1},v_i}), M_2(e_{v_{k-1},v_i}), M_3(e_{v_{k-1},v_i})\}$). In this case, we replace v_1 (resp. v_k) with any one vertex in $\mathcal{M}(e_{v_2,v_i})$ (resp. $\mathcal{M}(e_{v_{k-1},v_i})$) that is not on the path P. Observe that such a vertex must exist. Indeed, if $v_1 \in A_4$ (resp. $v_k \in A_4$), then $|\mathcal{M}(e_{v_2,v_i}) \setminus \{v_1, v_k\}| \geq 2$ (resp. $|\mathcal{M}(e_{v_{k-1},v_i}) \setminus \{v_1, v_k\}| \geq 2$). Thus, even in the previous replacements an inner vertex in the path P was replaced with a vertex in $\mathcal{M}(e_{v_2,v_i}) \setminus \{v_1, v_k\}$ (resp. $\mathcal{M}(e_{v_{k-1},v_i}) \setminus \{v_1, v_k\}$), there is still at least one vertex in $\mathcal{M}(e_{v_2,v_i}) \setminus \{v_1, v_k\}$ (resp. $\mathcal{M}(e_{v_{k-1},v_i}) \setminus \{v_1, v_k\}$) which is not on the path P at the moment.

After the above replacements, the path P does not contain any vertex in A_4, the set of vertices removed in the application of Crown Rule. Thus, P is a k-path for the reduced graph $\bar{G}$. $\qquad\square$

Theorem 2. *The k-PATH problem in split graphs has a $\frac{3}{2}k^2 + \frac{1}{2}k$ vertex-kernel.*

Proof. The kernelization first checks if the clique of the given split graph contains at least $k+1$ vertices. If so, any ordering of the vertices in the clique forms a k-path, and we are done. Otherwise, we apply the above reduction rules exhaustively. Let (C, I) be the split partition of the reduced split graph. Moreover, let J be the set of vertices in I that have at least two neighbors in C, and I' the set of vertices in I that have exactly one neighbor in C. According to Rule 1, we have that $|I'| \leq |C| \leq k$. According to Crown Rule, we have $|J| \leq \frac{3}{2}(|C|^2 - |C|)$. In total, the reduced graph contains at most $|J| + |I'| + |C| \leq \frac{3}{2}k^2 + \frac{1}{2}k$ vertices.

Due to the soundness of Rule 1 and Lemma 6, the original instance is a yes-instance if and only if the reduced instance is a yes-instance. Since each application of a reduction rule reduces the graph by at least one vertex and takes polynomial time, the kernelization takes polynomial time. $\qquad\square$

4   Conclusion

We have studied the kernelization of the k-VERTEX-DISJOINT PATHS problem and the k-PATH problem in split graphs. In particular, we derived a linear vertex-kernel for the former problem and a square vertex-kernel for the latter problem. A further research direction would be to investigate whether the k-PATH problem admits a linear vertex-kernel in split graphs.

References

1. Abu-Khzam, F.N.: A kernelization algorithm for d-hitting set. J. Comput. Syst. Sci. **76**(7), 524–531 (2010)
2. Abu-Khzam, F.N., Fellows, M.R., Langston, M.A., Suters, W.H.: Crown structures for vertex cover kernelization. Theor. Comput. Syst. **41**(3), 411–430 (2007)
3. Adcock, A.B., Demaine, E.D., Demaine, M.L., O'Brien, M.P., Reidl, F., Villaamil, F.S., Sullivan, B.D.: Zig-zag numberlink is NP-complete. JIP **23**(3), 239–245 (2015)
4. Bodlaender, H.L., Downey, R.G., Fellows, M.R., Hermelin, D.: On problems without polynomial kernels. J. Comput. Syst. Sci. **75**(8), 423–434 (2009)
5. Bodlaender, H.L., Thomassé, S., Yeo, A.: Kernel bounds for disjoint cycles and disjoint paths. Theor. Comput. Sci. **412**(35), 4570–4578 (2011)
6. Chor, B., Fellows, M., Juedes, D.W.: Linear kernels in linear time, or how to save k colors in $O(n^2)$ steps. In: Hromkovič, J., Nagl, M., Westfechtel, B. (eds.) WG 2004. LNCS, vol. 3353, pp. 257–269. Springer, Heidelberg (2004)
7. Frank, A.: Packing paths, circuits and cuts: a survey. In: Korte, B., Lovász, L., Prömel, H.J., Schrijver, A. (eds.) Paths, Flows, and VLSI-Layout, pp. 49–100. Springer, Berlin (1990)
8. Heggernes, P., Hof, P., van Leeuwen, E.J., Saei, R.: Finding disjoint paths in split graphs. Theor. Comput. Syst. **57**(1), 140–159 (2015)
9. Heggernes, P., Lokshtanov, D., Mihai, R., Papadopoulos, C.: Cutwidth of split graphs and threshold graphs. SIAM J. Discrete Math. **25**(3), 1418–1437 (2011)
10. Jansen, B.M.P.: Turing kernelization for finding long paths and cycles in restricted graph classes. In: Schulz, A.S., Wagner, D. (eds.) ESA 2014. LNCS, vol. 8737, pp. 579–591. Springer, Heidelberg (2014)
11. Karp, R.M., Leighton, F.T., Rivest, R.L., Thompson, C.D., Vazirani, U.V., Vazirani, V.V.: Global wire routing in two-dimensional arrays. Algorithmica **2**, 113–129 (1987)
12. Kramer, M.R., van Leeuwen, J.: The complexity of wire-routing and finding minimum area layouts for arbitrary VLSI circuits. Adv. Comput. Res. **2**, 129–146 (1984)
13. Lynch, J.F.: The equivalence of theorem proving and the interconnection problem. SIGDA Newsl. **5**(3), 31–36 (1975)
14. Merris, R.: Split graphs. Eur. J. Comb. **24**(4), 413–430 (2003)
15. Müller, H.: Hamiltonian circuits in chordal bipartite graphs. Discrete Math. **156**(1–3), 291–298 (1996)
16. Natarajan, S., Sprague, A.P.: Disjoint paths in circular arc graphs. Nord. J. Comput. **3**(3), 256–270 (1996)
17. Prieto, E., Sloper, C.: Looking at the stars. In: Downey, R.G., Fellows, M.R., Dehne, F. (eds.) IWPEC 2004. LNCS, vol. 3162, pp. 138–148. Springer, Heidelberg (2004)
18. Raman, V., Saurabh, S.: Short cycles make W-hard problems hard: FPT algorithms for W-hard problems in graphs with no short cycles. Algorithmica **52**(2), 203–225 (2008)
19. Robertson, N., Seymour, P.D.: An outline of a disjoint path algorithm. In: Korte, B., Lovász, L., Prömel, H.J., Schrijver, A. (eds.) Paths, Flows, and VLSI-Layout, pp. 267–292. Springer, Berlin (1990)
20. Robertson, N., Seymour, P.D.: Graph minors. XIII. The disjoint paths problem. J. Comb. Theor. Ser. B **63**(1), 65–110 (1995)

21. Wang, J., Yang, Y., Guo, J., Chen, J.: Planar graph vertex partition for linear problem kernels. J. Comput. Syst. Sci. **79**(5), 609–621 (2013)
22. Yang, Y.: Towards optimal kernel for edge-disjoint triangle packing. Inf. Process. Lett. **114**(7), 344–348 (2014)

The Bounded Batch Scheduling with Common Due Window and Non-identical Size Jobs

Hongluan Zhao[1(✉)], Guoyong Han[1,2], and Gongwen Xu[1,2]

[1] School of Computer Science and Technology,
Shandong Jianzhu University, Jinan 250101, China
hongluanzhao@163.com,
{hanguoyong,xugongwen}@sdjzu.edu.cn
[2] School of Information Science and Engineering,
Shandong Normal University, Jinan 250014, China

Abstract. The common due window scheduling problem with batching on a single machine is dealt with to minimize the total penalty of weighted earliness and tardiness. In this paper it is assumed that a job incurs no penalty as long as it is completed within the common due window. The problem is extended to the environment of non-identical job sizes. Then several optimal properties are given to schedule batches effectively. And by introducing the concept of PRN, it is proven that the PRN of each batch should be made as small as possible in order to minimize the objective. Based on these properties, an algorithm F-PRN for batch forming is proposed for the problem.

Keywords: Batch · Non-identical size · Due window · Earliness · Tardiness

1 Introduction

As the development of Just-In-Time (JIT) philosophy, the early-tardy job scheduling problem becomes a hot research point over years. JIT production assumes the existence of job due dates and the advantages are obvious, one of which is to eliminate inventory. If a job finishes before its due date, an early penalty will be incurred such as holding cost. And completing the job after it can result in such tardy cost as late charge, express delivery charge, or lost sale. So a JIT-schedule is to minimize the sum of these penalties. In this paper, the problem is extended. The due date term is generalized to the notion of due window, which is a time interval defined by an early due date and a window size.

On the other hand, there has been significant interest in scheduling problems that involve batching. The motivation for batching jobs is mainly for efficiency, since it may be cheaper or faster to process jobs in a batch than individually. Batching is encountered in many industries, such as semiconductor manufacturing and metal heat treatment. For example, in semiconductor manufacturing, a burn-in oven is regarded as a batch processing machine which can process a number of IC chips simultaneously. Further, the environment of non-identical job sizes is more practical.

A batch is a set of jobs processed simultaneously and completed together when the processing of all jobs in the batch is finished. Therefore, the processing time of a batch

© Springer International Publishing Switzerland 2016
D. Zhu and S. Bereg (Eds.): FAW 2016, LNCS 9711, pp. 250–256, 2016.
DOI: 10.1007/978-3-319-39817-4_24

is equal to the longest processing time of jobs assigned into the batch. Then, the batch scheduling problem involves assigning jobs into batches and sequencing the batches to achieve some objective.

In our paper, we combine the above areas of manufacture situation and study batch scheduling with non-identical job sizes and common due window. The objective is to minimize the sum of the earliness and tardiness penalties (E/T). The remainder of this paper is organized as follows: Sect. 2 gives a related literature review. In Sect. 3, a problem is described and some structural properties of an optimal schedule are proposed. One algorithm is presented in Sect. 4. Section 5 concludes the results.

2 Related Works

In the last several decades, many papers are studied about scheduling problems, but the majority is on traditional performance measures. Under the Just-In-Time conception, both earliness and tardiness should be discouraged. In the recent two decades, many results are concerned with earliness and tardiness penalty but most of them on due date constraint.

Articles on window scheduling problems are limited and almost about common due window. Kramer and Lee [1] firstly considered the problem of finding a schedule that makes the total weighted earliness and tardiness penalties (E/T) minimized. Liman et al. [2, 3], Weng and Ventura [4] analyzed the same problem as [1], where either the size or location of due window is given or to be determined. Zhao, Hu, and Li [5] gave a polynomial algorithm for a common due window scheduling problem with batching on a single machine to minimize total penalty of E/T.

The problems of scheduling on batch processing machines have received tremendous attention since it was first proposed by Ikura and Gimple [6]. Li and Lee [7] investigated the batch scheduling problems of minimizing the maximum tardiness and minimizing the number of tardy jobs and proved both are strongly NP-hard. Pan and Zhou [8] proposed a weighted cost rate heuristic (WCRH) algorithm for minimizing E/T with delivery restriction and distinct due dates. Yin, Cheng, Xu, and Wu [9] considered a problem with a common due date in a batch delivery system for the total cost of earliness, tardiness, inventory, and batch delivery. Xu, Chen, and Li [10] investigated the batch scheduling problem with dynamic job arrivals, using an ant colony optimization (ACO) meta-heuristic to minimize the makespan. All of the above studies examined problems with unit job size. Uzsoy [11] was the first to consider the importance of scheduling on burn-in ovens with non-identical job sizes and proposed several heuristics for problems of minimizing makespan and total completion time. After that, Sung and Choung [12] proposed a branch-and-bound algorithm and several heuristics for makespan minimization on a single burn-in oven. Damodaran, Manjeshwar, and Srihari [13] examined the same problem using a genetic algorithm (GA), and the results indicated GA outperformed previous algorithms. Chen, Du, and Huang [14] provided a novel insight into scheduling on a batch processing machine from a clustering perspective and developed a clustering algorithm.

3   Problem Description and Optical Properties

There are n jobs to be processed on a batch processing machine. The job set is denoted by J = {J_1, J_2, ..., J_n} and all of them are available at the same time. Let p_i and s_i be the processing time and the size of job $J_i \in J$.

The machine has a fixed capacity S. The total sizes of all the jobs in a batch cannot exceed the capacity of the machine. Once processing of a batch is initiated, it cannot be interrupted. And no jobs can be removed from or introduced into the batch until processing is completed.

The processing time of the batch is equal to the largest processing time of the jobs assigned to it. So these jobs have the same start time and the same completion time. For batch B_j, its processing time is denoted by T_j = max{p_i| $J_i \in B_j$} and the number of jobs in it as n_j. A common due window is defined by the earliest due date e and a latest due date d with the window size w = d-e. Without loss of generality, suppose that there are jobs out of [e,d] since otherwise, all jobs are contained in window set for an optimal schedule. Any job completed within the common due window has no penalty. If a job finishes out of the due window, it will incur an earliness or tardiness defined as the difference between the early or tardy due date and the completion time of this job, depending on whether it is completed before or after the due window. Our goal is to partition the jobs into batches and schedule the batches in a certain order.

Let S_i, C_i be the starting time and completion time of J_i, respectively. Thus for the batch B_j including J_i, its completion time is $C(B_j) = C_i$. The earliness and tardiness of job J_i are E_i = max{0, e−C_i} and T_i = max {0, C_i−d}, respectively. The objective function for schedule σ is defined as

$$\min F(\sigma) = \sum_{i=1}^{n} (E_i + T_i) \tag{1}$$

$$\text{s.t.} \quad \sum_{k=1}^{K} x_{ik} = 1, \ i = 1, \ldots, n \tag{2}$$

$$\sum_{i=1}^{n} s_i x_{ik} \leq S, k = 1, \ldots, K \tag{3}$$

$$x_{ik} \in \{0, 1\}, i = 1, \ldots, n; k = 1, \ldots, K \tag{4}$$

$$\left\lceil (\sum_{i=1}^{n} s_i)/S \right\rceil \leq K \leq n, i = 1, \ldots, n; k = 1, \ldots, K \tag{5}$$

where (1) is the objective function, (2) ensures each job J_i is arranged in one batch. (3) makes the total sizes of the jobs in one batch not exceeding the capacity S. x_{ik} = 1 indicates job J_i is arranged in B_k,otherwise x_{ik} = 0. K is the total batch numbers.

In a given schedule σ, the early set, window set, and tardy set are defined, respectively, as $E(\sigma) = \{J_i | C_i < e\}$; $W(\sigma) = \{J_i | e \leq C_i \leq d\}$; and $T(\sigma) = \{J_i | C_i > d\}$. They are denoted by E, W, and T, respectively, when it does not cause confusions. Since jobs in a batch have the same completion time, the batch is also assumed to be contained in the corresponding set of its jobs for simplicity. We assume that all parameters are positive integers.

From [1] for independent jobs, the scheduling problem with a common due window is NP-Complete to minimize the total weighted earliness and tardiness penalties. Extendedly, we have

Theorem 1. The batch scheduling problem, with non-identical job sizes, minimizing the total earliness and tardiness penalties, is NP-Complete.

Several dominant properties of an optimal schedule are presented as follow, which will be used to develop its fine algorithms.

Property 1 [5]. In an optimal schedule, no idle time is inserted between the starting time of the first batch and the completion time of the last batch.

Property 2 [5]. In an optimal schedule σ, there exists one batch B such that $C(B) = e$ or $C(B) = d$, unless the starting time of the first processed batch is zero.

Property 3. For a given set of formed batches, it is optimal to sequence the batches for set E in a batch weighted longest processing time (**BWLPT**) order such as $\frac{T_1}{n_1} \geq \frac{T_2}{n_2} \geq \ldots$; where T_j as the processing time of early batches. And for set T in a batch weighted shortest processing time (**BWSPT**) order such as $\frac{T'_1}{n'_1} \leq \frac{T'_2}{n'_2} \leq \ldots$; where T'_j as the processing time of tardy batches.

Property 3 is as a extended result from [15], where all jobs have identical sizes. Each batch has two attributes: processing time and number of jobs contained. The ratio of a batch processing time to job number in the batch is denoted by **PRN**. Suppose there was a batch B_k containing n_k jobs and its processing time is T_k. Then PRN $(B_k) = \frac{T_k}{n_k}$.

Property 4. If there are two batches in E where batches have been sequenced by BWLPT, any interchange of jobs between two batches, that decreases PRN of one batch and keep another's unchanged, will result in a smaller total penalty, under the capacity constraint without changing the number of jobs in each batch.

Proof. A feasible schedule with r batches where E has been sequenced by BWLPT. As for batches in E, we have $\frac{T_1}{n_1} \geq \frac{T_2}{n_2} \geq \ldots \frac{T_r}{n_r}$, where $n_1, n_2 \ldots n_r$ denote the number of jobs in each batch, respectively. Without loss of generality, we choose two batches B_i and B_j in E with assumption of $T_i \geq T_j$. If there exists a job J_k in batch B_i where $T_i \geq T_j > p_k$ and exchange the job J_k with the job having the longest processing time in batch B_j without violating the machine capacity constraint, then the processing time of batch B_i stays unchanged and that of batch B_j turns into T'_j, $T'_j < T_j$. Then $\Delta F = (n_1 + n_2 + \ldots + n_{j-1}) (T'_j - T_j) < 0$. So we get a smaller total penalty.   □

Moreover, the batches in set T have the similar result but in BWSPT.

4 Algorithm

There are three phases to solve our problem: (1) Assign the jobs into different sets, namely E, W and T; (2) assign the jobs into batches in each set, respectively; and (3) schedule the formed batches, respectively. As for the third phase, after the batches are formed, it is optimal to sequence the jobs by BWLPT and BWSPT in E and T respectively. Consequently, the more difficult remaining problem is how to form batches in E, W and T, respectively. It is critical to find an efficient batch forming algorithm, as the batch forming phase greatly affects our objective.

Through the analysis of our problem, the heuristic information of making PRN of each batch as small as possible is obtained. Thus, a new greedy algorithm **F-PRN** is proposed as follows.

Algorithm 1 (F-PRN)

Step 1. Index the jobs in SPT(shortest processing time) order, denoted as $J_1, J_2, \ldots, J_n$.

Step 2. Set $i = 1$, $j = 1$, $K = 1$. Put the first job J_1 in the first batch B_1.

Step 3. The remaining job set as $J' = \{J_{i+1}, \ldots, J_n\}$. Denote $B = \{B_1, \ldots, B_K\}$ be the batches presently and compute the ratios T_j/n_j, $j = 1, \ldots, K$.

Step 4. Set $i = i+1$. Put J_i in batch B_j satisfied $min\{j | \frac{p_i}{n_j+1} \leq \frac{T_j}{n_j}, s_i + \sum_{u \in Bj} s_u \leq S, Bj \in B\}$. If such j does not exist, push J_i into a new batch B_{k+1}. Set $K = K+1$.

Step 5 If $J' \neq \emptyset$, go to Step 3. Otherwise, stop.

The principle of the F-PRN algorithm is making PRN of each batch as small as possible, which has been proven in Property 4. The preconditions of adding a job into a given batch are (1) that the remaining sizes of the batch is large enough for the job and (2) PRN of the batch after merging this job is smaller than that of all other possibly added jobs. This procedure costs the time of $O(n^2)$.

The following property can be used from my previous one of reference [5]: there exists an optimal schedule, where the window set W contains the batches with smallest processing time.

Then, the remaining batches are in set E and T. Next, we can use the dynamic program in [1] of time $O(n^2 e + nd)$, as a more efficient algorithm. In it, the obtained batches are as ordinary jobs ignoring the size. But the processing time is treated as $n_j T_j$, because the contribution to the total penalty is $n_j T_j$, where T_j be the processing time and n_j be the number of jobs contained in batch B_j.

5 Conclusion

The common due window scheduling problem is investigated on a batch machine which can process many jobs simultaneously. In this paper, the bounded version is considered with batch capacity and non-identical job sizes. The objective is to minimize the weighted earliness and tardiness total penalty. Based on properties proposed, an efficient algorithm is presented.

As a future research direction, similar problems, where jobs have different penalty coefficient and due window, are challenging and worthy of investigation. When the release times of jobs are different or multiple machines are involved, the problems are more competitive.

Acknowledgments. This paper is supported by NSFC of Shandong Province (ZR2012GQ010), Development Projects of Science and Technology of Shandong Province (2015GGX101047, 2015GGX101018), and Ji'nan Higher Educational Independent Innovation Plan (201303001, 201401214).

References

1. Kramer, F.J., Lee, C.Y.: Common due-window scheduling. Prod. Oper. Manage. **2**(4), 262–275 (1993)
2. Liman, S.D., Panwalkar, S.S., Thong, S.: Determination of common due window location in a single machine scheduling problem. Eur. J. Oper. Res. **93**, 68–74 (1996)
3. Liman, S.D., Panwalkar, S.S., Thong, S.: Common due window size and location determination in a single machine scheduling. J. Oper. Res. Soc. **49**, 1007–1010 (1998)
4. Weng, M.X., Ventura, J.A.: A note on common due window scheduling. Prod. Oper. Manage. **5**, 194–200 (1995)
5. Hongluan, Z.H.A.O., Guojun, L.I.: Unbounded batch scheduling with a common due window on a single machine. J. Syst. Sci. Complexity **21**(2), 296–303 (2008)
6. Ikura, Y., Gimple, M.: Efficient scheduling algorithms for a single batch processing machine. Oper. Res. Lett. **5**(2), 61–65 (1986)
7. Li, C.L., Lee, C.Y.: Scheduling with agreeable release times and due dates on a batch processing machine. Eur. J. Oper. Res. **96**(3), 564–569 (1997)
8. Pan, Q., Zhou, B.: Dynamic scheduling algorithm for E/T problem in furnace district of semiconductor fab. Semicond. Technol. **33**(7), 639–643 (2008)
9. Yin, Y., Cheng, T., Xu, D., Wu, C.: Common due date assignment and scheduling with a rate-modifying activity to minimize the due date, earliness, tardiness, holding, and batch delivery cost. Comput. Ind. Eng. **63**(1), 223–234 (2012)
10. Xu, R., Chen, H., Li, X.: Makespan minimization on single batch-processing machine via ant colony optimization. Comput. Oper. Res. **39**(3), 582–593 (2012)
11. Uzsoy, R.: Scheduling a single batch processing machine with nonidentical job sizes. Int. J. Prod. Res. **32**(7), 1615–1635 (1994)
12. Sung, C.S., Choung, Y.I., Fowler, J.W.: Heuristic algorithm for minimizing earliness–tardiness on single burn-in oven in semiconductor manufacturing. In: Proceedings of MASM, pp. 217–222. Tempe, Arizona (2002)

13. Damodaran, P., Manjeshwar, P.K., Srihari, K.: Minimizing makespan on a batch processing machine with non-identical job sizes using genetic algorithms. Int. J. Prod. Econ. **103**(2), 882–891 (2006)
14. Chen, H., Du, B., Huang, G.Q.: Scheduling a batch processing machine with non-identical job sizes: A clustering perspective. Int. J. Prod. Res. **49**(19), 5755–5778 (2011)
15. Deng, X., Li, G., Feng, H., et al.: A PTAS for semiconductor burn-in scheduling. J. Comb. Optim. **9**(1), 5–17 (2005)

Parallel Identifying (l,d)-Motifs in Biosequences Using CPU and GPU Computing

Cheng Zhong[1,2]($\boxtimes$), Jing Zhang[1,2], Bei Hua[1,2], Feng Yang[1,2], and Zhengping Liu[1,2]

[1] School of Computer and Electronics and Information,
Guangxi University, Nanning, China
{chzhong,yf}@gxu.edu.cn

[2] Guangdong Key Laboratory of Popular High Performance Computers,
Shenzhen Key Laboratory of Service Computing and Applications, Shenzhen, China
zj1988129@126.com, huabei111@163.com, liu.zhengping@foxmail.com

Abstract. To accelerate cache access and reduce the access time, the large number of data produced with different combined positions and many candidate sequences are distributed to the texture memory in GPUs when the modeling computation is used to solve in parallel the (l,d)-motif identification problem. The size of thread blocks in GPUs is set according to the size of data in combined positions, the best number of running threads in a thread block is found, and a cache-efficient parallel algorithm for identifying (l,d)-motifs in biosequences is designed by CPU and GPUs cooperative computing. The experimental results show that the proposed parallel algorithm can solve some (l,d)-motif identification instances of large size in less computation time and obtain good speedup and scalability.

Keywords: Motif identification · Combinatorial computation · Parallel algorithm · Hybrid CPU and GPU architectures · Texture memory

1 Introduction

Identifying motifs from biological sequences is an important issue in biological information computing. For given n input sequences which the length of each sequence is L, each sequence contains a variant M_i of motif M of length l, $l < L$, M_i is evolved from M and the Hamming distance between M and M_i is not greater than d. If l and d are known, identifying motif M from the input sequences is called the implanted (l,d)-motif identification problem [1].

The (l,d)-motif identification problem is a well-known NP-Hard problem in computational biology. When the number n of the input sequences and length L of each sequence is large, identifying motifs process is very time consuming. Recently, people applied parallel computing to accelerate the solution of the implanted (l,d)-motif identification problem. Based on the unified projection approach and the voting algorithm, a parallel identifying motifs algorithm was

D. Zhu and S. Bereg (Eds.): FAW 2016, LNCS 9711, pp. 257–268, 2016.
DOI: 10.1007/978-3-319-39817-4_25

designed on heterogeneous cluster system of single-core processor [2]. The parallel algorithm [3] spent 6.9 h to solve successfully the (21,8)-motif identification instance on the multi-core computer having 16 cores. A modeling computation algorithm [4] was proposed to solve in parallel the planted motif identification problem on multi-processor system by combinatorial computation approach. The modeling computation algorithm produced large number of the data with different combined positions and many candidate sequences during identifying (l,d)-motifs. The modeling computation algorithm is difficult to solve the (l,d)-motif identification problem in an acceptable time when l is large. The parallel algorithm [5] solved the (25, 10) and (26, 11)-motif identification instances, and identified successfully the (50,12)-motif instance.

A brute force algorithm was presented to solve recursively the (l,d)-motif identification problem on multi-core processor [6]. The time complexity of this algorithm is reduced to the exponential level of d, where d is the number of mutation positions. By distributing large number of the data produced with different combined positions during the modeling computation to L3, L2 and L1 caches, a thread-level parallel algorithm called MLC-Modeling was designed to solve the (l,d)-motif identification problem [7]. An improved Gibbs sampling algorithm was presented to solve the (l,d)-motif identification problem on single GPU [8]. This parallel algorithm can solve the (l,d)-motif identification instance of larger l, but it was not cache-efficient. Based on CUDA programming model, a MEME algorithm was designed to solve the (l,d)-motif identification problem on single GPU in the two-level mode of parallel processing sequence and substrings [9]. To solve the planted motif finding problem, a parallelizable enumeration-based approach called BitBased [10] was proposed on CPU and GPU, which it was able to solve the (21,8)-motif identification instance. A GPUmotif method [11] was developed to accelerate the motif analysis, where the fragmentation technique was used to hide data transfer time between memories. A parallel projection algorithm for finding motifs [12] was implemented on GPUs. This algorithm can solve the instances of large input (600–1000 base pair per sequence) in an inordinate amount of time. To find DNA-binding motifs in ChIP-Seq and DNase-Seq data, EXTREME algorithm [13] was presented, which uses the expectation -maximization algorithm for motif discovery. The EXTREME algorithm can discover motifs in large datasets in a practical amount of time without discarding any sequences.

This paper designs a cache-efficient parallel algorithm for identifying (l,d)-motifs in biosequences on hybrid CPU and GPU architecture. The remainder is organized as follows. Section 2 describe and analyze the parallel algorithm. Section 3 reports the experimental results. Section 4 concludes the paper.

2   Algorithm

For the multi-core computer with single GPU, CPU converts the data types and transmits the data to the GPU, and allocates the space of video memory in GPU. Assume that there are r different combined positions when the modeling computation algorithm [4] is used. We let GPU execute the modeling computation for one position. Assume also that there are s threads in a thread block in GPU. Hence, the r/s thread blocks are set. When the kernel functions are run, position i of current thread in Grid can be computed by $blockDim.x \times blockIdx.x + threadIdx.x$. Each thread just deals with one position according to the computation mapping of i. The GPU returns the results to the main memory in CPU, and CPU identifies the motifs from the returned results.

Algorithm SGPU-Modeling describes parallel solving (l,d)-motif problem on single GPU.

Algorithm 1. SGPU-Modeling

Input: L, l, d, DNA sequence BS_i, $i=1,2,\ldots,n$;
Output: the motifs;
CPU side:
Begin
1. for $i=1$ to n do
2. for $j=1$ to $L-l+1$ do
3. If there is at least one subsequence of length l in each input DNA sequence except for BS_i and the Hamming distance between this subsequence and R_{ij} is smaller than or equal to d, R_{ij} is a subsequence of length l starting from position j in BS_i, then output motif R_{ij} and algorithm ends;
4. Construct corresponding sets C and CM for R_{ij}, C is the set of sequences whose the Hamming distance between each sequence of length l and R_{ij} is smaller than $2d$, and CM is the set of sequences whose the Hamming distance between each sequence of length l and R_{ij} is equal to $2d$;
5. for $k=1$ to $|CM|$ do
6. Record $2d$ different positions between R_{ij} and c_k and $r=\binom{2d}{d}$ combined positions produced by selecting randomly d ones from the $2d$ positions, and c_k is the k-th sequence in CM;
7. CPU sends R_{ij}, c_k, C and the array of combined positions to GPU, and distributes C and the array of combined positions to the texture memory in GPU;
8. Call kernel function $FindMotifkernel<<<num\_blocks, num\_threads>>>(input parameter, output parameter)$;
9. CPU identifies the motifs in the results from GPU.
End.
GPU side:
$FindMotifkernel<<<num\_blocks, num\_threads>>>(input parameter, output parameter)$
Begin
1. $i=blockDim.x*blockIdx.x+threadIdx.x$;
2. Thread i executes the modeling computation for R_{ij} and c_k in current mapping and obtains the result R'_{ij};
3. If R'_{ij} is a motif, its corresponding mark position is set to 1.
End.

In algorithm SGPU-Modeling, step 3 and step 4 require $O(n \times L)$ time respectively, step 6 requires $O(l \times r)$ time, step 7 requires $O(1)$ time, step 8 requires $O(d+|C|)$ time, and step 9 requires $O(r)$ time.

So, the time complexity of SGPU-Modeling algorithm is $= O(n \times L \times (n \times L + |CM| \times (|C| + r)))$, and the speedup is $O(\frac{n \times L + |CM| \times r \times |C|}{(n \times L + |CM| \times (|C| + r))})$.

Assume that the multi-core computer has ng GPUs and there are r different combined positions, each GPU executes the modeling computation for a position, and there are s threads in each thread block in each GPU. Hence, each GPU sets $r/(s \times ng)$ thread blocks. When the kernel function is run, position i of current thread in Grid is obtained by $ng_i \times gpuIndex + blockDim.x \times blockIdx.x + threadIdx.x$, where ng_i denotes the number of the i-th GPU. Each thread can just deal with a position according to computation mapping of i.

Algorithm 2. MGPU-Modeling

Input: L, l, d, DNA sequence BS_i, $i=1,2,\ldots,n$;
Output: motifs;
CPU side:
Begin
1. for $i=1$ to n do
2. for $j=1$ to $L-l+1$ do
3. If there is at least one subsequence of length l in each input sequence except for BS_i and the Hamming distance between the subsequence and R_{ij} is less than or equal to d, then output the motif R_{ij} and algorithm ends;
4. Construct corresponding sets C and CM for R_{ij};
5. for k=1 to $|CM|$ do
6. Record $2d$ different positions between R_{ij} and c_k and $r=\binom{2d}{d}$ combined positions produced by selecting randomly d ones from the $2d$ positions;
7. for $g=1$ to ng do in parallel
8. cudaSetDevice(g);
9. CPU sends R_{ij}, c_k, C and the array of combined positions to the g-th GPU, and distributes C and the array of combined positions to the texture memory in the g-th GPU;
10. Call kernel function $FindMotifkernel<<<num\_blocks, num\_threads>>>(input parameter, output parameter)$;
11. CPU identifies the motifs in results from GPUs;
End.
GPU side:
$FindMotifkernel<<<num\_block, snum\_threads >>>(input parameter, output parameter)$
Begin
1. $i=ng_i*gpuIndex + blockDim.x*blockIdx.x + threadIdx.x$;
2. Thread i execute the modeling computation for R_{ij} and c_k in current mapping and obtains the result R'_{ij};
3. If R'_{ij} is a motif, the corresponding mark position is set to 1;
End.

MGPU-Modeling is the parallel identifying (l,d)-motif algorithm on multiple GPUs. In algorithm MGPU-Modeling, step 3 requires $O(n \times L)$ time, step 4 requires $O(n \times L)$ time, step 6 requires $O(l \times r)$ time, both of step 8 and step 9 require O(1) time, step 10 requires $O(d+|C|)$ time, step 7 requires $O((d+|C|)/ng)$ time, and step 11 requires $O(r)$ time.

The time complexity of algorithm MGPU-Modeling is $O(n \times (L - l + 1) \times (2 \times (n \times L) + |CM| \times (l \times r + (d + |C|)/ng + r))) = O(n \times L \times (n \times L + |CM| \times (|C|/ng + r)))$, and the speedup is $O(\frac{n \times L + |CM| \times r \times |C|}{(n \times L + |CM| \times (r + |C|/ng)})$.

Now, we analyze the times of access to caches when algorithms SGPU-Modeling and MGPU-Modeling are executed. When the CPU transmits the data to the GPUs, the candidate sequence set C and the array of combined positions are bound to the texture memory in GPU, the sequence R_{ij} and c_k are stored into the global memory. In the global memory in GPU, each thread executes the modeling computation for R_{ij} and c_k, and it needs to execute d replacement operations. The times of comparing result sequences and the sequences in set C is $|C| \times l$ in the detection process. Therefore, each thread accesses to the global memory $(2 \times d + |C| \times l)$ times, and all threads in GPUs access to the global memory $r \times (2 \times d + |C| \times l)$ times. Each thread computes the data in a row of the array of combined positions, and it accesses to the texture memory d times. The times of comparing result sequences and the sequences in set C is also $|C| \times l$. Hence, each thread accesses to the texture memory $(d + |C| \times l)$ times, and all threads in GPUs access to the texture memory $r \times (d + |C| \times l)$ times. Although the global memory access frequency is almost the same as the texture memory access frequency, the data access to the texture memory can be accelerated by caching.

Therefore, SGPU-Modeling and MGPU-Modeling can make full use of the characteristics of GPU memory, and they are cache-efficient.

3 Experiment

3.1 Experimental Environment and Data

The computer contains 4 GPUs (4*Nvidia Tesla C2050 3 GB), 2 Intel Xeon E5620 2.4 GHz processors with each processor having 4 cores, and 12 GB main memory. The running operating system is Red Hat Enterprise Linux 5. The programming language and tools are C, OpenMP and CUDA respectively.

Here the experimental data are 20 simulated DNA sequences of length 600 and one motif sequence of length l. The bases (A,T,C,G) in each position of every sequence are generated independently with equal probability, and the variants of motifs are implanted in the random positions of the 20 DNA sequences.

3.2 Experimental Results

We first evaluate the execution time of SGUP-Modeling when multiple threads are run in a thread block in GPU. For solving the (24,6),(28,7), (32,8), (36,9),

Table 1. Execution time of SGPU-Modeling algorithm (/s)

(l,d)	Number of running threads in a thread block in GPU						
	64	96	128	192	224	256	512
(24,6)	29.68	27.75	26.67	27.75	27.73	27.74	29.69
(28,7)	30.97	28.95	27.82	28.96	28.64	27.9	28.9
(32,8)	37.07	32.23	32.32	32.65	32.79	34.22	34.37
(36,9)	58.29	56.04	53.22	56.35	56.07	55.1	56.79
(40,10)	131.09	124.84	115.04	119.35	119.39	121.27	121.15
(44,11)	695.01	646.65	615.2	626.28	635.4	629.4	637.83
(48,12)	1632.4	1538.6	1477.9	1497.9	1475.8	1480.2	1500.9

(40,10), (44,11) and (48,12)-motif identification instances, when the number of running threads is gradually increased, the execution time of SGUP-Modeling is shown in Table 1.

We can see from Table 1 that the required time of executing SGUP-Modeling is impacted by the number of running threads in a thread block. In addition, except for the (32, 8) and (48,12)-motif identification instances, when 128 threads are run in a thread block in GPU, the required time of executing SGPU-Modeling is the least.

Next, the required time of executing SGPU-Modeling on single GPU and the required time of executing MLC-Modeling on multi-core CPU are evaluated.

When 128 threads in a thread block are run in GPU, the execution time of SGPU-Modeling and MLC-Modeling is shown in Fig. 1.

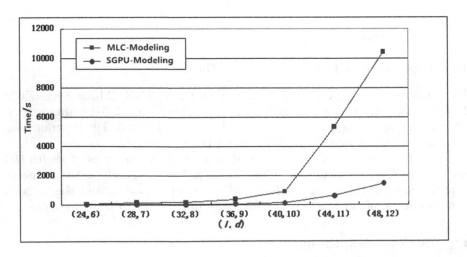

Fig. 1. Execution time of SGPU-Modeling and MLC-Modeling

As see from Fig. 1, when the (l,d)-motif identification instances of small size are solved, compared to algorithm MLC-Modeling on multi-core processor, the advantage of algorithm SGPU-Modeling is not obvious. But along with increase of l, the execution time of SGPU-Modeling is significantly less. This is because the use of thread blocks in GPU and parallel threads in thread block can greatly improve computational efficiency. In addition, the CPU is responsible for data pre-processing and serial computation, and the logical processing capability of CPU is fully utilized. It indicates that hybrid CPU and GPU computation can accelerate the (l,d)-motif identification.

Figure 2 shows the speedup of MLC-Modeling and SGPU-Modeling when 128 threads in a thread block in GPU are run.

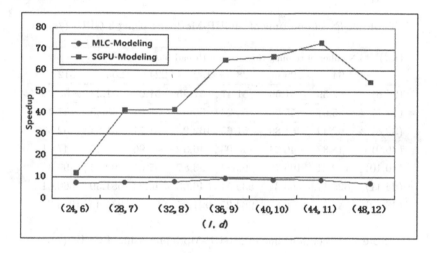

Fig. 2. Speedup of SGPU-Modeling and MLC-Modeling

From Fig. 2 we can see that the speedup of SGPU-Modeling begins to be decreased when $l > 44$ and $d > 11$, the speedup of MLC-Modeling begins to be decreased when $l > 36$ and $d > 9$, and the obtained speedup of SGPU-Modeling is obviously higher than that of MLC-Modeling. Furthermore, SGPU-Modeling can solve successfully the $(52,13)$-motif identification instance in 1 h and 50 min, but MLC-Modeling cannot solve this instance in the acceptable time.

When multiple GPUs are activated to solve in parallel the (l,d)-motif identification instances, the execution time of MGUP-Modeling is shown in Tables 2, 3 and 4 respectively.

From Tables 2, 3 and 4 we can see that the execution time of MGUP-Modeling is dependent on the number of GPUs activated, and the execution time is relatively less when the number of running threads in a thread block is 128 or 224 for a given (l,d)-motif identification instance. In addition, when three GPUs and four GPUs are activated to execute MGUP-Modeling, the required time for solving one instance of smaller size is longer than that for solving one instance

Table 2. Execution time of MGUP-Modeling using 2 GPUs (/s)

(l,d)	Number of running threads in a thread block in GPU						
	64	96	128	192	224	256	512
(24,6)	38.74	36.38	35.35	35.93	35.86	36.45	38.26
(28,7)	40.67	40.52	40.15	40.84	40.03	41.96	42.72
(32,8)	48.31	48.46	48.06	48.82	49.48	49.9	50.28
(36,9)	55.29	53.33	51.21	54.22	54.02	52.75	52.64
(40,10)	120.36	114.76	102.41	113.65	113.73	115.23	113.06
(44,11)	663.51	613.62	580.31	597.54	603.76	615.45	617.19
(48,12)	1463.7	1290.6	1183.5	1209.6	1221.3	1225.0	1311.4

Table 3. Execution time of MGUP-Modeling using 3 GPUs (/s)

(l,d)	Number of running threads in a thread block in GPU						
	64	96	128	192	224	256	512
(24,6)	52.59	51.59	51.47	51.95	51.08	51.28	53.92
(28,7)	55.67	55.61	53.24	55.55	53.51	55.35	57.47
(32,8)	62.44	61.89	61.65	62.24	61.71	61.84	62.35
(36,9)	48.37	46.24	45.03	46.32	46.96	47.85	47.61
(40,10)	109.21	105.76	90.34	93.87	95.61	99.14	99.38
(44,11)	629.17	563.34	539.51	546.93	546.24	551.39	560.61
(48,12)	1142.4	1036.8	964.25	991.35	968.28	975.42	1013.7

Table 4. Execution time of MGUP-Modeling using 4 GPUs (/s)

(l,d)	Number of running threads in a thread block in GPU						
	64	96	128	192	224	256	512
(24,6)	67.62	63.28	63.26	64.86	63.26	67.81	66.68
(28,7)	74.6	74.39	70.16	73.25	70.36	70.41	75.41
(32,8)	69.41	67.65	66.03	66.73	67.06	66.93	67.86
(36,9)	43.95	42.86	41.65	41.79	42.56	42.5	43.00
(40,10)	97.05	83.63	79.71	81.23	82.96	84.51	84.41
(44,11)	546.57	492.62	453.86	471.19	479.63	487.34	489.35
(48,12)	1034.6	925.86	793.54	817.47	804.15	811.03	856.71

of larger size. The reason is that when multiple GPUs are activated to solve in parallel the (l,d)-motif identification instances of smaller size, the time that CPU transmits data to multiple GPUs, the time of booting GPUs, and the waiting time of multiple threads synchronization are longer in this case. However, along with gradual increase of the size of the instances, hybrid CPU and multiple GPUs parallel computing can accelerate solving the (l,d)-motif identification instances and improve the efficiency of MGUP-Modeling.

When 128 threads in a thread block are run in GPU, one GPU, two GPUs, three GPUs and four GPUs are activated respectively, and the execution time of MGUP-Modeling is shown in Fig. 3.

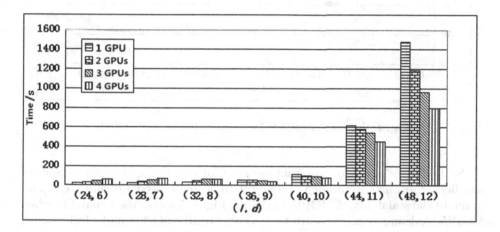

Fig. 3. Execution time of MGUP-Modeling using multiple GPUs

We can see from Fig. 3 that in the case of solving (l,d)-motif identification instances of small size, the execution time of MGUP-Modeling using multiple GPUs is little longer than that of MGUP-Modeling using single GPU. This is because at this time, the communication time between CPU and GPUs, the time of booting GPUs and the waiting time of multiple threads synchronization occupy a large portion of the required time for solving these instances, the advantage of multiple GPUs parallel computing is not played fully. However, along with gradual increase of the size of solving (l,d)-motif identification instances, the execution time of MGPU-Modeling on multiple GPUs is significantly less than that of MGPU-Modeling on single GPU. We can also see from Fig. 3 that the efficiency of MGPU-Modeling is the highest when four GPUs are activated. It illustrates that MGPU-Modeling is suit solving the (l,d)-motif identification instances of large size.

When 128 threads in a thread block are run in GPU, one, two, three and four GPUs are activated respectively, and the speedup of MGUP-Modeling is shown in Fig. 4.

Figure 4 shows that for solving (l,d)-motif identification instances of small size, the speedup of MGUP-Modeling using single GPU is relatively high, but with increase of size of the instances, the speedup of MGPU-Modeling using multiple GPUs is increased gradually.

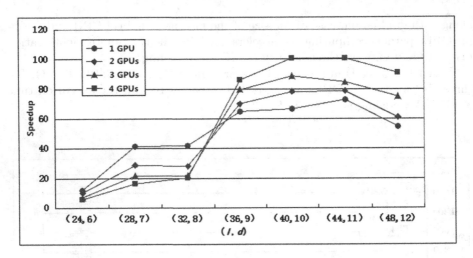

Fig. 4. Speedup of MGUP-Modeling using multiple GPUs

Scalability is also one important measure for evaluating the performance of parallel algorithms. According to the standard of equal speed measure [14,15], we evaluate the scalability of MGPU-Modeling. Figure 5 shows the scalability when MGPU-Modeling is run to solve the (36,9), (40,10), (44,11) and (48,12)-motif identification instances.

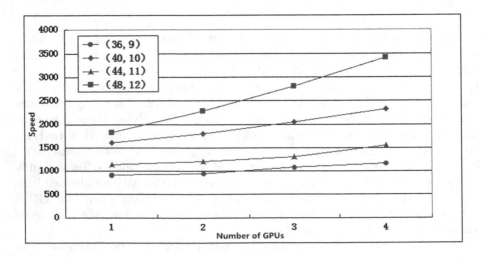

Fig. 5. Scalability of MGPU-Modeling

As can be seen from Fig. 5, with gradual increase of GPUs activated, the speed of MGPU-Modeling is also gradually increased. It illustrates that the

MGPU-Modeling has good scalability. We also see form Fig. 5 that when MGPU-Modeling solves the (l,d)-motif instances of larger size, it obtains higher speed in the case of the same number of GPUs activated.

4   Conclusions

The proposed parallel algorithm using hybrid CPU and GPU computing can solve efficiently the (l,d)-motif identification instances of large size in less computation time. The algorithm is cache-efficient and obtains good speedup and scalability. Further work will be focused on how to establish task scheduling model to balance the loads among compute nodes and solve the (l,d)-motif identification problem of larger l on heterogeneous hybrid CPU/GPU cluster system.

Acknowledgments. This work is supported in part by the National Nature Science Foundation of China under Grant No. 61462005, Nature Science Foundation of Guangxi under Grant No. 2014GXNSFAA118396 and 2014GXNSFAA118274, and Foundation of Guangdong Key Laboratory of Popular High Performance Computers, Shenzhen Key Laboratory of Service Computing and Applications under Grant No. SZU-GDPHPCL201414

References

1. Pevzner, P.A., Sze, S.H.: Combinatorial approaches to finding subtle signals in DNA sequences. In: The Eighth International Conference on Intelligent Systems for Molecular Biology, pp. 269–278. AAAI Press, Menlo Park (2000)
2. Li, J., Zhong, C.: Efficient and scalable parallel algorithm for motif finding on heterogeneous cluster systems (in Chinese). Comput. Sci. **39**(3), 279–282 (2012)
3. Dasari, N.S., Desh, R., Zubair, M.: An efficient multicore implementation of planted motif problem. In: 2010 International Conference on High Performance Computing and Simulation, pp. 9–15. IEEE Press, New York (2010)
4. Desaraju, S., Mukkamala, R.: Multiprocessor implementation of modeling method for planted motif problem. In: 2011 World Congress on Information and Communication Technologies, pp. 524–529. IEEE Press, New York (2011)
5. Nicolae, M., Rajasekaran, S.: Efficient sequential and parallel algorithms for planted motif search. BMC Bioinform. **15**, 34 (2014)
6. Marwa, A.R., Nawal, A.E.-F., Hossam, M.F.: Implementation of recursive brute force for solving motif finding problem on multi-core. Int. J. Syst. Biol. Biomed. Technol. **2**(3), 1–18 (2013)
7. Zhang, J., Zhong, C., Li, Z.: Cache-efficient parallel solving (l, d)-motif finding problem (in Chinese). Microelectron. Comput. **31**(6), 97–102 (2014)
8. Yu, L., Xu, Y.: A parallel gibbs sampling algorithm for motif finding on GPU. In: IEEE International Symposium on Parallel and Distributed Processing with Applications, pp. 555–558. IEEE Press, New York (2009)
9. Liu, Y., Schmidt, B., Liu, W., Maskell, D.L.: CUDA-MEME: accelerating motif discovery in biological sequences using CUDA-enabled graphics processing units. Pattern Recognit. Lett. **31**(14), 2170–2177 (2010)

10. Naga, S.D., Ranjan, D., Zubair, M.: High-performance implementation of planted motif problem on multicore and GPU. Concurrency Comput. Pract. Experience **25**(10), 1340–1355 (2013)
11. Zandevakili, P., Hu, M., Qin, Z.: GPUmotif: an ultra-fast and energy-efficient motif analysis program using graphics processing units. PLos ONE **7**(5), e36865 (2012)
12. Clemente, J.B., Cabarle, F.G.C., Adorna, H.N.: PROJECTION algorithm for motif finding on GPUs. In: Nishizaki, S., Numao, M., Caro, J., Suarez, M.T. (eds.) Theory and Practice of Computation. PICT, vol. 5, pp. 101–115. Springer, Heidelberg (2012)
13. Quang, D., Xie, X.: EXTREME: an online EM algorithm for motif discovery. Bioinformatics **30**(12), 1667–1673 (2014)
14. Sun, X.H., Rover, D.T.: Scalability of parallel algorithm-machine combinations. IEEE Trans. Parallel Distrib. Syst. **5**(6), 519–613 (1994)
15. Chen, G.: Parallel Computer Architectures (in Chinese). Higher Education Press, Beijing (2002)

On the Lower Bounds of Random Max 3 and 4-SAT

Guangyan Zhou$^{(\boxtimes)}$

Department of Mathematics, Beijing Technology
and Business University, Beijing 100048, China
`zhouguangyan@btbu.edu.cn`

Abstract. A k-CNF formula is said to be p-satisfiable if there exists a truth assignment satisfying a fraction of $1 - 2^{-k} + p2^{-k}$ of its clauses. We obtain better lower bounds for random 3 and 4-SAT to be p-satisfiable. The technique we use is a delicate weighting scheme of the second moment method, where for every clause we give appropriate weight to truth assignments according to their number of satisfied literal occurrences.

Keywords: Maximum satisfiability · The second moment method · Weighting scheme

1 Introduction

Maximum satisfiability (Max-SAT) is one of the central problems in theoretical computer science; it is the optimization version of the satisfiability problem. For a Boolean CNF formula F, Max-SAT is meant to determine whether there exists a truth assignment that satisfies a given number of clauses in F. The decision version of Max-SAT is NP-complete; however, Max-SAT can be approximated within a constant ratio. In recent years, a lot of results have been given by various approximation algorithms [BF, FK, GW, H1, H2]. Broder et al. [BFU] proved that the maximum number of satisfied clauses in a given CNF formula is tightly concentrated around its mean.

Recently, attention has been focused on the phase transition of random Max-SAT problems. A k-CNF formula with n variables and rn clauses, denoted by $F_k(n, rn)$, is said to be p-satisfiable if there exists a truth assignment satisfying a fraction of $1 - 2^{-k} + p2^{-k}$ of its clauses (note that $p = 1$ corresponds to the k-SAT problem; every k-CNF is 0-satisfiable). Say that a sequence of random events ξ_n occurs with high probability (w.h.p.) if $\lim_{n \to \infty} \mathbf{P}[\xi_n] = 1$. For every $k \geq 2$ and $p \in [0, 1]$, let

$$r_k(p) \equiv \sup\{r : F_k(n, rn) \text{ is } p\text{-satisfiable w.h.p.}\},$$
$$r_k^*(p) \equiv \inf\{r : F_k(n, rn) \text{ is } p\text{-unsatisfiable w.h.p.}\}.$$

Partially supported by NSFC 11301091.

D. Zhu and S. Bereg (Eds.): FAW 2016, LNCS 9711, pp. 269–278, 2016.
DOI: 10.1007/978-3-319-39817-4_26

Using the first moment method and algorithm analysis, Coppersmith et al. [CGHS] proved that for all $k \geq 2$ and sufficiently small p

$$\frac{k2^{k+2}}{\pi(k+1)^2} \times p^{-2} - O(p^{-1}) \leq r_k(p) \leq r_k^*(p) \leq 2(2^k - 1)\ln 2 \times p^{-2}.$$

Later, the bounds above were improved by Achlioptas et al. [ANP], who established that there exists a sequence $\delta_k = O(k2^{-k/2})$ such that for all $k \geq 2$ and $p \in (0,1)$

$$(1-\delta_k)\frac{2^k \ln 2}{p+(1-p)\ln(1-p)} < r_k(p) \leq r_k^*(p) \leq \frac{2^k \ln 2}{p+(1-p)\ln(1-p)}. \tag{1}$$

The approach to prove the lower bounds in (1) was by using a weighted version of the second moment method, which involves the application of the following Cauchy type inequality

$$\mathbf{P}[X > 0] \geq \frac{\mathbf{E}[X]^2}{\mathbf{E}[X^2]}, \tag{2}$$

where X represents any non-negative random variable. If X counts the number of p-satisfiable assignments, then (2) gives a lower bound on the probability of being p-satisfiable.

To state it precisely, let F be a random k-CNF formula on n variables and $m = rn$ clauses, and $S(F)$ be the set of satisfying truth assignments of F. For a fixed truth assignment $\sigma \in \{0,1\}^n$, let $H = H(\sigma, F)$ be the difference between the number of satisfied and unsatisfied literal occurrences in F under σ. Let $U = U(\sigma, F)$ be the number of unsatisfied clauses in F under σ. For fixed $0 < \gamma, \eta < 1$ and $u_0 = (1-p)2^{-k}$, Achlioptas et al. [ANP] adopted the following bivariate weighting scheme on the p-satisfiable assignments,

$$X_0 = X_0(\gamma, \eta) = \sum_\sigma \gamma^{H(\sigma,F)} \eta^{U(\sigma,F)-u_0 rn}. \tag{3}$$

In this way, "balanced" truth assignments (which mean the truth assignments that satisfy approximately half of all literal occurrences) were counted and assignments that violate more than $(1-p)2^{-k}rn$ clauses were suppressed exponentially. By (1) we see that, the ratio between the upper and lower bounds above tends to 1 as k increases, which greatly improves the previous results. However, there still exist big gaps between the lower and upper bounds for small k.

In this paper, we further improve the lower bounds of $r_k(p)$ for $k = 3$ and 4. Our method is a more general version of the weighting form (3), inspired by the works of [LGX, YV]. It's worth mentioning that by using different weighting schemes when applying the second moment method, the known best 2.68 lower bound for 3-SAT [AP] has been improved to 2.83 [LGX]; the known best 7.91 lower bound for 4-SAT [AP] has been improved to 8.09 [LGX, YV]. Specially, our lower bound $r_3(1)$ matches the lower bound in [LGX], and $r_4(1)$ matches the lower bound in [LGX, YV]. Finally, it is worth mentioning that, the computational complexity of this weighting scheme increases rapidly when k increases, but has little advantage over the one in [ANP] for large k.

2   Weighting and Estimating

In this section, we consider a refined transformation of the number of p-satisfiable truth assignments when applying the second moment method. For a random CNF formula F, appropriate weight will be given to every truth assignment $\sigma \in \{0,1\}^n$ depending on their number of satisfied literal occurrences in each clause. Later, we will give some necessary moment estimations which will be used to improve the bound of the probability of being p-satisfiable.

2.1   Our Weighting Scheme

For a random k-CNF formula F where the clauses are independently chosen, let $c = \ell_1 \vee \ell_2 \vee ... \vee \ell_k$ be a random clause, where $\ell_1, \ell_2, ..., \ell_k$ are i.i.d. uniformly distributed literals. Let $S_1(c)$ be the set of all truth assignments that satisfy exactly one of the k literals $\ell_1, \ell_2, ..., \ell_k$ in clause c. For some fixed $\gamma, \eta > 0$ and $u_0 = (1-p)2^{-k}$, we adopt the following weighting form

$$X = X(\gamma, \eta, \mu) = \sum_\sigma \prod_c \gamma^{H(\sigma,c)} \eta^{U(\sigma,c)-u_0}\left(1 + \mu \times 1_{\sigma \in S_1(c)}\right), \qquad (4)$$

where $\mu > -1$. For some fixed $A > 0$, we consider the following subset of truth assignments

$$S^* = \left\{\sigma \in \{0,1\}^n : H(\sigma, F) \geq 0 \text{ and } U(\sigma, F) \in \left[u_0 m, u_0 m + A\sqrt{m}\right]\right\}.$$

For fixed $\mu > -1$ and $u_0 = (1-p)2^{-k}$, let $\gamma_0 = \gamma_0(\mu)$ and $\eta_0 = \eta_0(\mu)$ be positive real values (in fact the existence of γ_0, η_0 can be seen by numerical calculation) satisfying the following equations

$$\begin{cases} 1 - \eta_0 = \left(1 - \gamma_0^2\right)\left(1 + \gamma_0^2\right)^{k-1} + (k-2)\mu\gamma_0^2 \\ u_0 = \dfrac{\eta_0}{(1 + \gamma_0^2)^k + k\mu\gamma_0^2 - (1 - \eta_0)}. \end{cases} \qquad (5)$$

Define

$$X_* = X_*(\gamma_0, \eta_0, \mu) = \sum_{\sigma \in S^*} \prod_c \gamma_0^{H(\sigma,c)} \eta_0^{U(\sigma,c)-u_0}\left(1 + \mu \times 1_{\sigma \in S_1(c)}\right). \qquad (6)$$

By definition, if $X_* > 0$ then at least one truth assignment must falsify at most $u_0 m + A\sqrt{m}$ clauses. Thus if we prove that there exists a constant $D > 0$ (which is independent of n) such that $\mathbf{E}[X_*^2] < D \times \mathbf{E}[X_*]^2$, then applying (2) we see that $F_k(n, m)$ is w.h.p. p'-satisfiable for all $p' < p$.

2.2   Moment Estimations and Truncation

We define two functions that will play an important role in our analysis.

$$\eta^{2u_0} f(\alpha, \gamma, \eta, \mu) =$$

$$\left(\alpha\left(\frac{\gamma^2 + \gamma^{-2}}{2}\right) + 1 - \alpha\right)^k - 2(1-\eta)\left(\frac{\alpha\gamma^{-2} + 1 - \alpha}{2}\right)^k + (1-\eta)^2\left(\frac{\alpha\gamma^{-2}}{2}\right)^k$$

$$+ 2k\mu\left[\left(\frac{\alpha\gamma^2 + 1 - \alpha}{2}\right)\left(\frac{\alpha\gamma^{-2} + 1 - \alpha}{2}\right)^{k-1} - (1-\eta)\left(\frac{1-\alpha}{2}\right)\left(\frac{\alpha\gamma^{-2}}{2}\right)^{k-1}\right]$$

$$+ k\mu^2\gamma^{-2(k-2)}\left[\left(\frac{\alpha}{2}\right)^k + (k-1)\left(\frac{1-\alpha}{2}\right)^2\left(\frac{\alpha}{2}\right)^{k-2}\right], \tag{7}$$

and

$$g_r(\alpha, \gamma, \eta, \mu) = \frac{f(\alpha, \gamma, \eta, \mu)^r}{\alpha^\alpha(1-\alpha)^{1-\alpha}}. \tag{8}$$

Lemma 1. *For every* $\gamma, \eta > 0$ *and* $\mu > -1$,

$$\mathbf{E}[X]^2 = \left(2g_r\left(\frac{1}{2}, \gamma, \eta, \mu\right)\right)^n.$$

Proof. By linearity of expectation and clause-independence we see that

$$\eta^{u_0 m}\mathbf{E}[X] = \sum_\sigma \mathbf{E}\left[\prod_c \gamma^{H(\sigma,c)}\eta^{U(\sigma,c)}\left(1 + \mu \times \mathbf{1}_{\sigma \in S_1(c_i)}\right)\right]$$

$$= 2^n\left(\mathbf{E}\left[\gamma^{H(\sigma,c)}\eta^{U(\sigma,c)}\left(1 + \mu \times \mathbf{1}_{\sigma \in S_1(c)}\right)\right]\right)^m.$$

It is easy to see that

$$\mathbf{E}\left[\gamma^{H(\sigma,c)}\eta^{U(\sigma,c)}\left(1 + \mu \times \mathbf{1}_{\sigma \in S_1(c)}\right)\right]$$

$$= \mathbf{E}\left[\gamma^{H(\sigma,c)}\right] - (1-\eta)\mathbf{E}\left[\gamma^{H(\sigma,c)}\mathbf{1}_{\sigma \notin S(c)}\right] + \mu\mathbf{E}\left[\gamma^{H(\sigma,c)}\mathbf{1}_{\sigma \in S_1(c)}\right].$$

Simple calculation yields that

$$\mathbf{E}\left[\gamma^{H(\sigma,c)}\right] = \left(\frac{\gamma + \gamma^{-1}}{2}\right)^k,$$

$$\mathbf{E}\left[\gamma^{H(\sigma,c)}\mathbf{1}_{\sigma \notin S(c)}\right] = \left(\frac{\gamma^{-1}}{2}\right)^k,$$

$$\mathbf{E}\left[\gamma^{H(\sigma,c)}\mathbf{1}_{\sigma \in S_1(c)}\right] = \binom{k}{1}\left(\frac{\gamma}{2}\right)\left(\frac{\gamma^{-1}}{2}\right)^{k-1}.$$

Therefore

$$\mathbf{E}\left[\gamma^{H(\sigma,c)}\eta^{U(\sigma,c)}\left(1 + \mu \times \mathbf{1}_{\sigma \in S_1(c)}\right)\right]$$

$$= \left(\frac{\gamma + \gamma^{-1}}{2}\right)^k - (1-\eta)\left(\frac{\gamma^{-1}}{2}\right)^k + k\mu\left(\frac{\gamma}{2}\right)\left(\frac{\gamma^{-1}}{2}\right)^{k-1}$$

$$\equiv Z(\gamma, \eta, \mu). \tag{9}$$

Note that $\eta^{-2u_0} Z(\gamma, \eta, \mu)^2 = f(1/2, \gamma, \eta, \mu)$, then

$$\mathbf{E}[X]^2 = (2g_r(1/2, \gamma, \eta, \mu))^2.$$

The following is to show that asymptotically $\mathbf{E}[X_*(\gamma_0, \eta_0, \mu)]$ is a constant fraction of $\mathbf{E}[X(\gamma_0, \eta_0, \mu)]$.

Lemma 2. *For every $u_0 \in [0, 2^{-k}]$ and $\mu > -1$, suppose that $\gamma_0, \eta_0 > 0$ satisfy Eq. (5), then there exists $\theta = \theta(k, u_0, \mu, \gamma_0, \eta_0, A) > 0$ such that*

$$\lim_{n \to \infty} \frac{\mathbf{E}[X_*(\gamma_0, \eta_0, \mu)]}{\mathbf{E}[X(\gamma_0, \eta_0, \mu)]} = \theta.$$

Proof. By linearity of expectation, it suffices to prove that there exists some $\theta = \theta(k, u_0, \mu, \gamma_0, \eta_0, A) > 0$ such that for the values of μ, γ_0 and η_0, and every truth assignment σ, it holds that

$$\lim_{n \to \infty} \frac{\mathbf{E}\left[\prod_c \gamma_0^{H(\sigma,c)} \eta_0^{U(\sigma,c)} \left(1 + \mu \times \mathbf{1}_{\sigma \in S_1(c)}\right) \mathbf{1}_{\sigma \in S^*(F)}\right]}{\mathbf{E}\left[\prod_c \gamma_0^{H(\sigma,c)} \eta_0^{U(\sigma,c)} \left(1 + \mu \times \mathbf{1}_{\sigma \in S_1(c)}\right)\right]} = \theta. \tag{10}$$

Recalling the formulas in our model are sequences of i.i.d. random literals $\ell_1, ..., \ell_{km}$, let $\mathbf{P}(\cdot)$ denote the probability assigned by our distribution to any such sequence, i.e., $(2n)^{-km}$. Now, fix any truth assignment σ and consider an auxiliary distribution $\mathbf{P}_\sigma$ on k-CNF formulas where the m clauses $c_1, ..., c_m$ are again i.i.d. among all $(2n)^k$ clauses, but where now for any fixed clause ω

$$\mathbf{P}_\sigma(c_i = \omega) = \frac{\gamma_0^{H(\sigma,\omega)} \eta_0^{U(\sigma,\omega)} \left(1 + \mu \times \mathbf{1}_{\sigma \in S_1(\omega)}\right) \mathbf{P}(\omega)}{Z(\gamma_0, \eta_0, \mu)}, \tag{11}$$

where the function $Z(\gamma, \eta, \mu)$ is given by (9), thus

$$Z(\gamma_0, \eta_0, \mu) = \mathbf{E}\left[\gamma_0^{H(\sigma,c)} \eta_0^{U(\sigma,c)} \left(1 + \mu \times \mathbf{1}_{\sigma \in S_1(c)}\right)\right].$$

Under $\mathbf{P}(\cdot)$ every k-CNF formula F with m clauses had the same probability $\mathbf{P}(F) = (2n)^{-km}$, but under $\mathbf{P}_\sigma$ its probability is

$$\mathbf{P}_\sigma(F) = \frac{\gamma_0^{H(\sigma,F)} \eta_0^{U(\sigma,F)} \mathbf{P}(F) \prod_{i=1}^m \left(1 + \mu \times \mathbf{1}_{\sigma \in S_1(c_i)}\right)}{Z(\gamma_0, \eta_0, \mu)^m}, \tag{12}$$

Let $\mathbf{E}_\sigma$ be the expectation operator corresponding to $\mathbf{P}_\sigma$. Keep the equations in (5) in mind, then

$$Z(\gamma_0, \eta_0, \mu) \mathbf{E}_\sigma[H(\sigma, c)] = Z(\gamma_0, \eta_0, \mu) \mathbf{E}_\sigma[H(\sigma, c) \mathbf{1}_{\sigma \in S(c)}] + Z(\gamma_0, \eta_0, \mu) \mathbf{E}_\sigma[H(\sigma, c) \mathbf{1}_{\sigma \notin S(c)}]$$

$$= \sum_{i=1}^k \binom{k}{i} (2i - k) \gamma_0^{2i-k} (1 + \mu \times \mathbf{1}_{i=1}) 2^{-k} - k\eta_0 \gamma_0^{-k} 2^{-k}$$

$$= k(2\gamma_0)^{-k} \left[(1 - \eta_0) - (1 - \gamma_0^2)(1 + \gamma_0^2)^{k-1} - (k-2)\mu\gamma_0^2\right]$$

$$= 0,$$

$$Z(\gamma_0, \eta_0, \mu) \mathbf{E}_\sigma[U(\sigma, c) - u_0] = (2\gamma_0)^{-k} \eta_0 - Z(\gamma_0, \eta_0, \mu) u_0 = 0.$$

Apply the multivariate central limit theorem to the i.i.d. mean-zero random vectors $(H(\sigma, c_i), U(\sigma, c_i) - u_0)$ for $i = 1, ..., m$. Observe that, since $k \geq 2$, the common law of these random vectors is not supported on a line. Then as $n \to \infty$

$$\mathbf{P}_\sigma[\sigma \in S^*(F)] = \mathbf{P}_\sigma\left[H(\sigma, F) \geq 0 \text{ and } U(\sigma, F) \in \left[mu_0, mu_0 + A\sqrt{m}\right]\right]$$
$$\to \theta(k, u_0, \mu, \gamma_0, \eta_0, A) > 0.$$

The right hand side is the probability that a certain non-degenerate bivariate normal law assigns to a certain open set, and its exact value is unimportant. By (11), this is equivalent to (10).

Next, we will give the estimation of the second moment of $X_*(\gamma_0, \eta_0, \mu)$.

Lemma 3.

$$\mathbf{E}[X_*^2] \leq 2^n \sum_{z=0}^{n} \binom{n}{z} \inf_{\gamma \geq \gamma_0, \eta \geq \eta_0} f(z/n, \gamma, \eta, \mu)^m. \tag{13}$$

Proof. Let σ and τ be any pair of truth assignments that agree on $z = \alpha n$ variables, and $\ell_1, \ell_2, ..., \ell_k$ be i.i.d. uniformly distributed literals and $c = \ell_1 \vee \ell_2 \vee ... \vee \ell_k$. Since $\sigma \in S^*$ implies $H(\sigma, F) \geq 0$ and $U(\sigma, F) \geq u_0 m$, then for any $\gamma \geq \gamma_0$ and $\eta \geq \eta_0$, we have

$$\mathbf{E}[X_*^2]$$
$$= \sum_{\sigma, \tau} \mathbf{E}\left[\gamma_0^{H(\sigma, F) + H(\tau, F)} \eta_0^{U(\sigma, F) + U(\tau, F) - 2u_0 m} \mathbf{1}_{\sigma, \tau \in S^*(\mathbf{F})} \prod_c (1 + \mu \times \mathbf{1}_{\sigma \in S_1(c)})(1 + \mu \times \mathbf{1}_{\tau \in S_1(c)})\right]$$
$$\leq \sum_{\sigma, \tau} \left(\mathbf{E}\left[\gamma^{H(\sigma, c) + H(\tau, c)} \eta^{U(\sigma, c) + U(\tau, c) - 2u_0} (1 + \mu \times \mathbf{1}_{\sigma \in S_1(c)})(1 + \mu \times \mathbf{1}_{\tau \in S_1(c)})\right]\right)^m. \tag{14}$$

Next, we will estimate (14). According to [ANP], it holds that

$$\mathbf{E}\left[\gamma^{H(\sigma, c) + H(\tau, c)} \eta^{U(\sigma, c) + U(\tau, c)}\right]$$
$$= \left(\alpha\left(\frac{\gamma^2 + \gamma^{-2}}{2}\right) + 1 - \alpha\right)^k - 2(1 - \eta)\left(\frac{\alpha\gamma^{-2} + 1 - \alpha}{2}\right)^k + (1 - \eta)^2\left(\frac{\alpha\gamma^{-2}}{2}\right)^k. \tag{15}$$

Note that $\mathbf{1}_{\sigma \in S_1(c), \tau \in S(c)} = \mathbf{1}_{\sigma \in S_1(c)} - \mathbf{1}_{\sigma \in S_1(c), \tau \notin S(c)}$, then

$$\mathbf{E}\left[\gamma^{H(\sigma, c) + H(\tau, c)} \eta^{U(\tau, c)} \mathbf{1}_{\sigma \in S_1(c)}\right]$$
$$= \mathbf{E}\left[\gamma^{H(\sigma, c) + H(\tau, c)} \mathbf{1}_{\sigma \in S_1(c)}\right] - (1 - \eta)\mathbf{E}\left[\gamma^{H(\sigma, c) + H(\tau, c)} \mathbf{1}_{\sigma \in S_1(c), \tau \notin S(c)}\right]$$
$$= k\left(\frac{\alpha\gamma^2 + 1 - \alpha}{2}\right)\left(\frac{\alpha\gamma^{-2} + 1 - \alpha}{2}\right)^{k-1} - k(1 - \eta)\left(\frac{1 - \alpha}{2}\right)\left(\frac{\alpha\gamma^{-2}}{2}\right)^{k-1} \tag{16}$$

If $\sigma, \tau \in S_1(c)$, we have two cases:

(1) Each of the k literals in clause c is assigned the same value in σ as that in τ;
(2) exactly two of the k literals in c are assigned the opposite value.

Then it follows that

$$\mathbf{E}\left[\gamma^{H(\sigma,c)+H(\tau,c)}\mathbf{1}_{\sigma\in S_1(c),\tau\in S_1(c)}\right]$$

$$= k\left(\frac{\alpha\gamma^2}{2}\right)\left(\frac{\alpha\gamma^{-2}}{2}\right)^{k-1} + k(k-1)\left(\frac{1-\alpha}{2}\right)^2\left(\frac{\alpha\gamma^{-2}}{2}\right)^{k-1}. \tag{17}$$

Combining (15)–(17), we can deduce that

$$\mathbf{E}\left[\gamma^{H(\sigma,c)+H(\tau,c)}\eta^{U(\sigma,c)+U(\tau,c)-2u_0}\left(1+\mu\times\mathbf{1}_{\sigma\in S_1(c)}\right)\left(1+\mu\times\mathbf{1}_{\tau\in S_1(c)}\right)\right]$$

$$= \eta^{-2u_0}\left(\mathbf{E}\left[\gamma^{H(\sigma,c)+H(\tau,c)}\eta^{U(\sigma,c)+U(\tau,c)}\right] + 2\mu\mathbf{E}\left[\gamma^{H(\sigma,c)+H(\tau,c)}\eta^{U(\tau,c)}\mathbf{1}_{\sigma\in S_1(c)}\right]\right.$$

$$\left. +\mu^2\mathbf{E}\left[\gamma^{H(\sigma,c)+H(\tau,c)}\mathbf{1}_{\sigma\in S_1(c),\tau\in S_1(c)}\right]\right)$$

$$= f(\alpha,\gamma,\eta,\mu). \tag{18}$$

Since the number of ordered pairs of assignments with overlap z is $2^n\binom{n}{z}$, (14) and (18) yield that

$$\mathbf{E}[X_*^2] \leq 2^n\sum_{z=0}^{n}\binom{n}{z}\inf_{\gamma\geq\gamma_0,\eta\geq\eta_0} f(z/n,\gamma,\eta,\mu)^m.$$

The proof of the following lemma follows by applying the Laplace method of asymptotic analysis [DB].

Lemma 4 [AM]. *Let ϕ be any real, positive, twice-differential function on $[0,1]$ and let*

$$S_n = \sum_{z=0}^{n}\binom{n}{z}\phi(z/n)^n.$$

Letting $0^0 \equiv 1$, define g on $[0,1]$ as

$$g(\alpha) = \frac{\phi(\alpha)}{\alpha^\alpha(1-\alpha)^{1-\alpha}}.$$

If there exists $\alpha_{\max} \in (0,1)$ such that $g(\alpha_{\max}) \equiv g_{\max} > g(\alpha)$ for all $\alpha \neq \alpha_{\max}$, and $g''(\alpha_{\max}) < 0$, then there exist constants $B,C > 0$ such that for all sufficiently large n

$$B \times g_{\max}^n \leq S_n \leq C \times g_{\max}^n.$$

The following result follows by combining Lemmas 1–4.

Corollary 1. *Let $\chi : [0,1] \rightarrow [\gamma_0,+\infty)$ and $\omega : [0,1] \rightarrow [\eta_0,+\infty)$ be arbitrary functions satisfying $\chi(1/2) = \gamma_0$, $\omega(1/2) = \eta_0$, and let $g_r^*(\alpha) = g_r(\alpha,\chi(\alpha),\omega(\alpha),\mu)$. If $g_r^*(1/2) > g_r^*(\alpha)$ for all $\alpha \neq 1/2$, and $(g_r^*)''(1/2) < 0$, then there exists a constant $D = D_{\chi,\omega}(k,r,p,\mu,A) > 0$ such that for all sufficiently large n*

$$\mathbf{E}[X_*^2] \leq D \times \mathbf{E}[X_*]^2.$$

Lemma 5. *For all* $0 < x \leq 1/2$, $g_r(1/2 + x, \gamma, \eta, \mu) > g_r(1/2 - x, \gamma, \eta, \mu)$.

Proof. Note that $\alpha^\alpha(1 - \alpha)^{1-\alpha}$ is symmetric around $1/2$, so it suffices to prove that for every $x \in (0, 1/2)$,

$$f\left(\frac{1}{2} + x, \gamma, \eta, \mu\right) > f\left(\frac{1}{2} - x, \gamma, \eta, \mu\right),$$

where $\gamma, \eta > 0$ and $\mu > -1$.

$$2^k \eta^{2u_0} f\left(\frac{1}{2} + x, \gamma, \eta, \mu\right)$$

$$= \frac{1}{2^k}\left[(\gamma^{-1} + \gamma)^k - (1 - \eta)\gamma^{-k} + k\mu\gamma^{-(k-2)}\right]^2$$

$$+ \frac{k}{2^{k-1}}\left[(\gamma^{-1} + \gamma)^{k-1} - (1 - \eta)\gamma^{-k} + (k - 2)\mu\gamma^{-(k-2)}\right]^2 x$$

$$+ \sum_{j=2}^{k}\binom{k}{j}\frac{1}{2^{k-j}}\left[(\gamma^{-1} + \gamma)^{k-j}(\gamma^{-1} - \gamma)^j - (1 - \eta)\gamma^{-k} + (k - 2j)\mu\gamma^{-(k-2)}\right]^2 x^j$$

$$= \sum_{j=0}^{k}\binom{k}{j}\frac{1}{2^{k-j}}\left[(\gamma^{-1} + \gamma)^{k-j}(\gamma^{-1} - \gamma)^j - (1 - \eta)\gamma^{-k} + (k - 2j)\mu\gamma^{-(k-2)}\right]^2 x^j$$

$$\equiv \sum_{j=0}^{k} a_j x^j.$$

Note that $a_j \geq 0$ ($j = 0, ..., k$) and at least one of the a_j's is non-zero. Moreover, for every $x > 0$ and odd j it holds that $x^j - (-x)^j > 0$, then the conclusion holds.

Remark. By Lemma 5, we know that

$$\inf_{\gamma \geq \gamma_0, \eta \geq \eta_0} g_r(1/2 + x, \gamma, \eta, \mu) \geq \inf_{\gamma \geq \gamma_0, \eta \geq \eta_0} g_r(1/2 - x, \gamma, \eta, \mu),$$

which simplifies the Corollary 1 to consider only $\alpha \in (1/2, 1]$.

3   Experimental Results and Performance Analysis

From Corollary 1 and Lemma 5 we see that, if for any given $p \in (0, 1]$ and $r > 0$ there exists constant $\mu > -1$, and functions $\chi : [0, 1] \to [\gamma_0, +\infty)$ and $\omega : [0, 1] \to [\eta_0, +\infty)$ such that for all $\alpha \in (1/2, 1]$, $g_r^*(1/2) > g_r^*(\alpha)$ and $(g_r^*)''(1/2) < 0$, then there exists a constant $D = D_{\chi,\omega}(k, r, p, \mu, A)$ such that for all sufficiently large n, $\mathbf{E}[X_*^2] \leq D \times \mathbf{E}[X_*]^2$. Thus, a random k-CNF formula $\mathbf{F}_k(n, rn)$ is p-satisfiable. For computing simplicity, we will let $0 < \gamma, \eta < 1$, and let $\mu \in (-1, 1)$ in this paper.

For $k = 3$ and $k = 4$, the lower bounds (shown in Figs. 1 and 3) can be determined numerically by using a refined adaptation of γ, η with respect to α. Specifically, for each value of $p \in [0, 1]$, we compute over all $\mu \in (-1, 1)$, and for every value of μ there exists a choice of functions χ, ω satisfying $g_r(1/2, \gamma_0, \eta_0, \mu) > g_r(\alpha, \chi(\alpha), \omega(\alpha), \mu)$ for all $\alpha \in (1/2, 1]$. Note that, to guarantee that $(g_r^*)''(\alpha)$ is twice differentiable at $1/2$, we will let

$$g_r^*(\alpha) = \begin{cases} g_r(\alpha, \gamma_0, \eta_0, \mu) & \text{if } \alpha \in (\frac{1}{2} - \varepsilon, \frac{1}{2} + \varepsilon) \\ \inf_{1 > \gamma \ge \gamma_0, 1 > \eta \ge \eta_0} g_r(\alpha, \gamma_0, \eta_0, \mu) & \text{otherwise} \end{cases} \tag{19}$$

where $\varepsilon > 0$ is a small enough constant. Let $r_k(p) \equiv \max_{\mu \in (-1,1)} r_k(p, \mu)$, we can then obtain the results in Figs. 1 and 3. The values of $r_k(p)$ for some specific p are shown in Tables 1 and 2.

Furthermore, we record an optimal value of μ for each p (as illustrated in Figs. 2 and 4). Note that, the value of μ may not be in one-to-one correspondence with p, which means that for each p there may exist several satisfying values of μ and we choose one of them arbitrarily.

Table 1. Comparison of lower bounds $r_3(p)$ for different p

p	1	0.9	0.8	0.7	0.6	0.5	0.4	0.3	0.2	0.1
Achlioptas et al.'s lower bound	2.68	3.25	4.04	5.2	6.9	9.89	15.29	26.82	60.01	238.93
Our lower bound	2.83	3.4	4.2	5.36	7.15	10.12	15.57	27.42	61.08	243.89

Table 2. Comparison of lower bounds $r_4(p)$ for different p

p	1	0.9	0.8	0.7	0.6	0.5	0.4	0.3	0.2	0.1
Achlioptas et al.'s lower bound	7.91	10.13	13.13	17.38	23.94	34.88	54.05	95.35	212.96	847.08
Our lower bound	8.09	10.59	13.92	18.75	26.1	38.46	60.85	110.22	245.67	922.46

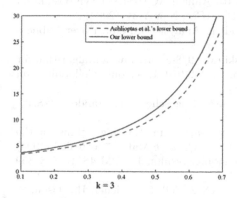

Fig. 1. Comparison of the lower bounds for the density r as a function of $q = 1 - p$ for $k = 3$.

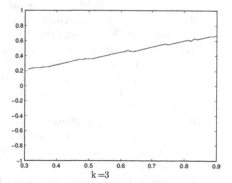

Fig. 2. The optimal choice of μ (with respect to $q = 1 - p$) for $k = 3$.

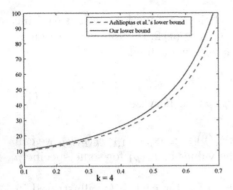

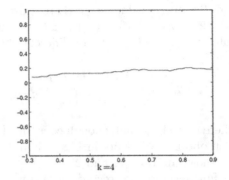

Fig. 3. Comparison of the lower bounds for the density r as a function of $q = 1 - p$ for $k = 4$.

Fig. 4. The optimal choice of μ (with respect to $q = 1 - p$) for $k = 4$.

References

[AM] Achlioptas, D., Moore, C.: The asymptotic order of the random k-SAT threshold. In: Proceedings of 43rd Annual Symposium on Foundations of Computer Science, pp. 126–127 (2002)

[ANP] Achlioptas, D., Naor, A., Peres, Y.: On the maximum satisfiability of random formulas. J. Assoc. Comput. Machinary **54**(2) (2007)

[AP] Achlioptas, D., Peres, Y.: The threshold for random k-SAT is $2^k \log 2 - O(k)$. J. Am. Math. Soc. **17**(4), 947–973 (2004)

[BF] Borchers, B., Furman, J.: A two-phase exact algorithm for MAX-SAT and weighted MAX-SAT problems. J. Comb. Optim. **2**(4), 299–306 (1998)

[BFU] Broder, A.Z., Frieze, A.M., Upfal, E.: On the satisfiability and maximum satisfiability of random 3-CNF formulas. In: Proceedings of 4th Annual ACM-SIAM Symposium on Discrete Algorithms, pp. 322–330 (1993)

[CGHS] Coppersmith, D., Gamarnik, D., Hajiaghayi, M.T., Sorkin, G.B.: Random MAX 2-SAT and MAX CUT. In: 14th Annual ACM-SIAM Symposium on Discrete Algorithms (Baltimore, MD). ACM, New York (2003)

[DB] de Bruijn, N.G.: Asymptotic Methods in Analysis, 3rd edn. Dover Publications Inc., New York (1981)

[FK] Fernandez de la Vega, W., Karpinski, M.: 9/8-approximation algorithm for random max 3-sat. Technical Report TR02-070, Electronic Colloquium on Computational Complexity (2002)

[LGX] Gao, Z., Liu, J., Xu, K.: A novel weighting scheme for random k-SAT. arXiv:1310.4303

[GW] Goemans, M., Williamson, D.: New 3/4-approximation algorithms for the maximum satisfiability problem. SIAM J. Discrete Math. **7**, 656–666 (1994)

[H1] Håstad, J.: Some optimal inapproximability results. J. ACM **48**(4), 798–859 (2001)

[H2] Hirsch, E.A.: A new algorithm for MAX-2-SAT. In: Reichel, H., Tison, S. (eds.) STACS 2000. LNCS, vol. 1770, p. 65. Springer, Heidelberg (2000)

[YV] Vorob'ev, F.Y.: A lower bound for the 4-satisfiability threshold. Discrete Math. Appl. **17**(3), 287–294 (2007)

Finding Disjoint Dense Clubs
in an Undirected Graph

Peng Zou[1], Hui Li[2], Chunlin Xin[3], Wencheng Wang[4], and Binhai Zhu[1(✉)]

[1] Department of Computer Science, Montana State University,
Bozeman, MT 59717-3880, USA
`peng.zou@msu.montana.edu, bhz@montana.edu`
[2] Department of Computer Science,
Beijing University of Chemical Technology, Beijing, China
`ray@mail.buct.edu.com`
[3] School of Economic and Management,
Beijing University of Chemical Technology, Beijing, China
`xinchl@buct.edu.com`
[4] State Key Laboratory of Computer Science, Institute of Software,
Chinese Academy of Sciences, Beijing, China
`whn@ios.ac.cn`

Abstract. In a social network, the trust among its members usually cannot be carried over many hops. So it is important to find disjoint clusters with a small diameter and with a decent size, formally called *dense clubs* in this paper. This paper focuses on handling this NP-complete problem. First, from the parameterized computational complexity point of view, we show that this problem does not admit a polynomial kernel (implying that it is unlikely to apply some reduction rules to obtain a practically small problem size). Then, we focus on the dual version of the problem, i.e., deleting d vertices to obtain some disjoint dense clubs. We show that this dual problem admits a simple FPT algorithm using a bounded search tree method (the running time is still too high for practical datasets). Finally, we combine a simple reduction rule together with some heuristic method to obtain a practical solution (verified by extensive testing on practical datasets).

1 Introduction

Social network gives people a new "world" where we can share everything that happens around us and social networks have grown enormously in recent years. It is full of data and has become an indispensable part of our life. Finding cohesive subgroups is vital to understanding the structure of the network. Clique is commonly used to describe a dense subgroup. However, the requirements of a clique are too restrictive in many situations, thus various relaxed cohesive subgroup structures based on clique have been proposed, such as s-club, s-clique, and s-plex, etc. [1,2].

Given an undirected graph $G = (V, E)$, a clique is a subset of vertices such that any pair of vertices in this subset form an edge in E. In fact, the maximum

© Springer International Publishing Switzerland 2016
D. Zhu and S. Bereg (Eds.): FAW 2016, LNCS 9711, pp. 279–288, 2016.
DOI: 10.1007/978-3-319-39817-4_27

clique problem is one of the most widely studied NP-complete problems [3]. Many algorithms for this problem are available in the literature [4–6]. Motivated by practical applications in social and biological networks, s-club is a diameter-based graph-theoretic generalization of clique, which was first introduced as an alternative approach to model a cohesive subgroup in the social network area [7,8]. A s-club is a subset of vertices $V' \subseteq V$, such that the diameter of the induced subgraph $G[V']$ is at most s.

With the development of social networks, the trust among its members has become a big issue. In a social network, the trust among its members usually cannot be carried over many hops. So it is important to find disjoint clusters with a small diameter and with a decent size, formally called *dense clubs* in this paper. Secondly, a complex social network is usually composed of several groups/communities, and this characterization of community structure means the appearance of densely connected groups of vertices, with only sparse connections between groups [9], see Fig. 1.

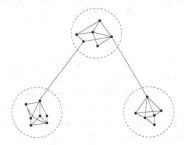

Fig. 1. A small example of the network structure. There are three groups denoted by the circle with many edges and only a small number of edges between the groups.

A s-clique is a subset of vertices $S \subseteq V$ if the shortest path distance $d_G(u, v) \le s$ for all $u, v \in S$. It is pretty obvious that a s-club is also a s-clique, but the converse is not true in general [10]. And for $s = 1$, a s-club is simply a clique. It is known that the maximum clique problem is a classical NP-complete problem [4,5], which is also hard to approximate [11]. The maximum s-club problem [12,13] is NP-complete for any fixed s, even when restricted to graphs of fixed diameter $s + 1$ [12]. In fact, testing whether a s-club is maximal is also NP-complete for any fixed integer s [14].

In reference [15], Bourjolly *et al.* posed three heuristic methods which are DROP, CONDTELLATION and s-DLIQUE-DROP for the maximum s-club problem. A variable neighborhood search (VNS) meta-heuristic algorithm was proposed by Shahinpour *et al.* [10]. In fact, they used the VNS heuristic method as the lower bound to develop an exact algorithm for the maximum s-club problem [10]. Recently, a new heuristic algorithm called IDROP for the largest s-club problem was given in the paper [16].

From the parameterized computational complexity point of view, two fixed-parameter tractable (FPT) algorithms for the maximum s-club problem were obtained [17]. The paper [18] extends the previous parameterized complexity study for 2-club and provides polynomial-size kernels for 2-club parameterized by "cluster editing set size of G" and "size of a cluster editing set of G". For the 2-club-editing problem, the paper [21] proposes an improved search tree algorithm with running time $O^*(3.31^k)$ based on two new branching cases, improving the trivial $O^*(4^k)$ bound.

In this paper, we first study the disjoint dense club problem. Specifically, we show that the problem does not have a polynomial kernel (unless the polynomial hierarchy collapses to the third level). This implies that it is unlikely to obtain any efficient FPT algorithm for the problem (and the related ones). Then we consider the dual problem of editing a graph (by deleting vertices) into disjoint s-clubs. Since the trivial bounded-degree search method takes $O^*((s+2)^d)$ time, which is not efficient for most real datasets, we propose three rules for build an efficient heuristic method. We then test this method with two real datasets.

The rest of this paper is organized as follows. We will introduce some necessary definitions and notations in Sect. 2. In Sect. 3, theoretical results are reported. All computational results are shown in Sect. 4. We conclude the paper in Sect. 5.

2   Preliminaries

In this section, we present the relevant definitions and review some useful notions. **FPT Algorithms and Kernels.** FPT (Fixed-Parameter-Tractable) algorithms are used to study the computational complexity of NP-hard problems [22–24]. Beside the input size n, we also consider a parameter k (or several parameters). An FPT algorithm is one which solves a parameterized problem L in $O(f(k)n^c) = O^*(f(k))$ time (i.e., decide whether $(x, k) \in L$, where $n = |x|$), where $f(-)$ is any computable function and c is a constant not related to n and k.

A parameterized problem L admits a problem *kernel* if there is a polynomial-time transformation of any instance (I, k) to an instance (I', k') such that (1) $(I, k) \in L$ iff $(I', k') \in L$; (2) $|I'| \leq g(k)$; and (3) $k' \leq k$. L has a polynomial kernel if $g(k)$ is a polynomial function. It is known that L admits an FPT algorithm iff it has a kernel (not necessarily polynomial). But if L has a polynomial kernel then usually it means L is relatively easier to solve.

Polynomial Parameter Transformations. A polynomial parameter reduction is used to reduce a problem known to be without a polynomial kernel to another problem B [19,20]. It is different from the traditional FPT-reductions.

Definition 1. Let P, Q be parameterized problem. P is polynomial parameter reducible to Q, written as $P \leq_{pp} Q$, if there exists a polynomial time computable function $f : \Sigma^* \times N \to \Sigma^* \times N$ and a polynomial p, such that for all $(x, k) \in \Sigma^* \times N$, (1)$(x, k) \in P$ if and only if $(x', k') = f(x, k) \in Q$, and (2) $|x'| \leq p(k)$. The function f is called a polynomial parameter transformation.

In [20], the following proposition was proved.

Proposition 1. Let P and Q be the parameterized problems and $\widetilde{P}$ and $\widetilde{Q}$ be the unparameterized versions of P and Q respectively. Suppose that $\widetilde{Q}$ is NP-complete and $\widetilde{P}$ is in NP. Furthermore, if there is a polynomial parameter transformation from P to Q, then if Q has a polynomial kernel, then P also has a polynomial kernel.

The above proposition can be used to prove kernelization lower bounds.

Graphs, Clubs, and Neighborhoods. We consider simple undirected graphs in this paper. Given a graph $G = (V, E)$, the distance $\delta(u, v)$ between two vertices $(u, v) \in V$ is the length of the shortest path between u and v. The *diameter* of G is the maximum of all δ-distances between pairs of nodes in V. The (open) i-neighborhood $N_i(v) = \{x | \delta(x, v) \leq i\}$ of v is the set of vertices that has distance at most i to v. The closed i-neighborhood of v is the set $N_i[v] = N_i(v) \cup \{v\}$. The exact i-neighborhood of v is the set $N_i^e(v) = \{x | \delta(x, v) = i\}$, which is the set of vertices have distance exactly i to v. For a vertex set T, $G[T]$ denotes the subgraph of G induced by T having edge set $E_T = \{(u, v) \in E | u, v \in T\}$.

A graph H is an s-club if the diameter of H is at most s. (A subset $S \subseteq V$ is an s-slub if $G[S]$ is an s-club.) And an s-club H is a (t, s)-club if the number of vertices in V is at least t, i.e., $V(H) \geq t$. Notice that when $s = 1$, an s-club is a clique. Throughout this paper, we assume that s is a small constant with $s \geq 2$. We next define the Maximum Club problem.

Problem (1): Maximum Club Problem.
Input: An undirected graph G and a constant integer $s \geq 2$, and integer t.
Question: Is there a (t, s)-club in G?

The Maximum Club problem is NP-complete [13]; in fact, it is NP-complete even in graphs of diameter $s + 1$ [12]. The parameterized version of the problem (with t being a parameter) is shown to be FPT but does not admit a polynomial kernel unless the polynomial hierarchy collapses to its third level (i.e., NP $\subseteq$ coNP/Poly) [17].

In a complex social network, finding a single club is probably not too interesting, so we define the following problems. As they are both NP-complete, we focus on the parameterized versions of the problems in this paper.

Problem (2): Disjoint Dense Clubs (DDC) Problem.
Input: An undirected graph G and a constant integer $s \geq 2$, and integers t, k.
Question: Are there k disjoint (t, s)-clubs in G?

Problem (3): Maximum Disjoint Dense Clubs (MDDC) Problem.
Input: An undirected graph G and a constant integer $s \geq 2$, and integer ℓ.
Question: Is there a set of disjoint s-clubs whose total size is at least ℓ in G?

3   Parameterized Results

In this section, we first present some theoretical results on the parameterized complexity for DDC and MDCC. Based on some of these results, we try to design a practical algorithm for them.

Theorem 1. *The disjoint dense clubs problem, parameterized by k and t, does not admit a polynomial kernel unless $NP \subseteq coNP/Poly$.*

Proof. We reduce the Maximum Club problem of finding a maximum s-club (with parameter p being its size) to the parameterized version of the DDC problem (parameterized by k and t), with a polynomial parameter reduction. Take the input $\langle G, s, p \rangle$, we construct p copies of G, i.e., $G_1, \ldots, G_p$, as the new graph $\mathcal{G}$. For $\mathcal{G}$ we set $k = p$ and $t = p$. Then G has an s-club of size p iff $\mathcal{G}$ has $k = p$ disjoint (p, s)-clubs.

"$\rightarrow$": If G has an s-club of size p, then each G_i has an s-club of size p (or, each G_i has a (p, s)-club). Obviously, $\mathcal{G}$ has p disjoint (p, s)-clubs.

"$\leftarrow$": If $\mathcal{G}$ has $k = p$ disjoint (p, s)-club, then each G_i must have a (p, s)-club. As G is known to be isomorphic to G_i, G must also have a (p, s)-club.

As deciding whether G has a (p, s)-club, parameterized by p, has no polynomial kernel unless $NP \subseteq coNP/Poly$, the DDC problem (parameterized by k, t) also has no polynomial kernel unless $NP \subseteq coNP/Poly$.   □

Theorem 2. *The maximum disjoint dense clubs problem, parameterized by ℓ, does not admit a polynomial kernel unless $NP \subseteq coNP/Poly$.*

Proof. The reduction is that same as before. Just set the parameter $\ell = p^2$.   □

The above two results, while simple, imply that it is unlikely to obtain efficient FPT algorithms to solve the DCC and MDCC problems. Next we look at the dual version of these problems.

Theorem 3. *The disjoint dense clubs problem, parameterized by $d = |V| - kt$, does admit an FPT algorithm running in $O^*((s + 2)^d)$ time.*

Proof. If between all pair of vertices u, v we have $\delta(u, v) \leq s$, then the whole graph G is an s-club. Then we are done. So to obtain disjoint s-club, we just need to branch over a pair of vertices u, v with $\delta(u, v) = s + 1$. Between and inclusive of u and v, there are $s + 2$ vertices. One of these $s + 2$ vertices must be deleted. Therefore, we have $s + 2$ choices at level-1. Then we repeat the above process for d rounds. Among the ($\leq (s+2)^d$) leaf nodes, if there are k (t, s)-clubs left, return YES; otherwise, return NO.   □

Theorem 4. *The maximum disjoint dense clubs problem, parameterized by $d = |V| - \ell$, does admit an FPT algorithm running in $O^*((s + 2)^d)$ time.*

Proof. The branching algorithm is the same as before. When we have a bound search tree of depth d, there are at most $\leq (s + 2)^d$ leaf nodes. We eliminate all those which are not an s-club, then check whether the (remaining) leaf nodes have a total size which is at least ℓ. The running time is $O^*((s + 2)^d)$.   □

Notice that the running time for the bounded search tree algorithm is too high for practical datasets. In fact, even if we could reduce the running time to roughly $O^*(3^d)$, similar to [21] for $s = 2$, it is still too high for practical datasets. So we need some practical method. We first present the following reduction rule.

Lemma 1. *(Reduction Rule I:) If $|N_s(v)| < t - 1$, then v cannot be in any (t, s)-club.*

Proof. For any two nodes u, v in the same club, we must have $|N_s[v] \cap N_s[u]| \geq t$. Hence we have $N_s(v) \geq t-1$. All the nodes in the club must satisfy this property, otherwise this node cannot be in any (t, s)-club. □

Note that we can repeatedly run this reduction rule. And when it is not possible to apply it, we have the following lemmas which do not necessarily help us reduce the problem size, but could help us reduce the solution search space.

Lemma 2. *(Branching Rule II:) Let $d(u, v) = 1$. If for all $w \in N_s(u) \cap N_s(v)$ we have $|N_s(w)| < t - 1$, then u, v cannot be in the same (t, s)-club.*

Proof. As $d(u, v) = 1$, if u, v are in the same (t, s)-club, then there must exist a $w \in N_s(u) \cap N_s(v)$ which is in the same (t, s)-club. However, by assumption, for all $w \in N_s(u) \cap N_s(v)$ we have $|N_s(w)| < t - 1$, i.e., there is not such a club containing u, v and w which is dense enough. Then we have a contradiction. Therefore, u, v cannot be in the same (t, s)-club. □

Lemma 3. *(Branching Rule III:) Let $d(u, v) = 1$. If for all $w \in N_s(u) \cap N_s(v)$, we have $|N_s[w] \cap N_s[u] \cap N_s[v]| < t$, then u, v cannot in the same (t, s)-club.*

Proof. As $d(u, v) = 1$, if u, v are in the same (t, s)-club, then there must exist a node $x \in N_s(u) \cap N_s(v)$ which is in the same (t, s)-club. (Here, $|N_s[x]|$ could be larger than t.) By assumption, for all $w \in N_s(u) \cap N_s(v)$ we have $|N_s[w] \cap N_s[u] \cap N_s[v]| < t$, i.e., there is no w connecting a club which contains both u and v, is completely in $N_s[u] \cap N_s[v]$ and is dense enough. (Intuitively, some nodes in $N_s[w]$ are more than s edges away from u or v.) Hence, we have a contradiction. Therefore, u, v cannot be in the same (t, s)-club. □

Based on Lemmas 1, 2 and 3, we design a new practical bounded search tree algorithm for DCC. (The algorithm also works for the MDCC problem.) The first step is to apply Lemma 1 on the input graph G. Lemma 1 can delete all nodes unlikely to be in any club. Then, we set $d(u, v) = 1$ to apply Lemmas 2 and 3. If we can find two nodes satisfying Lemma 2 or 3, one of these two nodes must be deleted. The running time of this algorithm is $O^*(2^d)$.

We would not be able to claim any theoretical result on the algorithm yet, as there are instances that the algorithm fails to handle (e.g., when we have two intersecting (t, s)-clubs). More investigation is needed to make the algorithm a true FPT algorithm for the dual version of DCC and MDCC with a running time of $O^*(2^d)$.

In the next section, we show some computational results based on this algorithm. The motivation is to avoid the $O^*((s + 2)^d)$ time FPT algorithm (Theorems 3 and 4), which is too high even for $s = 2, 3$.

4   Computational Results

We implemented the new bounded-search tree algorithm based on Lemmas 1, 2 and 3. We focused on $s = 2, 3$ for all of our empirical results. The algorithm was run on a laptop with Intel Core(TM) i7-3770 CPU, 3.40 GHz, 16 GB RAM. We tested our algorithm on two graphs. The first graph is the US Power Grid with 4941 vertices and 6594 edges. The diameter of the US Power Grid graph is 46. We downloaded it from the website:http://konect.uni-koblenz.de/networks/opsahl-powergrid. The second graph is "Autonomous System AS-733" with 6474 vertices and 12572 edges. The diameter of this graph is 9. We downloaded the graph from the website:http://snap.stanford.edu/data/as.html. It was shown on the webpage that the graph has 6474 vertices and 13895 edges. But there are self-loop edges which are meaningless to us. We preprocessed the graph and deleted all self-loop edges.

Table 1. The performance of Lemma 1 on the US Power Grid graph.

Diameter (s)	Size (t)	Nodes deleted	Time (ms)
2	10	4200	950
2	15	4857	935
2	20	4921	928
3	10	1236	595
3	15	3049	784
3	20	4240	936
3	25	4759	1011
3	30	4911	826

As we are using a bounded-search tree algorithm, we cannot expect to delete too many vertices from the graph (i.e., d should not be too large). The key contribution of Lemma 1 is to delete nodes unlikely to be a member of any t-s-club, i.e., before running the bounded-search step, we could already reduce the problem size by Lemma 1. We tested Lemma 1 on the US Power Grid and the results are listed in Table 1. From Table 1, we could find that Lemma 1 can delete more nodes with the increase of the size of clubs (e.g., t) when the diameter of clubs (e.g., s) is fixed. Moreover, Lemma 1 can delete more nodes with the decrease of the diameter of clubs (e.g., s) when the size of clubs (e.g., t) is fixed. In general, Lemma 1 is very sensitive to s and t.

We report the empirical results for the US Power Grid graph in Table 2. In the last column, 'No' means a solution has not been found after d steps and 'Yes' means a solution has been found. As the graph has a large diameter, there are in fact many small clubs.

We report the empirical results on the Autonomous System AS-733 graph in Table 3, where 'Yes/No' carry the same meaning as in Table 2. As the graph has a small diameter, it is reasonable to assume that there are not many clubs.

Table 2. Empirical results for the US Power Grid graph.

Diameter (s)	Size (t)	Steps (d)	Running Time (ms)	Result
2	10	10	24258	No
2	10	11	37874	No
2	10	12	65843	No
2	10	13	121148	No
2	10	14	232313	No
2	11	15	420401	No
2	14	15	521941	Yes
2	15	15	18543	Yes
3	11	10	81634	No
3	11	11	158064	No
3	12	11	142547	No
3	13	11	129704	No
3	14	11	119538	No
3	15	11	107597	No
3	16	12	179937	No
3	17	12	169046	No
3	18	12	178398	No
3	19	12	183689	No
3	20	12	140499	No
3	21	12	120817	No
3	22	13	235501	No
3	23	13	221456	No
3	24	13	157454	No
3	25	13	149456	No
3	26	14	266087	No
3	27	14	87690	Yes

The density of the US Power Grid graph is lower than the density of the Autonomous System AS-733 graph, and the diameter of the US Power Grid is much larger than that of the Autonomous System AS-733 graph. So the size of the clubs in the US Power Grid graph are smaller than the size of the clubs in the Autonomous System AS-733 graph. The running time of algorithm is faster when the input is the US Power Grid graph. For the same instance, the running time of this algorithm grows with the diameter of clubs and the size of clubs — which is reasonable.

Table 3. Empirical results for the Autonomous System AS-733 graph.

Diameter (s)	Size (t)	Steps (d)	Running Time (ms)	Result
2	1300	1	67848	Yes
2	700	5	30197324	No
2	750	4	10993184	No
2	760	5	1188221	No
2	770	5	1183132	No
2	770	8	1129920	No
2	780	8	1161247	No
2	790	8	114362	Yes
2	750	8	193337876	No
3	2800	1	8640385	No
3	2850	3	37265	Yes

5   Concluding Remarks

In this paper, we considered the Disjoint Dense Clubs problem which originates from social networks. While we proved that the problem is theoretically hard, we successfully designed a practical algorithm for its dual version (based on three rules). We tested this algorithm on two real graphs of 4941 and 6474 vertices respectively and the empirical results are promising. It would be interesting to add some extra rules to obtain an FPT algorithm, roughly running in $O^*(2^d)$ time, for this dual problem.

Acknowledgments. This research is partially supported by NSF of China under grant 61379087 and by the Opening Fund of Top Key Discipline of Computer Software and Theory in Zhejiang Provincial Colleges at Zhejiang Normal University.

References

1. Lee, V.E., Ruan, N., Jin, R., Aggarwal, C.: A survey of algorithms for dense subgraph discover. In: Aggarwal, C.C., Wang, H. (eds.) Managing and Mining Graph Data, pp. 303–336. Springer, New York (2010)
2. Seidman, S.B., Foster, B.L.: A graph-theoretic generalization a the clique concept. J. Math. Sociol. **6**(1), 139–154 (1978)
3. Karp, R.M.: Reducibility among combinatorial problems. In: Miller, R.E., Thatcher, J.W., Bohlinger, J.D. (eds.) Complexity of Computer Computations, pp. 85–103. Springer, New York (1972)
4. Pardalos, P.M., Xue, J.: The maximal clique problem. J. Glob. Optim. **4**(3), 301–328 (1994)
5. Bomze, I.M., Budinich, M.: The maximum clique problem. In: Du, D.-Z., Pardalos, P.M. (eds.) Handbook of Combinatorial Optimization, pp. 1–74. Springer, New York (1999)

6. Wu, Q., Hao, J.: A review on algorithms for maximum clique problem. Eur. J. Oper. Res. **242**(3), 693–709 (2015)

7. Alba, R.D.: A graph-theoretic definition of a sociometric clique. J. Math. Sociol. **3**(1), 113–126 (1973)

8. Mokken, R.J.: Clique clubs and clans. Qual. Quant. **13**, 161–173 (1979)

9. Newman, M.E.J., Girvan, M.: Finding and evaluating community structure in networks. Phys. Rev. **E69**, 026113 (2004)

10. Shahinpour, S., Butenko, S.: Distance-based clique relaxations in networks: s-Clique and s-Club. In: Goldengorin, B.I., Kalyagin, V.A., Pardalos, P.M. (eds.) Models, Algorithms and Technologies for Network Analysis, pp. 149–174. Springer, New York (2013)

11. Hastad, J.: Clique is hard to approximate with in $n^{1-\epsilon}$. In: Proceedings of the 37th Annual Symposium on Foundations of Computer Science, pp. 627–636. IEEE (1996)

12. Balasundaram, B., Butenko, B., Trukhanov, S.: Novel approaches for analyzing biological networks. J. Comb. Optim. **10**(1), 23–29 (2005)

13. Bourjolly, J., Laporte, G., Pesant, G.: An exact algorithm for the maximum k-club problem in an undirected graph. Eur. J. Oper. Res. **138**(1), 21–28 (2002)

14. Pajouh, F.M., Balasundaram, B.: On inclusionwise maximal and maximum cardinality k-clubs in graphs. Discrete Optim. **9**, 84–97 (2012)

15. Bourjolly, J., Laporte, G., Pesant, G.: Heuristics for finding k-clubs in an undirected graph. Comput. Oper. Res. **27**(6), 559–569 (2000)

16. Chang, M., Hung, L.: Finding large k-clubs in undirected graphs. Computing **95**(9), 739–758 (2013)

17. Schaefer, A., Komusiewicz, C., Moser, H., Niedermeier, R.: Parameterized computational complexity of finding small-diameter subgraphs. Optim. Lett. **6**(5), 883–891 (2012)

18. Hartung, S., Komusiewicz, C., Nichterlein, A.: Parameterized algorithmics and computational experiments for finding 2-clubs. In: Thilikos, D.M., Woeginger, G.J. (eds.) IPEC 2012. LNCS, vol. 7535, pp. 231–241. Springer, Heidelberg (2012)

19. Bodlaender, H., Downey, R., Fellows, M., Hermelin, D.: On problems without polynomial kernels. J. Comput. Syst. Sci. **75**(8), 423–434 (2009)

20. Bodlaender, H., Thomasse, S., Yeo, A.: Analysis of data reduction: transformations give evidence for non-existence of polynomial kernels. Technical report UU-CS–030, Dept. of Info. and Comput. Sciences, Utrecht University (2008)

21. Liu, H., Zhang, P., Zhu, D.: On editing graphs into 2-club clusters. In: Snoeyink, J., Lu, P., Su, K., Wang, L. (eds.) AAIM 2012 and FAW 2012. LNCS, vol. 7285, pp. 235–246. Springer, Heidelberg (2012)

22. Downey, R., Fellows, M.: Parameterized Complexity. Springer, New York (1999)

23. Flum, J., Grohe, M.: Parameterized Complexity Theory. Springer, Heidelberg (2006)

24. Niedermeier, R.: Invitation to Fixed-Parameter Algorithms. Oxford University Press, New York (2006)

Author Index

Printed in the United States
By Bookmasters